Remote Sensing Imagery

Remote Sensing Imagery

Edited by
Florence Tupin
Jordi Inglada
Jean-Marie Nicolas

iSTE

WILEY

First published 2014 in Great Britain and the United States by ISTE Ltd and John Wiley & Sons, Inc.

ISTE Ltd
27-37 St George's Road
London SW19 4EU
UK

www.iste.co.uk

John Wiley & Sons, Inc.
111 River Street
Hoboken, NJ 07030
USA

www.wiley.com

Library of Congress Control Number: 2013955375

British Library Cataloguing-in-Publication Data
A CIP record for this book is available from the British Library
ISBN 978-1-84821-508-5

FSC
www.fsc.org
MIX
Paper from responsible sources
FSC® C013604

Printed and bound in Great Britain by CPI Group (UK) Ltd., Croydon, Surrey CR0 4YY

Table of Contents

Preface

This book is aimed at students, engineers and researchers wishing to gain broad knowledge of remote sensing imagery. It is an introductory book that covers the entire chain from the acquisition of data to processing, and from physics and measuring principles to their resulting applications in the observation of planet Earth. We took it upon ourselves to present the different means of imagery together, be they optic, hyper-spectral or radar, since the user is increasingly acquainted with the joint use of this data. We have written this book for a readership that is acquainted with applied mathematics, signal processing or computer science, as taught in scientific professions and in engineering schools. It should thus enable engineers and researchers in image and signal processing, as well as geophysicians and mathematicians working in different applied fields of remote sensing: geographers, agronomists, urban planners, etc., to take advantage of the results. The book does not therefore specialize in the themes it approaches, but the references given will enable the curious readers to refer to other works.

This book was built as a coherent whole, giving a general view of remote sensing imagery, but it is equally possible to read the chapters independently. The book is organized into three parts. Part 1 is dedicated to acquisition systems, orbitography of satellites, sensors and the physics of measuring and observation, be it radar or optical. Part 2 presents the data and image processing tools, while emphasizing the specificity of remote sensing data. We will therefore discern between optical and multi-spectral signals, or radar data that show very different modelings. Part 3 is dedicated to the applications, the change detection and the analysis of multi-temporal series, as well as elevation measures, motion measures and data assimilation.

This book is a collective effort and several authors have had their input, which we hereby thank them for. We also wish to thank CNES (National Center of Space Studies), which has been the source of several studies which were indispensable for this book, and that has made the book more interesting through the numerous illustrations it has made available. We would also like to thank Henri Maitre, who initiated this book and also brought it to completion.

Florence TUPIN
Jordi INGLADA
Jean-Marie NICOLAS
December 2013

PART 1

Systems, Sensors and Acquisitions

Chapter 1

Systems and Constraints

1.1. Satellite systems

A remote sensing satellite placed in orbit around the Earth is subject to several gravitational forces that will define its trajectory and motion. We will see that orbit formalism dates as far as Kepler (1609), and the motion of satellites is modeled using Newton's laws. The Earth has specific properties, such as being flat at the poles; these specificities will introduce several changes to the Kepler model: quite strangely, as we will see, the consequences will turn out be extremely beneficial for remote sensing satellites, since they will allow us to have heliosynchronous sensors; this will enable them to acquire data at the same time as the solar hour, which in turn simplifies the comparisons of the respective data acquisitions.

The objective of this chapter is to briefly analyze orbital characteristics in order to draw some conclusions regarding the characteristics of imaging systems that orbit the Earth. For more details, readers can refer to the work of Capderou [CAP 03].

1.2. Kepler's and Newton's laws

By studying the appearance of the planets around the Sun (and, in particular, that of Mars), in 1609, Kepler proposed (in a purely phenomenological manner) the following three laws describing the motion of planets around the Sun:

– The planets' trajectory lies in a plane and is represented by an ellipse having the Sun as its focus.

Chapter written by Jean-Marie NICOLAS.

– The area swept out by the segment joining the Sun and the planet during a given period of time is constant.

– The square of the revolution period is proportional to the cube of the length of the major axis of the ellipse.

In 1687, Newton demonstrated these laws by giving us a model of the universal attraction. This model stipulates that two *punctual* masses m and M exercise a force F against each other, colinear to the line joining these two masses:

$$F = G\frac{mM}{r^2}$$

with $G = 6.672 \times 10^{-11}$ being the gravitational constant. Therefore, an interaction takes place between the two masses.

In the case of artificial satellites orbiting the Earth, it is obvious that the Earth's mass ($M_T = 5.5974 \times 10^{24}$ kg) is extremely large with respect to the mass of the satellite and we can easily assume that the center of the Earth may be mingled with the gravitational center of the Earth + satellite system. If we also assume that the Earth is a homogeneous sphere, we can consider it as a punctual mass by applying the Gauss theorem. The satellite with a mass m, located at a distance r from the Earth (with a mass M_T), is then subjected to a so-called attractive "central force":

$$\vec{F} = -\mu m\frac{\vec{r}}{r^3}$$

with $\mu = GM_T = 3.986 \times 10^{14}$ $\mathrm{m^3s^{-2}}$. Therefore, we can say that we have a central potential $U(r)$:

$$U(r) = -\mu\frac{m}{r} \qquad [1.1]$$

This force being the only one that can modify the motion of the satellite, we can therefore show that this motion verifies the following essential properties:

– The trajectory of the satellite lies in a plane, the "orbital plane". The distance r verifies, using polar coordinates, the equation of an ellipse:

$$r = \frac{p}{1 + e\cos(\theta - \theta_0)} \qquad [1.2]$$

described by two parameters: the eccentricity e and the parameter p of the ellipse. The Earth is at one of the foci of this ellipse: this is the first Kepler law. For an ellipse, a trajectory point is solely determined by the angle θ. Just like for $\theta' = \theta + 2\pi$ we find the same position values, we can say that the trajectory is closed in the orbital plane. The period of time that a satellite needs to pass from an angle θ to an angle $\theta + 2\pi$ is called a period: this is the period of time required for circling the Earth. An elliptic orbit has two main points:

– for $\theta - \theta_0 = 0$, we see that value r is at its minimum. We say that we are at the "perigee": the distance to Earth is denoted as r_P;

– for $\theta - \theta_0 = \pi$, we see that the value r is at its maximum. We say that we are at the "apogee": the distance to Earth is denoted as r_A.

We can easily deduce the relations:

$$\begin{cases} r_A = \frac{p}{1-e} \\ r_P = \frac{p}{1+e} \end{cases} \quad \text{and} \quad \frac{r_A}{r_P} = \frac{1+e}{1-e} \qquad [1.3]$$

– Since the attractive force is colinear to the distance vector $\vec{r}$, and there is no other force, the angular momentum $\vec{H}$

$$\vec{H} = \vec{r} \wedge m\vec{V} \qquad [1.4]$$

is conserved, so that:

$$H = mr^2 \frac{\mathrm{d}\theta}{\mathrm{dt}} = mC$$

where C is a constant that represents the law of equal areas, i.e. the second Kepler law.

– An ellipse can be characterized by its semimajor axis a defined by:

$$a = \frac{r_A + r_P}{2}$$

By applying the law of equal areas, we obtain the period T of the satellite:

$$T = 2\pi\sqrt{\frac{a^3}{\mu}} \qquad [1.5]$$

which is the expression of the third Kepler law. The parameters of this period T are, therefore, only a – the semimajor axis – and μ (related to the Earth's mass).

On an ellipse, the speed is not constant. We show that

$$V = \sqrt{\mu\left(\frac{2}{r} - \frac{1}{a}\right)} \qquad [1.6]$$

except when we have a perfectly circular trajectory, for which we have:

$$V = \sqrt{\frac{\mu}{a}} \qquad [1.7]$$

The speed of a satellite varies along its trajectory around the Earth. The speed is, therefore, higher as the distance r becomes smaller. More specifically, if v_P is the speed at the perigee (with $r_P = a(1 - e)$) and v_A is the speed at the apogee (with $r_A = a(1 + e)$), we get:

$$\begin{aligned} v_P &= \sqrt{\mu\left(\frac{2}{r_P} - \frac{1}{a}\right)} = \sqrt{\frac{\mu}{a}\frac{1+e}{1-e}} \\ v_A &= \sqrt{\mu\left(\frac{2}{r_A} - \frac{1}{a}\right)} = \sqrt{\frac{\mu}{a}\frac{1-e}{1+e}} \end{aligned}$$

From this, we may then deduce the following useful relation:

$$\frac{v_A}{v_P} = \frac{1-e}{1+e} \qquad [1.8]$$

which shows that the ratio of the speeds to the perigee and apogee depends only on the eccentricity and therefore on the shape of the ellipse.

To conclude on the general aspects of orbits, we must emphasize the fact that these ellipses only need two parameters to be described accurately. We often choose a, the semimajor axis, and e, the eccentricity.

1.3. The quasi-circular orbits of remote sensing satellites

The satellite era started with the launch of the first satellite Sputnik in 1957. Some numbers of civilian remote sensing satellites have since been placed in orbit around the Earth. These orbits, whose eccentricity is very low ($e < 0.001$), are quasi-circular and, therefore, described either by the semimajor axis a or by their altitude h, defined by the relation:

$$R_T + h = a$$

with $R_T = 6,378.137$ km being the radius of the Earth at the equator. We often speak of a circular orbit for this type of orbit.

Choosing an orbit for a remote sensing satellite needs to consider several things. First and foremost, since we can show that the energy of an orbit only depends on the semimajor axis a, we must note that the choice in altitude is restrictive in terms of launch: a high altitude requires a launcher that is both heavy and expensive. Therefore, it does not seem to be appropriate to choose a high altitude for an optical imaging system: the resolution being proportional to the distance, a low altitude will allow us – with an identical sensor – to distinguish more details than a high altitude. A last crucial point comes from the atmospheric drags (associated with the effects of solar winds). Friction, more significant as the altitude is lower, is difficult to model and can slowly decrease the altitude and lead the satellite to burn up in the Earth's low atmosphere. The gravity recovery and climate experiment (GRACE) satellite (a satellite devoted to geodesy), which was launched at an altitude of 485 km, was thus at only 235 km altitude in 5 years, or in other words, it had a loss in altitude of 250 km in 5 years. For the Satellite pour l'Observation de la Terre (SPOT) satellites (h = 820 km), the daily loss in altitude is of the order of 1 m, which leads to monthly orbit corrections. Thus, remote sensing satellites are regularly subject to orbital maneuvers seeking to return them to their nominal altitude, which, in turn, needs more fuel in order to be performed (RADARSAT-1 was thus carrying 57 kg of propellant). Therefore, these are the multiple reasons that cause the majority of civilian remote sensing satellites to be placed in orbit at altitudes varying between 450 km (QuickBird 2) and 910 km (Landsat 1).

In the case of a perfectly circular orbit, the orbital period T is thus written as (equation [1.5]):

$$T = 2\pi\sqrt{\frac{(R_T + h)^3}{\mu}} \qquad [1.9]$$

and the speed V is written as (constant on a circular orbit, relation [1.7]):

$$V = \sqrt{\frac{\mu}{R_T + h}}$$

which gives the following table for different circular orbits:

Altitude (km)	Period (h)	Number of orbits per day	Speed (km/s)
400	1.33	15.56	7.67
500	1.35	15.22	7.61
600	1.37	14.89	7.56
700	1.39	14.58	7.50
800	1.41	14.28	7.45
900	1.43	13.98	7.40

It is important to remember the orders of magnitude: for a remote sensing satellite, the orbital period is of the order of 100 min, the speed is of 7.5 km/s and the number of orbits per day is around 15. More precisely, SPOT is at an altitude of 822 km: its period is 101.46 min, its speed is 7.424 km/s and it orbits 14.19 times a day.

We have seen that for an elliptic orbit, the distance to the focus and the speed are not constant. If we analyze the values of the eccentricities of the remote sensing satellite orbits, we notice that they are extremely small: for example, for SPOT, we have $e = 1.14 \times 10^{-3}$. In this case, the elliptical trajectories present some particularities:

– First, the shape of the orbit is nearly circular. If we use the Cartesian formula (described by the semimajor axis a and the semiminor axis b), we obtain $b = a\sqrt{1-e^2} \sim a - ae^2/2$. For SPOT ($a = 7,200$ km), we find a 4 m difference between a and b: the ellipse, therefore, has a semicircular shape.

– If the orbit is elliptic, it is not by its shape that we know this, but because the Earth is not at the center of the orbit but at its focus. The distance d between the center and the focus is given by the relation $d = ae$, which gives SPOT a value of 8.21 km. Between the perigee and the apogee, the altitude will therefore vary by approximately 16.4 km, which will have consequences on the sensors whose performances depend on the altitude. It is interesting to compare the deviation from the circle (several meters) and the variations in altitude (10 km).

– Finally, the speed of the satellite will not be constant. Using relation [1.8], we show that for SPOT, the speed varies between 7.416 km/s at the apogee and 7.432 km/s at the perigee: the deviation seems very small, but we must note that at the end of 1 s the difference in the trajectory is approximately 17 m, i.e. nearly 2 pixels.

1.3.1. *The orbit in the terrestrial referential: the recurrence cycle*

In the hypothesis that the Earth can be modeled by a punctual mass, the trajectory of a remote sensing satellite lies in a plane and its orbit is closed and quasi-circular. The orbital plane has a well-defined orientation that we determine using a unitary vector $\vec{n}$ positioned at the focus (the Earth center) and perpendicular to this plane: in this orbital plane, the period T of the satellite is determined by its altitude.

The Earth is also characterized by a daily rotation around an axis Oz (by definition, the north–south axis) defined by a unitary vector $\vec{k}_z$. It is, therefore, in relation to this axis that we define i, the "tilt angle" of the orbital plane (also called orbit inclination):

$$\cos i = \vec{n}.\vec{k}_z$$

For a terrestrial observer, the problem becomes more complex because the Earth turns around its axis. In a first approximation, we may notice that the observer belongs to a non-Galilean referential that completes one rotation per day. The relative motion of the satellite is no longer reduced to a simple circular motion, but it associates the proper motion of the satellite (rotation about $\vec{n}$) with the rotation of the Earth (rotation about $\vec{k}_z$).

Let us consider a satellite with a T period whose orbital plane tilt is different from zero. The orbit of this satellite will cross the equatorial plane at two points: it will then have two points at the equator covered by the satellite in one revolution. Let us consider one of these two points of the equator: at the end of a satellite revolution, this point, related to the Earth, will have traveled a distance $V_E T$ toward east, with $V_E = 463$ m/s being the speed of the Earth at the equator. If the period is of the order of 100 min we find a distance of 2,800 km. Therefore, when a revolution starts, the satellite flies over another point of the equator, that has not been previously flown over and that is located at approximately 2,800 km west.

The orbit of a satellite is usually described by its "track", defined by the points on Earth that have been flown over by the satellite. If the Earth did not have its own rotation motion, the track would be the intersection between the orbital plane and the terrestrial sphere: we would, therefore, have a circle whose center would be the center of the Earth. However, because the Earth has its own rotation motion, the track has a specific appearance that results from the combination of the rotation motion of the satellite and that of the Earth. Figure 1.1 shows a SPOT orbit on Earth, which illustrates the equatorial lag at the end of a period.

The purpose of a remote sensing satellite is to acquire information from the surface of the terrestrial globe, if possible in its entirety. To do this, let us consider a rotation of the equatorial plane so that the normal vector in the equatorial plane $\vec{n}$ is perpendicular to the Earth's axis: in this case study, the satellite flies over the north and south poles and we therefore speak of a "polar orbit". Because of the Earth's own rotation, it will be made to fly over different points of the equator with each revolution.

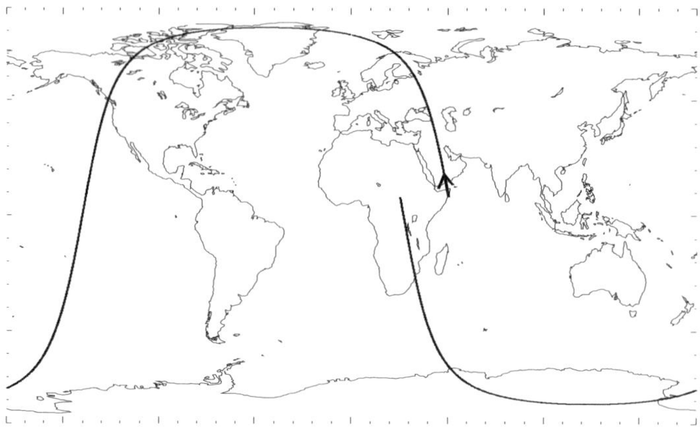

Figure 1.1. *A SPOT orbit represented in an Earth-bound referential*

We have seen that the T period of the satellite only depends on the h altitude: the designers of satellite missions, therefore, have the possibility of choosing the period by fine-tuning the altitude h. One of the essential points in the choice of altitude lies with p, the number of orbits per day (of a duration T_J):

$$p = \frac{T}{T_J}$$

which is, *a priori*, a real number. In the case where p is expressed as a rational fraction:

$$p = p_0 + \frac{r}{q} p_0 \in \mathbf{N} q \in \mathbf{N} r \in \mathbf{N} \qquad [1.10]$$

with r and q being coprime integers, we note that at the end of q days, the number of orbits N will be:

$$N = qp_0 + r$$

and then N is an integer. For a given ground point, exactly q Earth rotations will have passed and this point can then be found in the exact same configuration as before. As for the satellite, at the end of N orbits (N is an integer), it is found exactly on the same position on its orbital plane as before. Therefore, the satellite flies over exactly the same point; thus, there is a "cycle" in the relative motion of a satellite in relation to Earth, as soon as the number of orbits per day is written as a rational fraction (equation [1.10]). We say that the satellite is "phased" with a cycle of q days. In q days, the satellite will have flown over N points of the equator. These points are evenly spaced: if $\mathcal{C}$ is the terrestrial circumference ($\mathcal{C} = 40{,}000$ km by historic definition), we therefore have a distance $\frac{\mathcal{C}}{N}$ between these points.

The aim of a phased orbit is to be able to predict the satellite's position in the future (in order to be able to task data acquisitions) as much as in the past (in order to analyze the archives). This is the reason why space agencies choose this type of orbit that allows for efficient data management. Using the formula of circular orbits [1.9], let us calculate the usual altitudes for which $r = 0$ (i.e. the number of orbits in a day is an integer: therefore, the cycle of this satellite is 1 day). Thus, we obtain the altitudes of 561 km (15 orbits/day) and 891 km (14 orbits/day). We will see that there is a remote sensing satellite, Formosat-2, whose altitude is 891 km and cycle is 1 day, and which does 14 orbits/day. Table 1.1 gives some examples of orbital parameters.

Mission	Altitude (km)	Orbits by a day	Cycle in days	Orbits by cycle	Range between orbits at equator (km)	Daily lag
Landsat-1	917	$p = 13 + \frac{17}{18}$	18	251	159	159 km
Formosat-2	891	$p = 14$	1	14	2,857	0 km
SPOT	822	$p = 14 + \frac{5}{26}$	26	369	108	540 km
TerraSAR-X	515	$p = 15 + \frac{2}{11}$	11	167	240	480 km

Table 1.1. *Orbital parameters of some remote sensing satellites*

If q defines the cycle, r defines the "daily lag". To understand the importance of this, let us take the example of Landsat-1 ($h = 917$ km), for which $p = 13 + \frac{17}{18}$; its cycle is therefore 18 days. In 1 day, it completes a little less than 14 orbits: therefore, it needs more than an entire day to fulfill 14 orbits. Once the 14th orbit has been circled, the satellite will fly over a point that is slightly to the west of the one it had initially flown over. Slowly by slowly, this shift west becomes increasingly significant and at the end of 18 days, this shift will bring the satellite exactly above a track that had already been traveled.

Let us now take the example of SPOT ($h = 822$ km), for which $p = 14 + \frac{5}{26}$ ($p \sim 14.19$); its cycle is therefore 26 days. In 1 day, it completes a little more than 14 orbits: at the end of 14 orbits, a little less than a day will have passed. Once the 15th orbit has been circled, a little bit more than a day will have passed; the satellite will then fly over a point that is slightly to the west from the one it had flown over the previous day. At the end of $r = 5$ days, it will have completed $70 + \frac{25}{26}$ orbits and therefore its track at that moment will be very close to its initial track. SPOT is an "agile" platform: it can point its acquisition around the vertical axis so that the area acquired can be estimated on neighboring tracks. Because of this property, it can complete an acquisition on a given point on the Earth every 5 days, except that the acquisition mode is different from the one used initially.

1.3.2. *The effects of the Earth's flattening: the precession of the orbits*

The simplified model used thus far has supposed that the Earth was spherical and homogeneous and that, using the Gauss theorem, it could be considered as a punctual mass located in its gravity center. The reality, however, is completely different: the Earth is neither perfectly spherical nor is it homogeneous. Thus, since the beginning of astronautics, we have been able to observe that the trajectories were not Kepler's ideal trajectories and that we had to, therefore, introduce slight modifications into the elliptical orbits in order to better predict the positions of satellites.

The first modification that had to be considered was the Earth's flattening at the poles. A well-known fact for geodesists is that this flattening can be characterized through ground measurements; we know the Earth's radius is 6,378.137 km at the equator and 6,356.752 km at the poles. The Earth displays somewhat of an equatorial "roll". This being the case, we can no longer speak of a potential field in $\frac{1}{r}$ because we can tell from the latitude that the gravitational field will be slightly modified because of this "roll", so that the gravitational force is not necessarily directed toward the center of the Earth.

Without going into the theoretical modifications generated in the gravitational field U in detail, we can, however, say that, since the potential is no longer central, the trajectory of the satellite is no longer plane. In other words, the angular momentum $\vec{H}$ [1.4] no longer has a fixed direction. Therefore, we can show the following properties:

– The tilt of the orbit remains constant: we always have $\vec{n}.\vec{k}_z = Cste = \cos i$.

– $\vec{n}$ completes a precession motion around the north–south axis of the Earth, which we can study by considering the vector $\vec{l} = \vec{n} \wedge \vec{k}_z$. This vector $\vec{l}$ completes a uniform rotation $\dot{\Omega}$ around the north–south axis of the Earth, characterized by the relation:

$$\dot{\Omega} = \frac{\mathrm{d}\Omega}{\mathrm{dt}} = -\frac{3}{2}\sqrt{\frac{\mu}{a^3}}\frac{R_T^2}{r^2}J_2 \cos i \qquad [1.11]$$

where J_2 is a parameter defined on the basis of the Earth's flattening. This expression shows that if we fixed a (to have a well-defined period yielding a phased orbit), we could adjust $\dot{\Omega}$ due to this tilting of the orbit. Let us remark that for $i = 90^{\mathrm{o}}$, $\dot{\Omega} = 0$: symmetrically, the terrestrial roll effect can only modify the value of the angular momentum of the satellite $\vec{H}$ and not its direction.

Because of the Earth's flattening, we therefore have a precession of the orbit, the magnitude order of this precession is small *a priori*, because we find a value of the order of degrees per day for the quasi-polar low orbits ($i \simeq 98^{\circ}$).

1.3.3. *Heliosynchronous orbits*

Up to now, we have only considered the Earth as a sphere that rotates around its axis (period = 86,160 s). However, the Earth moves around the Sun in an orbit that is also slightly elliptic, with a period of 365.24 days, given that the duration of a day is 86,400 s. If we consider a satellite placed on a polar orbit whose rotation vector in an absolute referential has a fixed direction $\vec{n}$ (always pointing toward the same star in the sky), since the referential connected to the Earth completes a revolution in 1 year, the vector $\vec{n}$ describing the satellite will, therefore, also complete a revolution in 1 year (in this referential connected to the Earth).

When we introduced phasing for remote sensing satellites, it was so that we could easily use the archives of a satellite, or be able to define acquisition schedules. If we wish that all acquisition parameters be identical at the end of a cycle, the acquisition time must also remain the same. To do this, the vector $\vec{n}$ describing the satellite has a direction fixed in the referential related to the Earth: therefore, it goes around the Earth once per year in an absolute referential. We will then say that the satellite is "heliosynchronous": regardless of the day it passes over the equator, it will always be the same solar time.

One revolution per year becomes one rotation of the value $K = 0.9856^{\circ}/\text{days}$. This value is accessible due to the flattening of the Earth and the specific value of its coefficient J_2. By reversing relation [1.11], we obtain the "ideal" tilting value of the orbit for a quasi-circular orbit with a radius a or with an altitude h:

$$i = \text{Acos}\left(-\frac{a^{\frac{7}{2}} 2K}{3\sqrt{\mu}R_T^2 J_2}\right) = \text{Acos}\left(-\frac{(R_T + h)^{\frac{7}{2}}\, 2K}{3\sqrt{\mu}R_T^2 J_2}\right)$$

The tilting must be slightly larger than the value 90°, which causes this type of orbit to be retrograde. Table 1.1 gives values of orbit tilting for different heliosynchronous satellites. This tilt is close to 90° and the orbits remain quasi-polar (we sometimes call them *near polar orbits* (NPO)). Therefore, they allow for a good Earth coverage outside of the polar caps.

1.3.4. *Tracking the orbits*

Because of the flattening of the Earth, Kepler's laws cannot be applied. However, the effects of this flattening are weak and the orbit of a remote sensing satellite can be approximated at any time by an ellipse described by a focus (the Earth), an orbital plane (with a normal vector $\vec{n}$) and the parameters of the Kepler ellipse. The major axis of the ellipse is called the "line of apsides": this line obviously goes through the center of the Earth. We must now enhance this description to be able to position this ellipse in relation to the Earth.

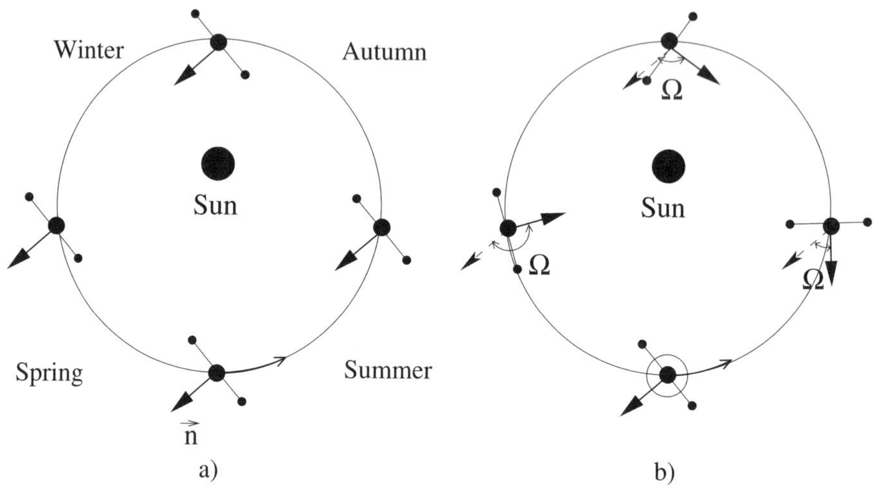

Figure 1.2. *Evolution of the orientation of the orbit of a polar satellite having an orbital plane with a direction given at the summer solstice: the orbital plane is represented by the ascending and descending nodes. a) The Earth is supposed to be perfectly spherical and homogeneous and the orientation of the orbit remains identical throughout the year. b) The Earth is flattened, modifying the orientation of the orbit throughout the year. In this example, the orbit is not heliosynchronous and it does not complete an entire revolution during the year*

To do this, we must define a frame related to the Earth and first characterized by a privileged direction: the north–south axis (with a directing vector $\vec{k_z}$) and in the equatorial plane (the perpendicular plane with $\vec{k_z}$ that the center of the Earth belongs to as well as the equator). The angle formed by a point on the Earth (described by the vector $\vec{r}$) and $\vec{k_z}$ allows us to define the latitude. The equatorial plane is then enhanced with a conventional tracking method based on the Greenwich meridian: the angle formed by the position of the meridian and a point of the equator gives us the longitude.

In this terrestrial frame, we will also have the following definitions:

– The "line of nodes" (ascending node and descending node) that is at the intersection of the orbital plane and the equatorial plane. The node line passes, by definition, through the center of the Earth. Conventionally, when it goes through the ascending node, the satellite flies toward the northern hemisphere.

– The “inclination” i of the orbit with respect to the equatorial plane, so that $\cos i = \vec{n}.\vec{k}_z$. Conventionally, the orientation of this angle is defined by the line of the nodes and the ascending node: if the inclination is equal to zero, the satellite turns toward the east.

– The “right ascension of the ascending node” Ω that expresses, in the equatorial plane, the angle between the fixed reference axis of the equatorial plane (given by the Greenwich meridian) and the line of nodes.

– Finally, the “argument of the perigee” ω that expresses, in the plane of the orbit, the angle between the line of nodes and the line of apsides (i.e. the line joining the perigee and the apogee). A more in-depth analysis of the field $U(r)$ will show that this angle is not constant, except for some kinds of very specific orbit called “frozen orbits” (SPOT, for example, has such an orbit).

For remote sensing heliosynchronous satellites, we often provide the time the satellite passes through the ascending node instead of the straight ascension of the ascending node (which is the same thing).

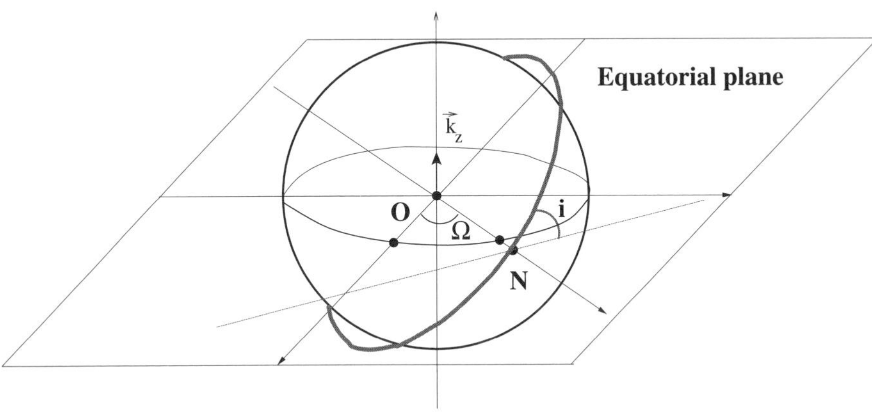

Figure 1.3. *Orbit of a satellite with inclination i and intersecting the equatorial plane at the ascending node N, in the Earth frame, which is defined by a reference meridian (Greenwich)*

A heliosynchronous orbit can, therefore, be described using five parameters: a, e, i, Ω and ω. When the orbit is circular, the description is limited to three parameters: h, i and Ω with $R_T + h = a$ (R_T being the Earth radius).

Even if the orbit is not strictly a Keplerian orbit, these descriptions are enough because the perturbations are weak and we can keep in mind the concept of an orbit that is locally plane and elliptical. In the case of a heliosynchronous orbit, we can have an idea of the non-flatness of the orbit. Indeed, let us consider an orbit with a period of 100 min: in 1 year, the satellite will have circled approximately 5,260 orbits. Since it is heliosynchronous, the ascending node will go round the Earth on the equator (40,000 km) in a year, which amounts to a shift toward the east of approximately 7.6 km/orbit.

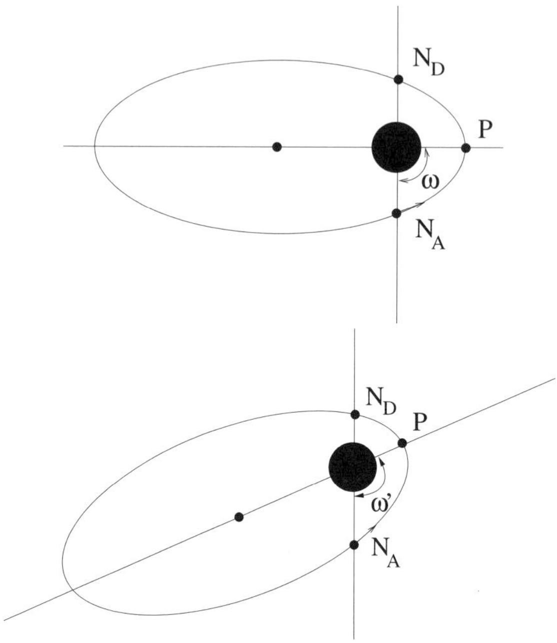

Figure 1.4. *Two examples of an orbit having the same orbital plane, even ascending and descending nodes, but whose arguments of the perigee ω and ω' are different. The figures are represented in the orbital plane. In these figures, only the node line belongs to the equatorial plane*

1.3.5. *Usual orbits for remote sensing satellite*

To summarize, we see that the space mechanics allow us to have around the Earth satellites on orbits that we can choose to be near circular, cyclical (when we choose specifically the altitude) and heliosynchronous (when we choose the tilting of the orbit). Table 1.2 gives several examples of the most usual satellite orbits, as well as the order of magnitude of the pixel size of corresponding images (for the satellites having several acquisition modes, the given value corresponds to the most current configurations).

These orbits display a weak eccentricity that has, nevertheless, several consequences on the altitude and the speed that slightly vary along the orbit.

Therefore, we have the following relations for the altitudes and the speeds at the apogee and the perigee:

$$\frac{R_T + h_A}{R_T + h_P} = \frac{v_P}{v_A} = \frac{1+e}{1-e} \sim 1 + 2e \qquad [1.12]$$

Satellite			Altitude (km)	Orbits by a day	Cycle (days)	Inclination	Pixel size
Landsat-1	1972	O	917	13+17/18	18	99.09°	30 m
Landsat-7, Terra	1999, 1999	O	705	14+9/16	16	98.21°	15 m
SPOT-1	1986	O	822	14+5/26	26	98.72°	10 m
SPOT-5	2002	O	822	14+5/26	26	98.72°	2.5 m
Ikonos	1999	O	680	14+9/14	14	98.11°	1 m
QuickBird	2001	O	450	15+7/8	8	97.21°	60 cm
Formosat-2	2004	O	891	14	1	99.00°	2 m
WorldView-1	2007	O	496			97.50°	50 cm
GeoEye	2008	O	684			98.10°	50 cm
Pléiades	2011	O	695	14+15/26	26	98.15°	50 cm
ERS, ENVISAT	1991, 2002	R	780	14+11/35	35	98.55°	12.5 m
RADARSAT-1	1995	R	798	14+7/24	24	98.58°	8 m
TerraSAR-X	2007	R	514	15+2/11	11	97.45°	2 m
Cosmo-Skymed	2007	R	619	14+13/16	16	97.86°	2 m

Table 1.2. *Orbital parameters and image characteristics of several remote sensing satellites*

1.4. Image acquisition and sensors

The objective of this section is to provide the basis for the essential concepts in satellite imagery, that is perspective ray, resolution and ground swath. These concepts will allow us to understand how linear sensors build satellite images because of the satellite's own motion.

Two sensor families are used in satellite imagery: optical sensors and radar "SAR" sensors (Synthetic Array Radar). The first sensors are passive sensors: they measure the backscattering of the solar light on the ground (or for the so-called thermal sensors the terrestrial ground's own radiation). The second sensors are active sensors: they emit an electromagnetic wave using an antenna, then they use the same antenna to receive the backscattering of this wave on the ground. Both of them are directional: they mainly process the information that comes from a given direction of observation called line of sight (LOS).

1.4.1. *Perspective ray in optical imagery for a vertical viewing*

In the case of optical sensors, the elementary sensor (a bolometer, a part of photosensitive film, a Charged Coupled Device (CCD) element, etc.) is a passive system that captures the photons coming from the LOS. This capture is done in a limited period of time T_{int} called the integration time (or exposure).

Since the propagation medium can be, upon a first approximation, considered homogeneous, the celerity of the light is then constant and the light is propagated in a straight line. To each given LOS corresponds a straight line coming from the sensor. For an ideal elementary sensor, only the objects that are located on this straight line contribute to the response of the sensor: we will call this straight line a "perspective ray". Depending on the wavelength analyzed, the contribution of the objects can be the backscattering of the sunlight (i.e. for the wavelengths ranging from visible to middle infrared) or the object's heat radiation (in thermal infrared). The solar illumination being incoherent (as is the heat radiation), the intensity measured by a sensor is the sum of these elementary contributions during a given duration called "integration time" T_{int}.

Actually, the optical system (made of lenses and mirrors) is subject to diffraction laws that define a cone around this perspective ray; this way, any object in this cone contributes to the response being measured. An elementary sensor is, therefore, characterized by a perspective ray (defined by a direction) and an angular aperture $\delta\omega$ (called instantaneous field of view (IFOV)). This angular aperture can usually be defined by a relation that links the aperture L of the optical system (the dimension of the mirror or of the lens) and the analyzed wavelength λ. For circular optics, the angular aperture is expressed as the main lobe of a circular aperture (whose impulse response – or *point spread function* (PSF) – is expressed as a Bessel function). We can choose the following relation as an expression of the angular aperture, which is valid for a monochromatic radiation:

$$\delta\omega = 1.22\frac{\lambda}{L}$$

LOS and IFOV define a cone that characterizes the "resolution"[1] δr of the system in any point of the perspective ray. The optical system points toward the ground of the Earth: the cone associated with the perspective ray then intersects the ground on a small surface, called a "footprint" (FP).

---

1 Let us recall a definition of resolution: it is a measure of the ability of the instrument to separate images of two identical neighboring object points.

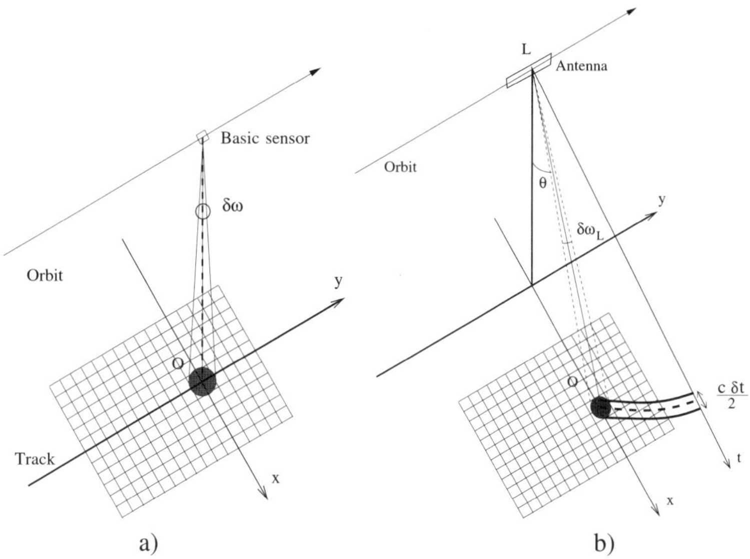

Figure 1.5. *a) Optical monosensor viewing the ground vertically: the perspective ray (the dotted line) and the LOS correspond to the same line. The area intersected on the ground has a surface that is proportional to* $(R\delta\omega)^2$*. b) Radar monosensor viewing the ground with an angle* θ*: the perspective ray (dotted line) is perpendicular to the LOS because it corresponds to an isochronous line for the wave coming from the sensor; the area intersected on the ground, i.e. the footprint, has a surface that is proportional to* $\frac{R\delta\omega_L c}{2\delta t}$

It is worth noting that given the celerity of the light, the flight time between the objects belonging to the cone associated with the perspective ray and the sensor is much shorter than the time of the sensor's integration. Therefore, all of these objects will contribute to the same output value, regardless of their range from the sensor.

One last important point remains, however, to be noted: in the most general cases, light rays cannot go through objects. Thus, only the backscattering object closest to the sensor will be observed: the other objects located on the perspective ray will then be hidden and will have no contribution.

1.4.2. *Perspective ray in radar imaging*

A radar imaging system is an active system based on a completely different principle: echolocation. The antenna emits a wave of the shortest possible duration and then acts as a receiver to capture the echoes produced by elementary reflectors (punctual objects, surfaces and volumes). As the celerity of electromagnetic waves through the atmosphere can be assumed to be constant, it is easy to convert a delay measured upon the received signal into a range. Thus, spatial positions that belong to a isosynchronous surface, defined as the sphere whose center is the radar antenna, correspond to a delay given by the real signal (reference time).

This isochronous surface has a range bounded by diffraction laws. In the most general cases, the antenna is rectangular, with a size $l \times L$: the LOS is perpendicular to the antenna and defines the "range axis". In this chapter, we will assume that the side of the antenna corresponding to the size L is aligned with the trajectory of the satellite; thus, the axis defined is called the "azimuth axis". Thus, the aperture defined is expressed as the main lobe of a rectangular aperture (whose PSF is expressed as the product of two cardinal sinuses). It has an extension $\delta\omega_L$ along the azimuth direction, and $\delta\omega_l$ along the third direction (that is perpendicular to the range axis and the azimuth axis). We then have the relations:

$$\delta\omega_l = \frac{\lambda}{l}\delta\omega_L = \frac{\lambda}{L}$$

defining the "main lobe" of the antenna. Given the orientation of the antenna in relation to the track, this main lobe points perpendicularly to the track.

On the isosynchronous surface, every backscattering object illuminated by the radar will then re-emit the incidental wave. The contribution of all backscattering targets belonging to this isochronous surface defines the nature of the signal received by the antenna: thus, we can call this isochronous surface a "perspective surface". We must also note that the illumination is coherent (single, quasi-monochrome source): so the received signal must be seen as the sum in amplitude (and not the sum in energy) of the elementary signals backscattered by each elementary target.

In the plane that is perpendicular to the track, this perspective surface is reduced to a curve that we can call a perspective ray: all the points located on this perspective ray will have a contribution, arriving at the same time on the antenna, therefore, without being resolved. The spatial limitation of this perspective ray is due to the diffraction laws applied to the antenna. If we place ourselves far enough from the antenna, the perspective surface can be locally approximated by a plane, which we can call the "perspective plane": the perspective ray then becomes a perspective straight line, perpendicular to the LOS of the antenna.

Now we can see the essential difference between a radar sensor and an optical sensor. For the optical sensor, the perspective ray comes from the sensor and is mingled with the LOS; as soon as a diffusing object is located on the perspective ray, it covers the other objects located further on the perspective ray. For a radar sensor, the perspective sensor is perpendicular to the LOS. All the objects located on this perspective ray can contribute to the received signal.

A real radar system cannot distinguish, from a temporal point of view, two events that are very close to each other: it has a temporal resolution δt. We will see that it is the shape of the emitted wave that dictates temporal resolution. Knowing the celerity of electromagnetic waves, we deduce the spatial resolution along the range axis $\delta r = \frac{c\delta t}{2}$.

Thus, instead of an isochronous surface, we have, at time t, corresponding to a range $R = \frac{ct}{2}$, a small volume around this isochronous surface, with an extent $\delta r = \frac{c\delta t}{2}$ along the range axis and the extension $R\delta\omega_L$ along the azimuth axis. All the backscattering objects belonging to this small volume will be mixed in the same output value.

The antenna points toward the Earth and the intersection of this small volume with the ground defines the footprint: it is worth noting that along the range axis, the spatial area depends on the shape of the wave and the angle of incidence and that along the azimuth axis, the spatial area depends on the dimension L of the antenna.

As in the optical case, we can note that the matter absorbs electromagnetic waves; thus, masking elements can appear as soon as absorbing objects are in the LOS of the antenna.

Upon a first analysis, it is interesting to note that a perpective ray can be defined for an optical system as much as for a radar system, but with a fundamental difference in relation to the LOS:

– the optical perspective ray comes from the sensor and is therefore identical to the LOS;

– the radar perspective ray is perpendicular to the LOS.

Thus, analyzed in the plane perpendicular to the track, the perspective ray of an optical system that regards a ground point along an incidence θ and the perspective ray of a radar system whose LOS is $\pi/2 - \theta$ are identical (Figure 1.6).

In the two cases, the masking effects will appear along the LOS, which indeed corresponds to the optical perspective ray, but does not to the radar perspective ray.

1.4.3. *Resolution and footprint*

1.4.3.1. *The case of optical imagery*

Let us consider an optical system in a vertical LOS (Figure 1.7): it is characterized by the perspective cone defined by the perspective ray and the angular aperture $\delta\omega$ (due to the diffraction of the optical system). At any given range R, two objects belonging to this cone cannot be resolved: we can choose as the definition of the resolution δx, the "footprint", which is a function of the range R:

$$\delta x = FP_{optical} = R\delta\omega \quad [1.13]$$

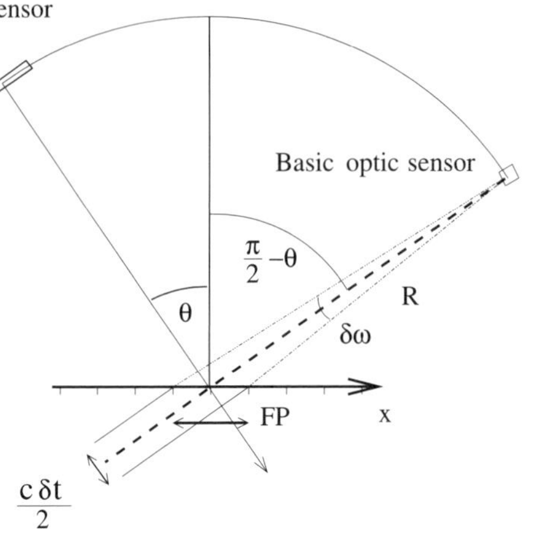

Figure 1.6. *Optical monosensor and radar monosensor sharing the same footprint and the same perspective ray. These two footprints can be similar if the incidence angle is θ for the radar and the incidence angle is $\pi/2 - \theta$ for the optical sensor. The figure belongs to the plane perpendicular to the track*

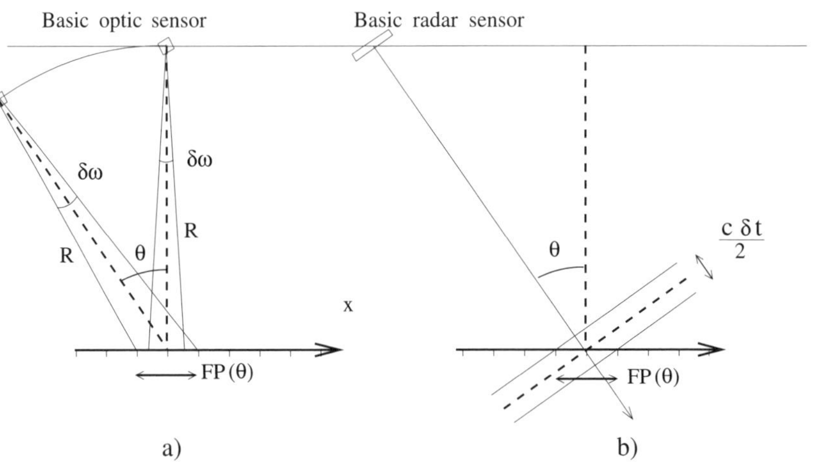

Figure 1.7. *a) Footprint in optical imagery. b) Footprint in radar imagery*

1.4.3.2. *The case of radar imagery*

Similarly, the perspective ray of a radar system of a temporal resolution δt viewing a ground point with a local incidence θ will have a footprint FP equal to (Figure 1.7):

$$FP_{radar}(\theta) = \frac{\delta r}{2\sin\theta} = \frac{c\delta t}{2\sin\theta}\theta \neq 0 \qquad [1.14]$$

If several objects belong to the same radar FP, their individual contributions will add up coherently (real parts and imaginary parts will add together independently) and their contributions will be mingled. These objects retransmit data toward the antenna in the same way as the footprint was a "ground antenna" emitting a wave: however, unlike usual antennas, the amplitude and phase laws on this "ground antenna" do not verify any simple law, which prevents the calculus of the diagram of this ground antenna.

As in the optical case, we can choose as definition of the resolution δx along the direction Ox (swath), the footprint, which depends on the incidence angle θ:

$$\delta x = FP_{radar}(\theta) = \frac{c\delta t}{2\sin\theta}\theta \in]-\frac{\pi}{2};\frac{\pi}{2}$$

1.4.3.3. *Relations verified by the footprint: the optical case*

We have seen that an optical system observing the Earth is characterized by a footprint whose dimensions are deduced from the resolution of the system. We will now see that the geometry of the footprint depends on the range between the sensor and the ground, as well as the local incidence θ, i.e. the angle between the LOS and the ground.

– If we consider the same system at two different altitudes and viewing the Earth with a null incidence (the perspective ray is then perpendicular to the surface of the Earth), the optical laws will then easily demonstrate that the size of the footprint varies linearly with the range. For two ranges R and R', the footprints $FP(R)$ and $FP(R')$, verify:

$$FP(R') = FP(R)\frac{R'}{R} \qquad [1.15]$$

which is shown in Figure 1.8.

– Let us now model how the footprint varies with the local incidence. Let us assume that the ground presents a local slope α: the local incidence is then no longer null. Since in this configuration the range R remains the same, it is easy to show that in this case we have:

$$FP(\alpha) = \frac{1}{\cos\alpha}FP(\alpha = 0) \qquad [1.16]$$

which is shown in Figure 1.9.

– Let us now consider the case where the sensor is at a given altitude $H = R(\theta = 0)$ and where we modify the LOS of the sensor. On the one hand, by assimilating the Earth to a plane, the range between the sensor and the Earth is written as:

$$R'(\theta) = \frac{R(\theta = 0)}{\cos\theta}$$

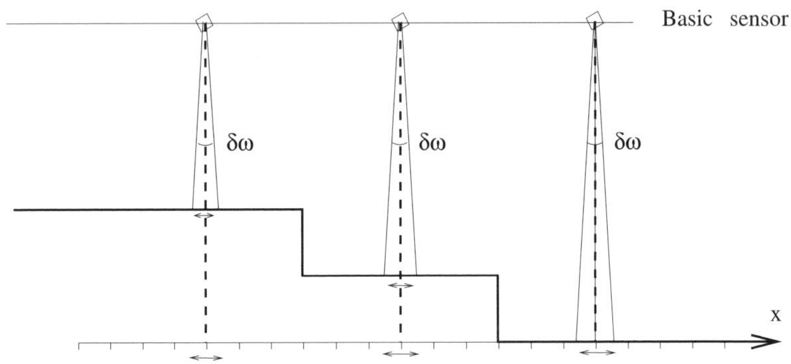

Figure 1.8. *Optical monosensor viewing the ground vertically from different altitudes: the footprint varies linearly with the altitude (equation [1.15])*

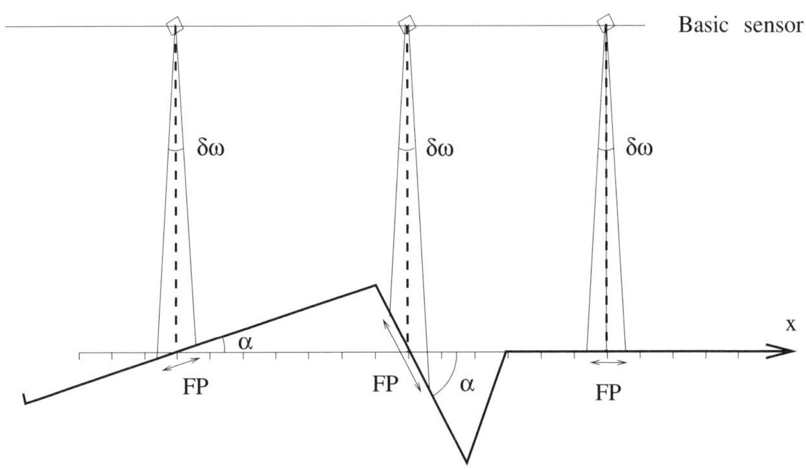

Figure 1.9. *Optical monosensor viewing the ground vertically. As the ground is not horizontal, the footprint depends on the local slope α (equation [1.16])*

On the other hand, the perspective ray will have a local incidence θ. By then applying the two previous formulas, we deduce:

$$FP(\theta) = \frac{1}{\cos^2\theta} FP(\theta = 0) \qquad [1.17]$$

We see that, for an optical system, the closer the incidence is to the vertical and the lower the altitude of the satellite, the more the FP is reduced.

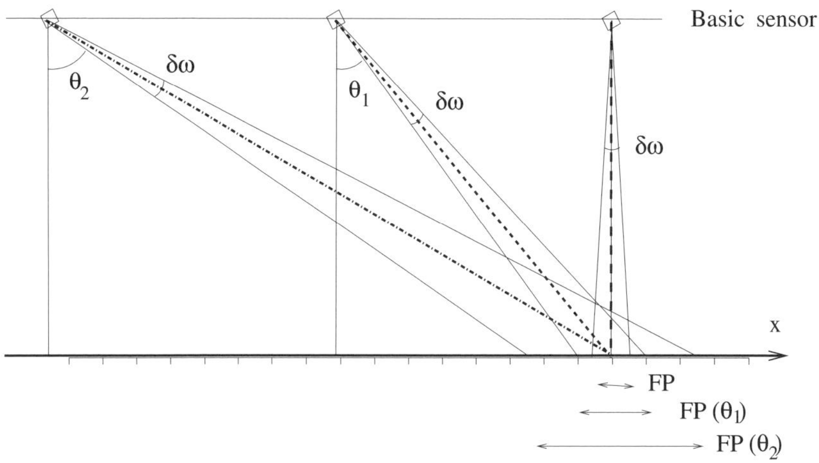

Figure 1.10. *Optical monosensors of the same angular aperture $\delta\omega$, at the same altitude and viewing the same point. The local incidence angle θ depends on the LOS. The footprint is proportional to $\frac{1}{(\cos\theta)^2}$ (equation [1.17])*

1.4.3.4. *Relations verified by the footprint: the radar case*

Similarly, the footprint of radar systems verifies some specific relations. However, these relations differ from the specific relations of optical imagery and it is important to emphasize them beforehand.

The echolocation signal is a temporal signal of a duration δt. For a system viewing a ground point with the local incidence θ, the radar footprint along the range axis is written as (relation [1.14]):

$$FP_{radar} = \frac{c\delta t}{2\sin\theta}$$

It is important to emphasize that it does not depend on the range between the ground point and the sensor.

In the presence of local slopes, the footprint verifies the following relation:

$$FP_{radar} = \frac{c\delta t}{2\sin(\theta - \alpha)}$$

where α is the local slope of the ground. The footprint is so much larger that the local slope orients itself in the direction of the sensor. If the footprint is larger, its backscattering will be so much higher as a large number of elementary targets belong to this footprint. Figure 1.11 illustrates this relation.

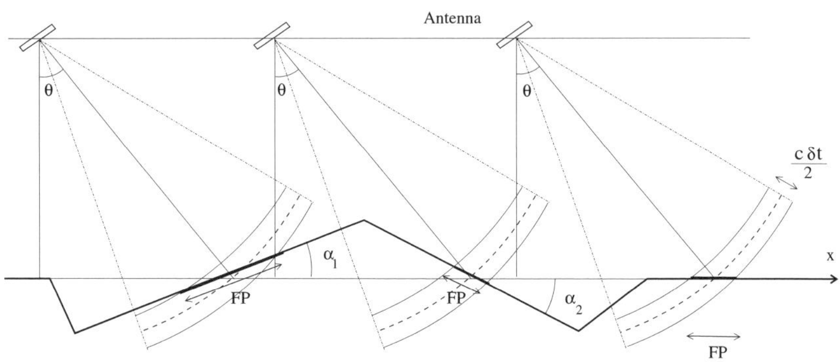

Figure 1.11. *Radar antennas with the same temporal resolution, the same altitude and the same emission angle, viewing Earth areas with various local slopes α. The footprint is larger as the local slope orients itself in the direction of the sensor*

Finally, Figure 1.12 shows how the footprint varies depending on the local incidence angle, in the case of antennas placed at an identical altitude. We then have the relation:

$$FP(\theta_2) = \frac{\sin\theta_1}{\sin\theta_2} FP(\theta_1) \qquad [1.18]$$

1.4.4. *The swath in satellite imagery*

An elementary sensor will not be of much use in satellite imagery since it only gives information on a very small area of the Earth: the footprint. Thus, a system of satellite imagery has some abilities allowing information acquisition on a wider area, and eventually the construction of an image. The first stage that must be overcome is passing a piece of information acquired on a point to a piece of information acquired on a line.

For an optical sensor, the acquisition of a line of data is carried out by using several LOS, which in turn allows us to acquire information that comes from different perspective rays. This set of perspective rays will be called the "perspective bundle": the area on the ground intersected by this perspective bundle is called the "swath".

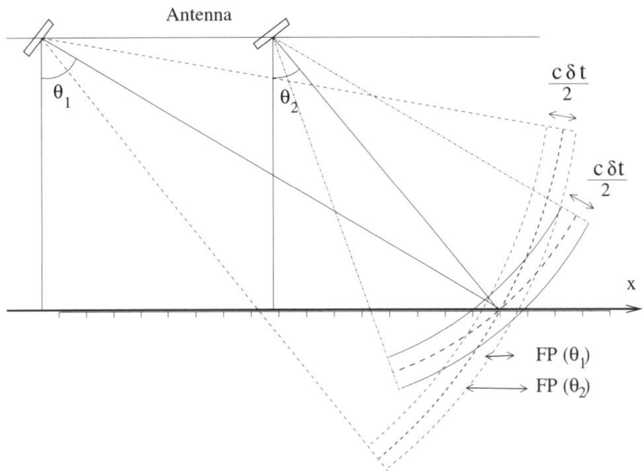

Figure 1.12. *Radar antennas with the same temporal resolution and the same altitude, viewing the same area with two incidence angles θ_1 and θ_2. The law followed by the footprint is in* $\frac{1}{\sin\theta}$

Historically, the first satellite sensors were based on a unique sensor (for example a microbolometer) and the LOS was modified using a circular scanning obtained by the rotation of a mirror. Thus, we construct the perspective bundle step by step, this bundle being a plane perpendicular to the mirror rotation axis. Let us note that this acquisition mode was very well adapted to analog signals used both for acquisition and transmission to the ground reception stations.

The introduction of numeric sensors (CCD sensors) allowed us to increase the number of elementary sensors. Historically, the first CCD system was the "whiskbroom" where the perspective bundle is always obtained by the rotation of a mirror but where a small number of elementary sensors are aligned according to the mirror rotation axis; thus, we have the possibility to acquire several lines at once, during the rotation of the mirror. For example, the MODIS sensor (for example on board the Terra satellite) is fitted with 30 CCD elements allowing us to acquire 30 image lines in approximately 4 s, i.e. half a mirror rotation.

Nowadays, optical systems are generally based on a linear sensor made up of a large number of CCD elements (several thousand). This CCD array belongs to the focal plane of the optical system and in a perpendicular plane to the motion of the satellite (that is the plane that is perpendicular to the rotation axis of the first systems): we then speak of a "pushbroom" sensor. The number of elements has increased considerably in recent years: if SPOT-1 had a CCD array with 6,000 elements, QuickBird has six CCD arrays, each with 4,736 CCD elements. To each CCD element corresponds a different LOS and therefore a different perspective ray. All of these perspective rays make up the perspective bundle, defined by an angular aperture Ω. In the case where, in a first approximation, we consider the Earth as a flat surface, the swath, defined by the intersection of the perspective bundle and the ground, has a length $L_S \simeq R\Omega$.

For the radar imagery, there is an unique antenna. After each emission, the received signal is sampled with a time step δt and analyzed on a duration T_W much larger than the time step δt. The time swath is based on these time samples. The ground swath is obtained by projecting this time swath on the ground. The nearest point of the ground swath is called the near range and the farthest point, the far range. By assuming the Earth is flat, the swath can be expressed as:

$$L_S = \frac{cT_W}{2 \sin \hat{\theta}}$$

$\hat{\theta}$ being the average incidence angle.

Let us note that the diffraction laws define a main lobe to every radar antenna. Only the area of the Earth belonging to the main antenna lobe can contribute to the backscattered signal so that there is a physical limit to the duration of the time swath T_W.

1.4.5. *Images and motion*

We have seen in section 1.3 that a remote sensing satellite belonging to a circular orbit moves at a constant speed of $V_S \simeq 7.5$ km/s. This motion specific to the so-called non-synchronous satellites depends on the orbital parameters and allows them to fly over part of the Earth or the entire Earth. This motion can thus be used to build the images.

Indeed, let us consider, at the time T, a non-synchronous satellite fitted with a CCD array (Figure 1.14): it acquires a line (corresponding to the swath at the time T) during the integration time T_{int}. Let a time interval $\Delta t \geq T_{int}$. The system can then acquire a new line at the time $T + \Delta t$. Since, on its orbit, the satellite will have moved a distance equal to $V_S \Delta t$, this new swath will be at a distance $V_{S,ground} \Delta t$ from the previous one, $V_{S,ground}$ being the relative speed of the satellite on the Earth.

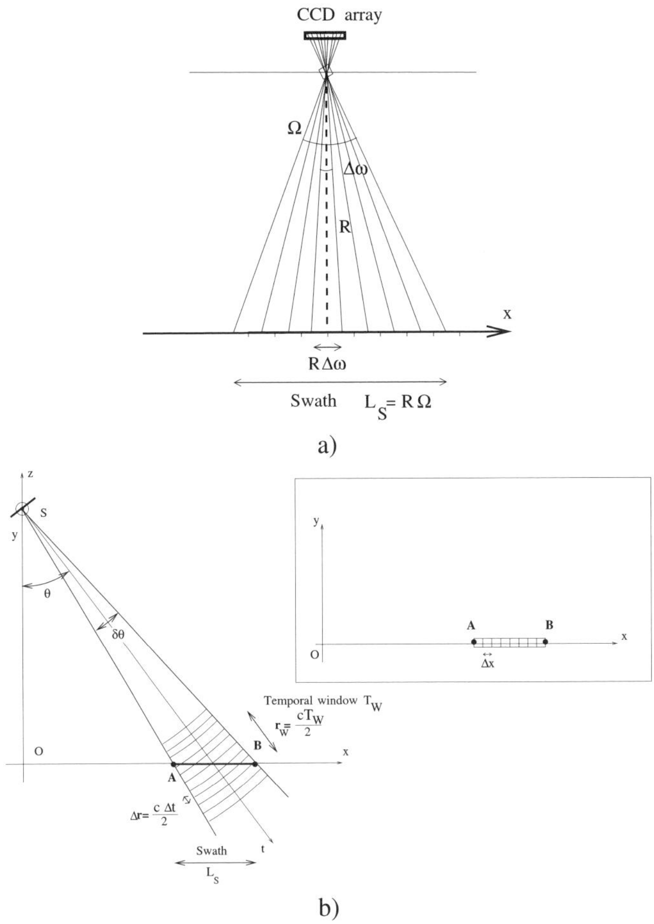

Figure 1.13. *Definition of the swath in a plane perpendicular to the satellite motion. a) An optical sensor based on a CCD array and viewing the ground vertically. The aperture Ω defines the swath ($R\Omega$). b) Radar antenna aiming along the incidence θ with a sidelobe $\delta\theta$. The window of temporal analysis is T_W: it corresponds, on the LOS, to a range r_W which, projected on the ground, results in a swath AB with a length L_S*

We can, therefore, build an image starting from these lines acquired at different times and by using a simple linear sensor: this is the reason why the CCD 2D matrices are only marginally used in remote sensing.

In the case of a radar sensor, if the duration T_F corresponds to the length of the swath L_S corresponding to the points on the Earth backscattering the emitted wave, we can emit these waves at a specific frequency called the pulse repetition frequency (PRF). One of the conditions is that the duration associated with the PRF, T_{PRF}, should be higher than T_F (if not, the signals coming from some points of the swath would be mixed from one emission to another). Thus, we find ourselves with a configuration analogous to the optical case: the satellite having moved a distance equal to $V_S T_{PRF}$ between each radar emission, we can thus build an image starting from these consecutive emissions. Each line is then at a distance $V_{S,ground} T_{PRF}$ from the previous one, $V_{S,ground}$ being the relative speed of the satellite with respect to the Earth. This radar image construction is called "stripmap mode".

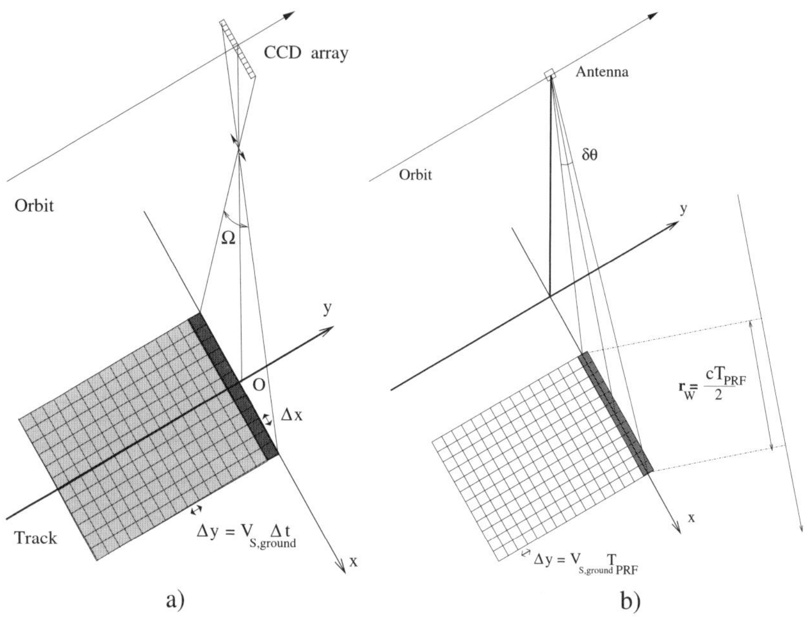

Figure 1.14. *a) Principle of the optical pushbroom. The CCD array acquires pixel lines periodically (period Δt), spatially sampled along the direction of the swath. b) Principle of the construction of a radar "strip map" image, assuming that the wave fronts are parallel. The radar acquires pixel lines periodically (period $\Delta t = 1/F_{PRF}$), time sampled at the frequency F_e*

1.5. Spectral resolution

1.5.1. *Introduction*

Here, we purposely introduce only a few details on the notion of spectral resolution, as it will be developed in more detail in Chapter 3.

Optical sensors measure the quantity of light reflected by the surface observed in a given range of wavelengths. For example, the wavelengths of the visible light range between 400 and 700 nm. If these wavelengths are observed together, we speak of a panchromatic image (containing all the colors). If this range of wavelengths is divided into sub-bands in order to be able to measure the light intensity separately in each band, we speak of a multispectral image. Often, in the case of the visible spectrum, the chosen bands correspond to the colors blue, green and red, which allows us to rebuild images in "natural colors" by composing these three channels. Other choices can be made, as we will see, for example, in Chapter 3 for the Formosat-2 sensor (whose five bandwidths will be detailed in Figure 3.1).

The multispectral image thus allows us to better characterize the observed surfaces, because it allows us to separate the contributions in different wavelength ranges. Thus, the majority of multispectral sensors contain a near infrared band, since the vegetation has a strong response in these wavelengths.

The wavelengths observed in optical do not stop at the visible and the near infrared (NIR). We can, thus, implement sensitive sensors in the short wave infrared (SWIR) and in the thermal infrared (TIR). Thus, the number of bands can go over tens or even hundreds: then, we speak of super and hyperspectral sensors, respectively, the limit between the different categories not being very well defined. We then obtain a very precise signature of the observed materials.

1.5.2. *Technological constraints*

The technology used for building sensors depends on the wavelength we are interested in. The sensitivity of the materials is indeed different in the different wavelengths and the difficulty of implementing sensors is not the same in all areas of the electromagnetic spectrum.

To simplify, the ability of a sensor to measure light with a good signal-to-noise ratio depends on the flux of received light. This flux is proportional to the surface of the sensor and to the bandwidth. The superspectral and hyperspectral sensors require a thin division of the spectrum. The price to pay is a loss in the spatial resolution, since the footprint must be of a larger size to allow a reasonable number of received photons to be detected.

1.5.3. *Calibration and corrections*

Calibration is the process of converting the magnitude provided by the sensor (a voltage) into a measure of the observed physical phenomenon. If we are interested in the light flux, a first interesting transform consists of the transformation of digital numbers into radiances (normalized flux related to the surface of the sensor). The light flux measured by the sensor contains contributions from several sources. There is the sunlight, the different atmospheric reflexions and, finally, the quantity of light emitted by the observed scene.

To gain access to this last magnitude, the measure issued from the sensor must suffer a set of transformations and corrections. The effects that are most difficult to correct are those induced by the light traveling through the atmosphere. Beyond the fact that the terrestrial atmosphere is opaque for some wavelengths – we are not interested in these spectral bands – the optical depth of the atmosphere varies with time, space and wavelength.

The calibration and adjustment processing is detailed in Chapter 3, section 3.6 for the optical data, and in Chapter 4, section 4.4 for the radar data.

1.5.4. *Image transmission*

The images acquired by the satellites must then be transmitted to the ground in order to be exploited. This transmission is done via a radio link (often in X band or Ka band) whose bandwidth (data transmission rate) is limited.

The volume of images acquired depends on the following factors: the number of pixels (field of view and spatial resolution), the number of spectral bands and the radiometric resolution of the number of quantification levels used to encode the values of pixels.

From these elements, we can deduce the size of a rough image. Often, these images are compressed on the satellite, before transmitting them on the ground. This compression allows us to reduce the volume of data to be transmitted, while minimizing information loss.

Transmitting images to the ground needs a radio link to be established between the satellite and a reception station on the ground. The majority of the remote sensing satellites being on polar orbits, it is impossible to always have ground stations that can be seen from these moving satellites.

This is why the images are often stored in a memory on board the satellite while waiting for the passage above a reception station. The size of this memory limits the number of images that can be acquired by the satellite during each orbital period.

Another parameter that limits the acquisition capacity of the satellite is the data transmission to the ground. It is indeed necessary for the satellite to be able to empty its memory during the visibility duration of the ground stations.

Chapter 2

Image Geometry and Registration

In Chapter 1, we outlined the background concepts and the constraints of orbital imaging systems. In this chapter, we will analyze how an image is built, how the space agencies distribute this type of data to end users and what kind of processing is required in order to ensure an optimal exploitation.

2.1. The digital image and its sampling

2.1.1. *Swath sampling*

2.1.1.1. *The case of optical images*

We have seen that the swath is defined by an angular aperture Ω intersecting the ground where a row of the numerical image is acquired. This acquisition requires a spatial sampling step that is carried out because of the sampling produced by the elements of the CCD matrix.

The CCD matrix is, therefore, composed of a large number of elementary sensors, spaced from one another at a constant distance of d. Knowing the focal distance of the instrument f and the distance to the ground R, by approximating $f << R$, we then obtain a sampling of the swath in "ground cells" of the size Δx:

$$\Delta x = \frac{f\,R}{d}$$

Chapter written by Jean-Marie NICOLAS and Jordi INGLADA.

corresponding to an angular aperture $\Delta\omega$:

$$\Delta\omega = \frac{\Delta x}{R} = \frac{f}{d}$$

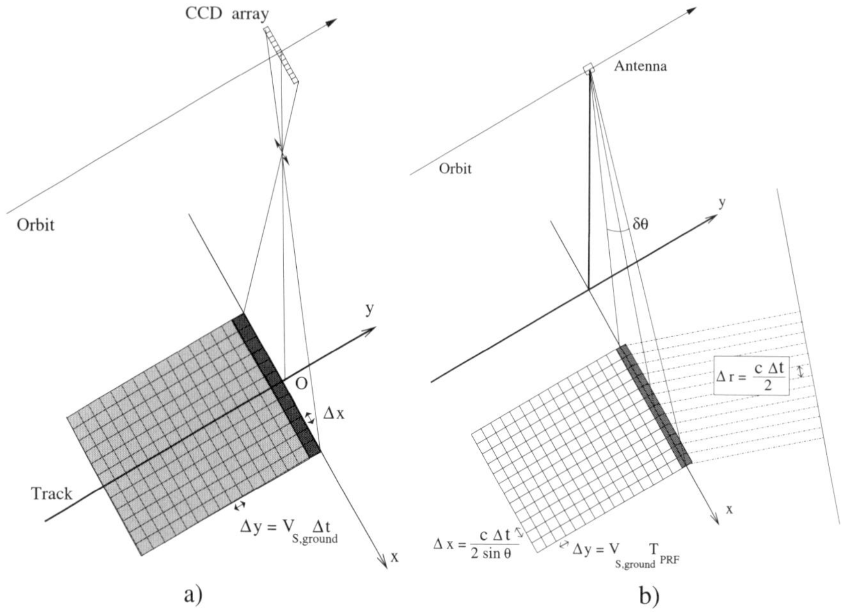

Figure 2.1. *a) Optical pushbroom principles. The CCD linear detector acquires, at a period Δt, pixel rows sampled with a Δx spacing along the direction of the swath. The concatenation of the acquired rows allows us to generate an image, the interval between ground rows being given by $\Delta y = V_{S,ground}\Delta t$, $V_{S,ground}$ being the projection of the speed of the satellite on the ground along the track. b) Radar image construction principles in* strip map *mode (assuming that the wavefronts are parallel). At a period $\Delta t = 1/PRF$, the radar acquires elementary rows, which, when concatenated, will give the final image. These elementary rows are the result of a sampling of the time signal received at a given frequency F_e, which leads to a time cell Δt and to a ground cell Δx*

We have previously seen that the physical characteristics of the optical system lead to defining an angular aperture $\delta\omega$ and the ground resolution δx. The Shannon criterion will, therefore, result in:

$$\Delta x \leq \delta x$$

For a swath L_S, the number of CCD elements N must verify:

$$N\Delta x = L_S$$

2.1.1.2. *The case of radar images*

In the case of radar data, we have seen that the useful received signal had a duration T_W characteristic of the swath. This signal is sampled at a given frequency F_e defining a time cell Δt:

$$\Delta t = \frac{1}{F_e}$$

A radar system is defined by a time resolution δt. To verify the Shannon criterion, we must have:

$$\Delta t \leq \delta t$$

As the resolution of an SAR system depends only on its bandwidth B (see, for example, [MAS 08]), the Shannon criterion is then verified by the relation:

$$F_e \geq B$$

Since the radar is an echolocation system, we associate a "range cell" Δr with this time cell, via the relation:

$$\Delta r = \frac{c\Delta t}{2}$$

Through ground projection, we obtain the following corresponding "ground cell":

$$\Delta x = \frac{c\Delta t}{2 \sin \theta}$$

This only depends on the local incidence angle and on the sampling frequency (we must note that, as opposed to optical systems, it does not depend on the distance R).

For a swath L_S, the number of ground cells N must verify the relation:

$$N\Delta x = L_S$$

2.1.2. *The pixels in optical imagery and in radar imagery*

A pixel of the image is, thus, defined by the sampling of the swath, but also by a sampling along the track. Both for optical system and radar, the size of the pixel along the track is induced by the acquisition rate of the data and the speed of the satellite. We obtain the relation:

$$\Delta y = V_{S,ground} \Delta t$$

$V_{S,ground}$ being the projection of the speed of the satellite on Earth. This step in the sampling process must also verify the Shannon criterion:

– For an optical sensor (the lens aperture is generally circular), we therefore must have:

$$\Delta y \leq R \delta \omega$$

– For a radar sensor, we must also have:

$$\Delta y \leq \frac{L}{2}$$

The latter relation is the result of a specific kind of processing, called SAR focusing, that allows us to overcome the constraints related to the limited size of the physical antenna (see, for example, [MAS 08]) and that gives an azimuthal resolution to such a system that is equal to $L/2$.

2.2. Sensor agility and incidence angle

2.2.1. *Agility of optical sensors*

2.2.1.1. *First agility concepts: the SPOT case*

The first remote sensing optical satellites were equipped with sensors viewing the ground vertically: this is the case of the Landsat systems whose swath (around 180 km) allowed for every point of the equator to be revisited in 16 days (in this case, the revisiting time is the same as the cycle time). This value being too high for some remote sensing applications, SPOT-1 was innovative in 1986, being equipped with an agile sensor: using a mirror, the pointing of the sensor can be done with an angle that varies between $-27°$ and $27°$. In comparison with Landsat (60 km), if the swath is reduced, it can be relocated inside an "agility corridor" of around 800 km length (Figure 2.2, bottom). Because of this agility, every point of the equator can be imaged in approximately 5 days: this revisit time is then much lower than the duration of the cycle (26 days).

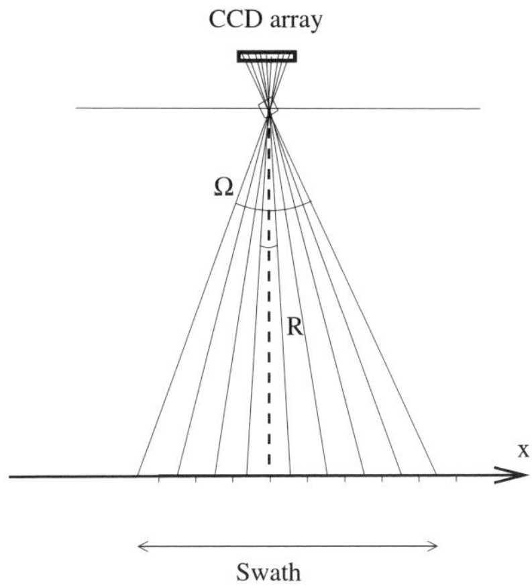

a)

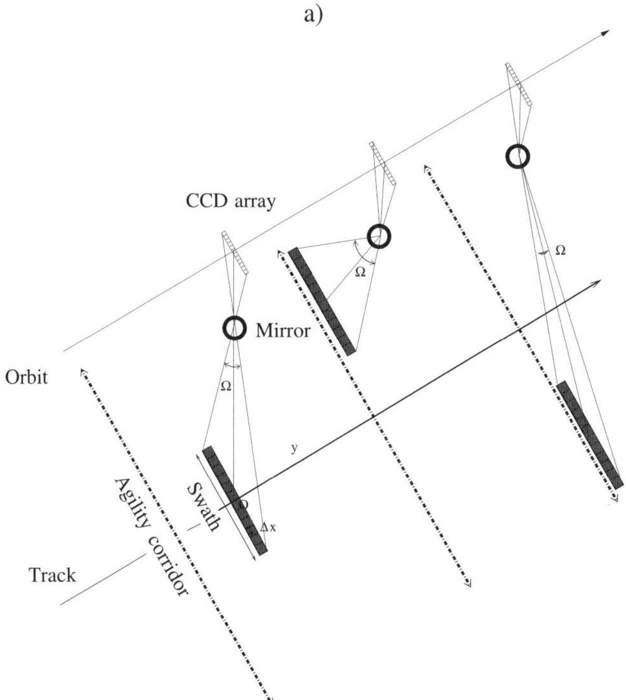

b)

Figure 2.2. *a) Static optical sensor. The line of sight is always vertical. b) Agility of the SPOT sensor: using a mirror, the swath can move between* $-27°$ *and* $27°$*,thus increasing the area that will be accessible for a given position of the satellite*

2.2.1.2. *The agility of optical VHR sensors: the example of Pleiades*

This concept has been extended to every direction of sight on the latest Very High Resolution (VHR) optical satellites (such as Pleiades): the direction of sight can be chosen according to a variable angle not only from left to right, but also in a forward and backward direction. To guarantee this agility, there is no longer a mirror: it is the entire satellite that swings over in the required direction, with a $\pm 45°$ viewing angle.

2.2.2. *Agility of radar sensors*

Similarly, we find, in the radar world, satellites whose viewing angle (i.e. the ground incidence angle if we assume that the Earth is flat) has a constant value: this is the case of the European Remote Sensing Satellite (ERS) that had a swath of 100 km for an average incidence angle of 23, the antenna always points toward the same side of the satellite (the right side, to view the Atlantic rather than the Artic). Since 1995 with RADARSAT-1, then in 2002 with ENVISAT, the use of an antenna with an electronic focusing has allowed us to extend the localization of this swath in an agility corridor of around 500 km (for an average local incidence between approximately 20° and 45°, Figure 2.3). The year 2000 witnessed the emergence of sensors allowing us to aim both sides of the satellite. One such example is the RADARSAT-2 system, which, due to an antenna pointing at the nadir and equipped with electronic focusing modulus antenna, allows a right sight viewing as well as a left sight viewing. Another example is TerraSAR-X, which, as with Pleiades, can also modify the viewing direction by a rolling motion of the satellite. Because of this agility, the revisiting time can be improved: so the Cosmo-Skymed constellation allows for a daily revisit time with its four satellites (whereas the cycle of one of these satellites is of 16 days).

2.2.3. *The effects of the incidence variation on the ground cell size*

The spatial extension of a ground cell depends on the same factors as the spatial extension of the footprint.

For an optical sensor, the ground cell therefore depends on the distance and the incidence. We have the following relation in the case of vertical incidence and a variable altitude (deduced from [1.15]):

$$\Delta x_{optic}(R') = \Delta x_{optic}(R)\frac{R'}{R}$$

and for a sensor at a given altitude depending on an incidence θ (deduced from [1.17]):

$$\Delta x_{optic}(\theta) = \frac{1}{\cos^2\theta}\Delta x_{optic}(\theta = 0)$$

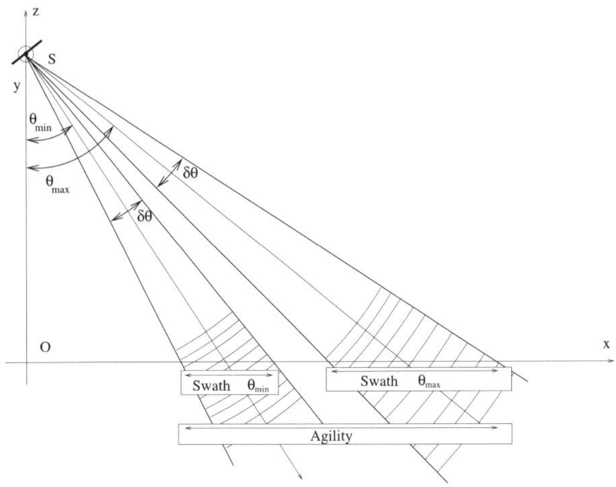

Figure 2.3. *Radar sensor with electronic focusing. The lateral sight incidence angle can be chosen between two values* θ_{min} *and* θ_{max}

For an optical system in an off-nadir viewing (that is so that the local incidence θ is not null), the variation of the size of the ground cell along the swath has to be emphasized, since the local incidence varies along the swath as well. If the first ground cell, corresponding to a local incidence θ, has a size of $\Delta x(\theta)$, a ground cell inside the swath and corresponding to a local incidence θ' has the size:

$$\Delta x_{optic}(\theta') = \frac{\cos^2\theta}{\cos^2\theta'}\Delta x_{optic}(\theta)$$

this expression, therefore, gives the law of the variation of the ground cell along the swath.

Similarly, for a radar system, we characterize the "range cell" Δr with the help of the time sampling Δt:

$$\Delta r = \frac{c\Delta t}{2}$$

This range cell is equal to the ground cell if the incidence is equal to 90°:

$$\Delta x_{radar}(\theta = \pi/2) = \Delta r = \frac{c\Delta t}{2} = \frac{c}{2F_e \sin\theta}$$

For a given incidence angle θ, the ground cell can then be written as:

$$\Delta x_{radar}(\theta) = \frac{c\Delta t}{2\sin\theta} = \frac{1}{\sin\theta}\Delta x_{radar}(\theta = \pi/2) \qquad [2.1]$$

The ground cell will be reduced even more since the incidence angle will be close to $\pi/2$. However, this situation is unrealistic because every object having some elevation will be able to mask the objects that will be behind it.

If we analyze the size of the ground cells along the swath, two ground cells corresponding to two local incidence angles θ and θ' will be connected by the relation:

$$\Delta x_{radar}(\theta') = \frac{\sin\theta}{\sin\theta'}\Delta x_{radar}(\theta)$$

It is interesting to see that we find relations that are analogue to the optical case, without the squaring of the trigonometric function (the variations in distance do not affect the size of the radar ground cell).

2.2.4. *The consequences of agility*

A last point that must be analyzed is the orientation of the track, which depends on the local incidence. If we decide to acquire images at different local incidences, this will impose a different position on the orbit. Since the orbit tracks are not parallel, the acquired data will have different orientations.

Upon a first analysis, we see the counterpart to agility: the image data acquired do not have the same ground cell size and do not have the same orientation in relation to the geographical north. As soon as we decide to use several images to analyze an area of the Earth, a registration step is required for correcting these effects of geometrical distortion.

2.3. Georeferencing of remote sensing images

2.3.1. *From an image to an orthoimage*

An imaging satellite sensor provides an elementary image (that will, hereon, be called the raw image) that can be considered as a two-dimensional (2D) array of the values connected to the local physical phenomena such as the backscattering of the ground in relation to solar radiation or to radar emission. As an image analyst has to be able to use this properly, these values must be linked to ground points described through their georeferenced coordinates (longitude, latitude and elevation). It is, therefore, necessary to build an "orthoimage", that is a 2D array of values so that the geometry of this array corresponds to a particular cartographic geometry (standard if it locally preserves the angles, equivalent if it preserves the surfaces and "aphylactic" if it preserves the distances). This step is almost unavoidable as soon as we use several images that can come from different sources (same sensor but different acquisition positions, different sensors, etc.). It requires different levels of modeling that will be detailed later on.

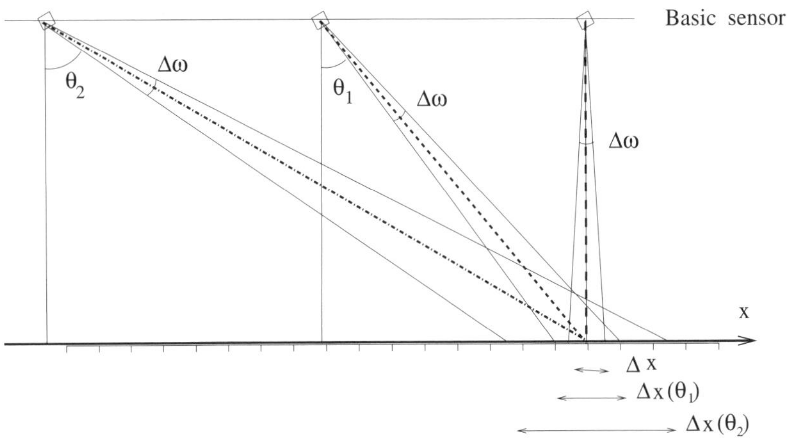

Figure 2.4. *Effects of the local incidence variation for an optical sensor at a given altitude*

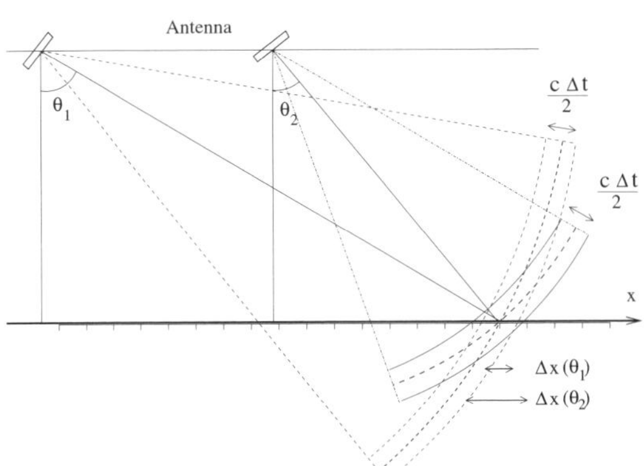

Figure 2.5. *Effects of the local incidence variation for a radar sensor at a given altitude and with the same sampling frequency (i.e. the same time cell Δt)*

We will call "georeferencing" the process that allows us to assign a latitude and a longitude on the Earth's surface to each element of the pixel array. To build an orthoimage, the model of the flat Earth that helped us to pose the problem of satellite images must be rethought, to consider, on the one hand, the non-spherical aspect of the Earth and its relief and, on the other hand, a model that is as precise as possible in terms of acquisition geometry, which needs very good knowledge of the sensor as well as the representation of the Earth. There are two possible representation models:

– The forward model, which, for each pixel of the image, assigns a longitude value, a latitude value and an elevation value to the surface of the Earth. If $\vec{\theta}$ is a set of parameters that describes the sensor (focal distance, ground distance, speed of the platform, altitude of the platform), it is then possible to express the longitude λ and the latitude φ as a function of the pixel described by (r, c) (row and column positions) and also as a function of the altitude:

$$\begin{aligned} \lambda &= f_\lambda(r, c, h, \vec{\theta}) \\ \varphi &= f_\varphi(r, c, h, \vec{\theta}) \end{aligned}$$

– The inverse model, which assigns a pixel to each latitude, longitude and elevation values, described by (r, c):

$$\begin{aligned} r &= g_r(\lambda, \varphi, h, \vec{\theta}) \\ c &= g_c(\lambda, \varphi, h, \vec{\theta}) \end{aligned}$$

Implementing these models can be a delicate task. These models can be more or less complex and also more or less accurate. We can choose to perfectly describe the characteristics of the sensor to reach models, where each parameter has a physical signification. In these cases, we have very accurate models, called physical models, but whose equations are complex. Furthermore, in this context, it is often easier to describe the inverse model, whereas it is very difficult to express the forward model analytically.

We can choose to approach the "true" model of the sensor through a simplified expression: for instance, a polynomial transform or a rational function transform (thus called the *rational polynomial model* (RPM)) described by its coefficients, called rational polynomial coefficients (RPC). In general, the coefficients of these polynomes are obtained with the help of ground control points that are uniformly spread across the image. These particular pixels can be selected by hand by a human operator with the help of digital maps, or automatically determined on reference images that would themselves be orthorectified. There are local similarity measures that can be applied to find control points that we will then call homologous points.

2.3.2. *The metaparameters of VHR optical images*

At the present time, many VHR optical images are provided with the RPCs of their inverse model, which allows us, to some degree, to find out the value of a pixel knowing the geographical parameters of a ground point. The models used usually have 20 coefficients for the numerator and 20 coefficients for the denominator. Other parameters associated with the study area are, thus, provided:

– offsets for longitude (Lo_{off}), latitude (La_{off}) and elevation (El_{off});

– scaling factors for longitude (Lo_{sc}), latitude (La_{sc}) and elevation (El_{sc});

– image row offset (r_{off}) and image column offset (c_{off});

– scaling factors of the image rows (r_{sc}) and the image columns (c_{sc}).

These allow us to define normalized values for the longitude (Lo), the latitude (La) and the elevation (El):

$$\begin{aligned} Lo &= \frac{\lambda - Lo_{off}}{Lo_{sc}} \\ La &= \frac{\varphi - La_{off}}{La_{sc}} \\ El &= \frac{h - El_{off}}{El_{sc}} \end{aligned}$$

as well as normalized values for the column index c and the row index r (let us note that these indices are not necessarily whole numbers):

$$\begin{aligned} r_n &= \frac{r - r_{off}}{r_{sc}} \\ c_n &= \frac{c - c_{off}}{c_{sc}} \end{aligned}$$

From this set of parameters, the pixel position corresponding to the given latitude, longitude and elevation can be written in normalized values:

$$\begin{aligned} r_n &= \frac{\sum_{i=1}^{20} C_{r,i} p_i \left(Lo, La, El\right)}{\sum_{i=1}^{20} D_{r,i} p_i \left(Lo, La, El\right)} \\ c_n &= \frac{\sum_{i=1}^{20} C_{c,i} p_i \left(Lo, La, El\right)}{\sum_{i=1}^{20} D_{c,i} p_i \left(Lo, La, El\right)} \end{aligned}$$

The polynomes p_i are specific to space agencies. For example, for the DigitalGlobe data as well as for the Pleiades data, the provided model follows the RPC00B specification, which is expressed for a coefficient set R_i through the relation:

$$\begin{array}{lllll} \multicolumn{5}{l}{\sum_{i=1}^{20} R_i p_i\,(Lo, La, El) =} \\ R_1 & +R_2 Lo & +R_3 La & +R_4 El & +R_5 Lo\, La \\ +R_6 Lo\, El & +R_7 La\, El & +R_8 Lo^2 & +R_9 La^2 & +R_{10} El^2 \\ +R_{11} Lo\, La\, El & +R_{12} Lo^3 & +R_{13} Lo\, La^2 & +R_{14} Lo\, El^2 & +R_{15} Lo^2\, La \\ +R_{16} La^3 & +R_{17} La\, El^2 & +R_{18} Lo^2\, El & +R_{19} La^2\, El & +R_{20} El^3 \end{array} \quad [2.2]$$

However, these orthoimages thus obtained are not perfect: indeed, some characteristics of the platform (roll, pitch and yaw motions) are sometimes not known with sufficient precision and the digital terrain models (DTMs) may contain errors or imprecisions. Thus, a residual registration can be required; this stage will be discussed in section 2.4.

2.3.3. *The levels of the images*

We have seen that the acquisition of image-type data was bound by different elements of geometric distortions that make the image comparison difficult, if not impossible, and all the more so given that the acquisition characteristics (local incidence) were different. Therefore, the space agencies have always proposed their images along different "levels", all corresponding to different steps of corrections.

This processing is applied almost all the time in the same order, which allows us to define successive levels of corrections. For example, Centre National d'Etudes Spatiales (CNES) has defined the following levels for SPOT images:

– Raw: matrix of digital numbers. The data did not suffer any corrections, neither geometric nor radiometric.

– Level 1A: radiometric correction of the distortions due to sensitivity discrepancies between the elementary detectors of the imaging instrument.

– Level 1B: geometric correction of the global effects (panoramic effects, Earth rotation and curvature). The internal distortions of the image are corrected, which authorizes the distance, angle and surface measurements.

– Level 2A: geometric correction carried out in the standard cartographic projection without taking ground control points. This allows us, taking into account the possible deviation in the localization, to combine the image with geographic information of different types (vectors, raster maps and other satellite images).

– Level 2B: the cartographic projection by taking ground control points either taken over from the map or from on-site measurement, using global positioning system (GPS). The image is corrected to an average elevation in a projection and a standardized cartographic tiling and used when the deformations due to relief are not strong (flat terrain, etc.).

– Level 3: cartographic projection using control points and a digital elevation model (DEM) issued from Reference3D data set[1] in order to eliminate the distortions due to the relief.

This process can also be found in the case of VHR sensors. For example, Pleiades data have three levels of distribution:

– Primary product, corresponding to data processed for radiometric and optical corrections. The multispectral and panchromatic channels are registered.

– Standard ortho, corresponding to georeferenced data (WGS84).

– Tailored ortho, upon demand and when the user provides a DEM of a higher quality.

Similarly, DigitalGlobe proposes different products for its sub-metric images (WorldView-2):

– Basic imagery: images corrected of the sensor distortions. The camera model is provided so that the users can themselves carry out the photogrammetric processing of this type of data.

– Georeferenced: images so that a geographic coordinate can be linked to a given pixel in a cartographic referential.

According to the projected application, it is indeed possible to choose one's level. For example, the 1B SPOT level allows for topographic information extraction. However, we must note that for the most recent sensors, the most usual data are orthorectified (image associated with a RPM). We must not forget that this type of image has, therefore, suffered some postprocessing steps that can have damaging effects on the final quality of the image and on the quality of the underlying information. Furthermore, some aspects of the acquisition remain omitted: for example, the Pleiades images are provided with a ground cell of 50 cm, whatever the local incidence of the acquisition (we know, however, that the resolution of this kind of image is multiplied by 2 between a vertical viewing and a 45° viewing).

1 Reference 3D is a DEM in a DTED2 format, i.e. with 1 s arc step, built on the basis of SPOT5/HRS data by Astrium Services and the IGN.

Owing to all these reasons, we must look closely at the revisiting characteristics of the most modern satellites: the values of the guaranteed resolution (hence related to the extreme values of the local incidence) are associated with the revisiting time. Thus, we can note that for WorldView-2 (for which the pixel is always of 50 cm), the revisiting time is 3.7 days for a resolution better than 59 cm and 1.1 days for a resolution better than 1 m.

2.3.4. *SAR image specificities*

In the case of radar images, the space agencies do not provide the RPCs for the basic data (e.g. the *single-look complex* data). However, one of the greatest turning points in the technology of current satellites lies in the exceptionally good knowledge of their orbital trajectory. Because of the excellent performances of inertial platforms and the very precise knowledge of the GPS position, it is becoming possible to locate pixels with an accuracy of several meters.

More particularly, the interferometric processing is simplified because we can, without any major problems, register the slave image in the geometry of the master image (see Chapter 9). Without DTM, the result will still have a lot of fringes related to the relief (let us remark that there are fringe-based methods to reconstruct the local relief). If we have a DTM, it will then be possible to detect the changes in the position of the Earth's crust (Chapter 10).

2.4. Image registration

2.4.1. *The need for image registration*

We have seen in section 2.3.1 that the use of the geometric model of acquisition allows us to project the images in a cartographic reference. However, this procedure has its limits due to a lot of possible errors (model parameters, accuracy of the DTM, etc.).

If some applications can cope with the obtained accuracy, these residual orthorectification errors can be too significant for other applications, as, for example, when several images are conjointly used. When these errors are about a tenth of the pixel's resolution, the performances of classification and change detection algorithms can decrease.

Image registration, thus called "image matching", tries to palliate this problem by estimating sub-pixel deviations between the images. We will only discuss here the registration of image pairs in the case of scalar images (single-band). In the case of multi-images, these same techniques can be used, but there are some approaches based on the redundancy between a large set of images that are better suited.

Let us also point out that there are image registration errors that do not have any solution, such as the presence of hidden faces or the projections of several points of the scene on the same set of pixels (layover, etc.) Thus, in the multisensor case, where the physics of the measure of the systems used is very different, images correspond to observations of different objects (for example, the case of ground penetration by the radar wave).

2.4.2. *Modeling the problem*

The objective of image registration is, therefore, to find a geometric transformation between the two images, which maximizes their degree of similarity. This formulation of the problem allows us to implement optimization strategies whose main elements are as follows:

– a measure of similarity;

– a geometric transformation;

– an optimization technique.

2.4.2.1. *The similarity measure*

The similarity measure is a function that allows us to evaluate the degree of similarity between two images. In the case of image registration, a first image is taken as a reference. A geometrical transformation is applied on the second image so that it is possible to measure the similarity between this modified image and the reference. If the measured degree of similarity is high, we may say that the images have been registered. Otherwise, we must find another geometric transformation.

The simplest way of assessing the similarity between two images is to calculate the difference between them pixel by pixel:

$$\mathcal{D}^{diff}_{(I_1,I_2)}(m,n) = I_1(m,n) - I_2(m,n) \qquad [2.3]$$

and to make an average for every pixel inside a given window: thus, we obtain the similarity: $\tilde{\mathcal{D}}_{(I_1,I_2)}(m,n)$:

$$\tilde{\mathcal{D}}^{diff}_{(I_1,I_2)}(m,n) = \frac{1}{KL}\sum_{k=-K/2}^{K/2}\sum_{l=-L/2}^{L/2}\left(I_1(m-k,n-l) - I_2(m-k,n-l)\right) \quad [2.4]$$

Several options are possible. More precisely, in order to avoid the compensation effects between the positive and negative differences, we choose to calculate the mean square error (MSE), which gives the similarity $\tilde{\mathcal{D}}^{MSE}_{(I_1,I_2)}$:

$$\tilde{\mathcal{D}}^{MSE}_{(I_1,I_2)}(m,n) = \frac{1}{KL}\sum_{k=-K/2}^{K/2}\sum_{l=-L/2}^{L/2}\left(\mathcal{D}^{diff}_{(I_1,I_2)}(m-k,n-l)\right)^2 \qquad [2.5]$$

If we wish to obtain a global measure, we can choose the pixel at the center of the image and use the values K and L allowing a full covering of the image.

We can also eliminate, in equation [2.5], the constant terms only depending on one of the two images, yielding the cross-correlation measure:

$$\tilde{\mathcal{D}}^{prod}_{(I_1,I_2)}(m,n) = \frac{1}{KL}\sum_{k=-K/2}^{K/2}\sum_{l=-L/2}^{L/2} I_1(m-k,n-l)I_2(m-k,n-l) \qquad [2.6]$$

This last measure is sensitive to lighting conditions. In the case of an affine grayscale transformation with $I_1(m,n) = \alpha I_2(m,n) + \beta$, the MSE can be very high, whereas the images represent the same scene and are therefore perfectly registered. This is why one of the most common similarity measures is the normalized correlation coefficient:

$$\tilde{\mathcal{D}}^{corr}_{(I_1,I_2)}(m,n) = \frac{1}{KL}\sum_{k}\sum_{l}\frac{(I_1(m-k,n-l)-\mu_1)(I_2(n-k,m-l)-\mu_2)}{\sigma_1\sigma_2} \qquad [2.7]$$

with μ_i and σ_i the average and the standard deviation of the image i calculated locally or globally. This normalization with respect to the average μ_i and the standard deviation σ_i allows us to surpass lighting changes between the two observations of the same scene.

In the case where the images that must be registered are acquired with very different sensors (different optical spectral bands, different radar polarizations, etc.), the correlation coefficient becomes less useful, because it cannot account for the differences in the value of the pixels corresponding to the same geographic position, but representing different physical measures (for example reflectance in blue and in near infrared). In these cases, we assume an underlying link due to the observed landscape and we measure the statistical dependence between the images.

A criterion that has often been used is mutual information [ING 04]. Initially proposed for the registration of medical multimodal images [VIO 95, MAE 97], this measure seeks to maximize the entropy of the images, and, at the same time, to minimize the joint entropy. Let us note by f_{I_i} the distribution of the gray levels of the image and I_i and $f_{(I_1,I_2)}$ the joint distribution of the two images. Actually, these values are quantified, so we are dealing with probabilities. They are estimated by calculating the histograms and the joint histogram of images (see Chapter 5). Thus, the entropy of a variable that follows a probability density function f_{I_i} is written as:

$$H_{I_i} = -\sum_{q=1}^{q=Q} f_{I_i}(q) \log_2(f_{I_i}(q)) \qquad [2.8]$$

where Q represents the number of gray levels. The entropy of the joint probability distribution $H_{(I_1,I_2)}$ is obtained from the previous equation by replacing f_{I_i} with the joint probability $f_{(I_1,I_2)}$. It can be considered as the information that one image brings with respect to the other. The more the two images resemble each other, the lower this measurement.

The mutual information between the images is defined by the following formula:

$$IM_{(I_1,I_2)} = H_{I_1} + H_{I_2} - H_{(I_1,I_2)} \qquad [2.9]$$

During the registration operation, maximizing the mutual information comes to seek the joint entropy minimization and thus to find a deformation making the two images as similar as possible.

We will return to these similarity measures in Chapters 8–10 since they are used for change detection, elevation calculation and motion estimation.

The similarity measure can be calculated on the whole set of the image pixels that must be registered or on a subset of the image pixels. Besides reducing the computation time, the assessment on a limited number of pixels can have other motivations as well. We can, for example, try to exclude given areas from the images because of the presence of clouds or changes previously known and that could lead to measurement errors.

Another reason for working on a limited number of points is to make sure that the geometric transformation will be retained, will be precise on points of particular interest (ridge lines, road network and coast lines) and, at the same time, will avoid too significant errors on other points. Indeed, the geometric transformation used is often a simplification of the true deformation that exists between the images and it is therefore impossible to achieve a good superposition throughout the entire image.

When we have no prior knowledge about the regions that must be privileged or excluded, we assess the similarity on a standardized grid of points defined on the reference image (for example one point out of 10 points).

2.4.2.2. *The geometric transformation*

The geometric transformation is a reference change that allows us to move from the coordinates of the pixels of the reference image (m, n) to those of the pixels of the image that needs to be registered (m', n'):

$$m' = f(m, n), n' = g(m, n)$$

where (m', n') can be non-integer.

In light of this, it plays a role that is similar to that of the geometrical acquisition models.

There are two main families of geometric transformation that are currently used: analytical functions and grids.

– Analytical functions, as suggested by their name, are analytical expressions of the transform, which depend on a limited number of parameters. Some examples of analytical transformations are translations, affine transformations or polynomial transformations. These transformations are often simplifications of the true deformation between the images. This means that they rarely provide an accurate transform for all the points in the images, but rather it means that the global superposition is satisfactory. The main advantages of the analytical transformations is the small number of parameters that need to be determined.

– The grids are lists of points in the reference image with the coordinates of their homologous points in the image that needs to be registered. These grids can be considered as a sampling (in the sense of signal processing) of the true deformation between the images that have to be registered. They can be extremely accurate for the points in the grid, but do not give any information on the rest of the points. For these other points, an interpolation of the coordinates has to be done with the help of the nearest points in the grid. If the precision makes up the main advantage of the grids, their major drawback is the computational cost needed to determine the homologous points in the grid.

The choice of the type of geometric transformation depends on the data and the cartographic projections of the considered application. In the case of image fusion or change detection at the pixel level, a very accurate registration is needed and we prefer to use grids. In the case of visual analyses (photo interpretation), we will be able to choose simple analytical transformations.

2.4.2.3. *The optimization*

The goal of the optimization will be to find a set of parameters of the geometric transformation that maximize the similarity measure chosen for assessing the quality of the superposition of the images that we wish to register.

This optimization will be local or global depending on the measure of the similarity chosen and the type of geometric transformation used.

There are many available optimization techniques, but they are not specific to image registration. These techniques are part of the scientists's or engineer's toolbox, which is why we will not discuss this subject here.

However, it is interesting to know the main existing families in order to be able to make an informed choice. This choice will depend on the complexity of the solution space. This solution space is defined by the images to be registered, by the similarity measure and by the geometric transformation. The space complexity depends on its monotoniticity (convexity, the presence of local minima) and on the noise level.

2.4.2.3.1. Image complexity

The images to be registered can create a range of solutions that is extremely complex, depending not only on their noise level – allow us to think particularly of the case of radar images with the speckle phenomenon described in Chapter 6 – but also on the type of landscape observed. One example that illustrates this dependence with respect to the landscape are the regions with pseudo-periodic structures (agricultural parcel and urban structures) that can be confused with a translation.

2.4.2.3.2. Complexity related to the similarity measure

The similarity measure can be a complex function whose variables are, for example, the values of the pixels of the images to be registered. This function can generate a highly nonlinear and non-convex space, whose global minimum – or an acceptable local minimum – can be difficult to find.

2.4.2.3.3. Complexity related to the geometric transformation

The decisive aspect of the geometric transformation as regarding the complexity of the solution range is the number of parameters and their dynamic and variation scales. For example, a rigid translation is determined by two parameters (translations in X and Y), which have similar dynamics and equivalent variation scales. However, even if a 2D similitude only has two parameters (rotation and scale), their variation dynamics are different (for example from 0.5 to 2 for the scale and from 0 to 2π for the rotation). Thus, a factor 2 on the scale generates a deformation that is completely different to a deformation generated by a factor 2 in rotation.

In the case of geometric transformations with more parameters, the solution range can be very difficult to optimize.

2.5. Conclusion

Because of the advances of sensors, the digital systems and the storage capacities, satellite imaging allows access to large quantities of georeferenced information almost everywhere on the Earth. We must not omit some essential points that are well known by signal and image specialists:

– Resolution and sampling are two fundamentally different magnitudes. The information available in an image depends on the global resolution of the acquisition and processing chain. The sampling is simply connected to the representation mode of the data, which, in the case of orthoimages, is connected to a positioning in a georeferenced frame. To make the processing of a data set acquired on a given area easy to use, some steps leading to oversampling the data become unavoidable; the image can then appear to be smooth, although that does not mean the actual information has been modified (as long as the Shannon criterion has been observed).

– The complementarity of the current sensors opens up new possibilities for methods of data fusion using all kinds of data: this point will be discussed in more detail in the chapter devoted to data assimilation (Chapter 11).

– Finally, the current sensors warrant a very regular coverage, i.e. a daily coverage, of some parts of the globe. We should introduce, besides space resolution, the notion of temporal resolution: this massively influences the change detection methods that will be discussed in Chapter 8.

Chapter 3

The Physics of Optical Remote Sensing

An optical remote sensing instrument is sensitive to the energy of the light it receives during the exposure time of the instrument. Having undergone several stages of processing, the measurements of the light energy by optical remote sensors are finally used to determine properties of the vegetation, the atmosphere or the oceans. The physics of optical measurements provides the theoretical framework for implementing this processing and understanding the inherent errors and uncertainties.

3.1. Radiometry

The detectors of optical remote sensing instruments are sensitive to the radiant energy (the energy brought by the light) that they receive during an exposure time. On a camera, the exposure time corresponds to the interval between the opening and the closing of the shutter. On a digital observation instrument, the exposure time is often set by the electronics, and is usually named "integration time".

The received radiant energy is related to properties of the observed scene, but it also depends on a large number of other parameters related to the instrument itself, such as the detector surface, the instrument sensitivity or the optics aperture. Therefore, the radiant energy is not a convenient unit, for example when we wish to compare the measures taken by different instruments. In the following sections, we will have to successively define physical quantities that will enable us to progressively bypass the parameters that depend on the instrument, in order to obtain an inherent characteristic of the surface that has been observed. This will require

---

Chapter written by Olivier HAGOLLE.

defining an entire set of physical quantities: describing this series of quantities can seem a little off-putting for the reader (and indeed, for the writer as well), but it will finally prove useful in understanding what is being observed.

3.1.1. *Radiant energy, spectral energy, spectral sensitivity and equivalent energy*

A sensor is sensitive to a radiant energy ε that has been received during the integration time t_i. The unit of the radiant energy is joules (J). The radiant energy is brought by photons of different wavelengths and the spectral (radiant) energy $\varepsilon(\lambda)$ received in each wavelength λ depends on the observed scene; it is expressed in joules per wavelength unit (often in $J/\mu m$). Furthermore, we must consider the fact that the instrument's spectral sensitivity $S(\lambda)$ changes with the wavelength; $S(\lambda)$ is a unitless quantity (see Figure 3.1).

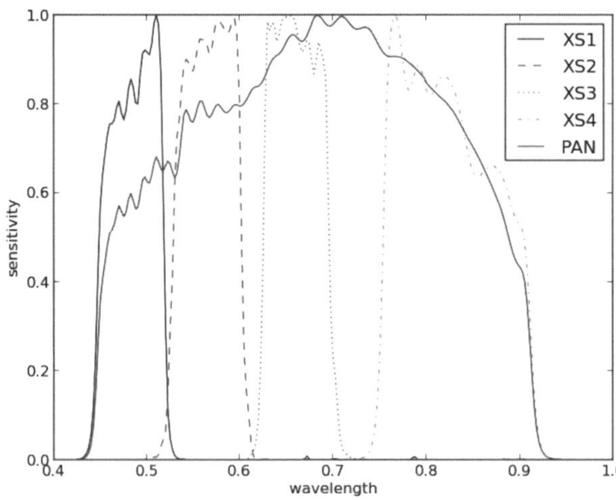

Figure 3.1. *Spectral responses of the five bands of the Formosat-2 satellite. The panchromatic band spreads all over the range of the visible wavelengths. The unit of the abscissas is the micrometer (micron)*

The radiant energy measured during the integration time by an instrument whose sensitivity is not null in the wavelength interval $[\lambda_1, \lambda_2]$ is, therefore, expressed as follows:

$$\varepsilon = \int_{\lambda_1}^{\lambda_2} S(\lambda)\varepsilon(\lambda)\, d\lambda \qquad [3.1]$$

For a given instrument, the larger the spectral band $[\lambda_1, \lambda_2]$, the greater the radiant energy collected. To escape this effect, it is useful to define the "equivalent spectral energy", as the average of the spectral energy in the band $[\lambda_1, \lambda_2]$, weighted by the sensitivity of the sensor in the band $[\lambda_1, \lambda_2]$: $S_{[\lambda_1,\lambda_2]} = \int_{\lambda_1}^{\lambda_2} S(\lambda)\, d\lambda$:

$$\varepsilon_{eq\,[\lambda_1,\lambda_2]} = \frac{\int_{\lambda_1}^{\lambda_2} S(\lambda)\varepsilon(\lambda)\, d\lambda}{\int_{\lambda_1}^{\lambda_2} S(\lambda)\, d\lambda} = \frac{\int_{\lambda_1}^{\lambda_2} S(\lambda)\varepsilon(\lambda)\, d\lambda}{S_{[\lambda_1,\lambda_2]}} \qquad [3.2]$$

The equivalent spectral energy does not depend on the bandwidth anymore, and it keeps the same unit as the spectral radiant energy.

3.1.2. *The flux*

For a given wavelength, the spectral energy is the integral of the light power that reaches the sensor during the integration time; the light power is called the "spectral flux" and is expressed in Watts per wavelength (often in W/μm). It is, thus, the derivative of the spectral energy versus time:

$$\phi(\lambda) = \frac{d\varepsilon(\lambda)}{dt} \qquad [3.3]$$

Just like in the case of energy, the "equivalent spectral flux" can be defined as follows:

$$\phi_{eq\,[\lambda_1,\lambda_2]} = \frac{\int_{\lambda_1}^{\lambda_2} S(\lambda)\phi(\lambda)\, d\lambda}{\int_{\lambda_1}^{\lambda_2} S(\lambda)\, d\lambda} = \frac{\int_{\lambda_1}^{\lambda_2} S(\lambda)\phi(\lambda)\, d\lambda}{S_{[\lambda_1,\lambda_2]}} \qquad [3.4]$$

and by developing this formula we can verify that the equivalent quantities are interconnected in the same way as spectral quantities:

$$\phi_{eq\,[\lambda_1,\lambda_2]} = \frac{d\varepsilon_{eq\,[\lambda_1,\lambda_2]}}{dt} \qquad [3.5]$$

The flux that reaches the instrument, however, remains a quantity that is not directly related to the optical properties of the scene that we seek to measure, since it depends on the sensor surface. To bypass this parameter, it is necessary to introduce a new quantity: the irradiance.

3.1.3. *The irradiance*

For a given source, the spectral irradiance is the spectral flux that reaches the sensor per surface unit ds of the sensor. It is therefore defined as:

$$E(\lambda) = \frac{d\phi(\lambda)}{ds} \qquad [3.6]$$

Similarly, the equivalent irradiance is defined by:

$$E_{eq\,[\lambda_1,\lambda_2]} = \frac{\int_{\lambda_1}^{\lambda_2} S(\lambda)E(\lambda)\,d\lambda}{\int_{\lambda_1}^{\lambda_2} S(\lambda)\,d\lambda} = \frac{\int_{\lambda_1}^{\lambda_2} S(\lambda)E(\lambda)\,d\lambda}{S_{[\lambda_1,\lambda_2]}} = \frac{d\phi_{eq\,[\lambda_1,\lambda_2]}}{ds} \qquad [3.7]$$

The irradiance allows us, for example, to characterize the light power per surface unit that reaches a surface that is perpendicular to the source. In the visible or near infrared range, since it is the Sun that lights up the Earth, we will consider the solar irradiance. It varies with the Earth–Sun distance, which itself varies throughout the year. Thus, the solar irradiance integrated over the whole spectrum in day j for a surface that is perpendicular to the Sun is equal to $Es(j) = E_0 \frac{d^2 j}{d_0^2}$, with: $E_0 = 1{,}367\,W/m^2$ and:

$$\frac{d^2 j}{d_0^2} = 1 - 0.01673 \cos\left[\frac{2\pi}{365.3}(j - j_0 - 2)\right]$$

and j_0 is January 1st, 1950. The spectral solar irradiance strongly depends on the wavelength (as shown in Figure 3.2). We will see in section 3.1.5 that it corresponds to a black body at 5,800 K.

Irradiance is a very useful quantity, but it still depends on the field of view of the sensor. The thinner the field of view, the lower the irradiance. Irradiance also depends on the orientation of the sensor with regard to the direction of the incoming light. For example, if the sensor is parallel to the direction of the photons, it will not receive any light. Conversely, when the sensor is perpendicular to the direction of the light, it receives a maximum irradiance. To obtain a unit that is independent of the characteristics of the instrument, we have yet to define another quantity.

3.1.4. *The radiance*

The spectral radiance is the spectral flux that reaches the instrument per surface unit ds and per unit of solid angle $d\Omega_s$, perpendicularly to the surface of the instrument. The spectral radiance unit is W/m^2/sr/μm:

$$L(\lambda) = \frac{d^2\phi(\lambda)}{ds\,d\Omega_s\,\cos\theta_s} \qquad [3.8]$$

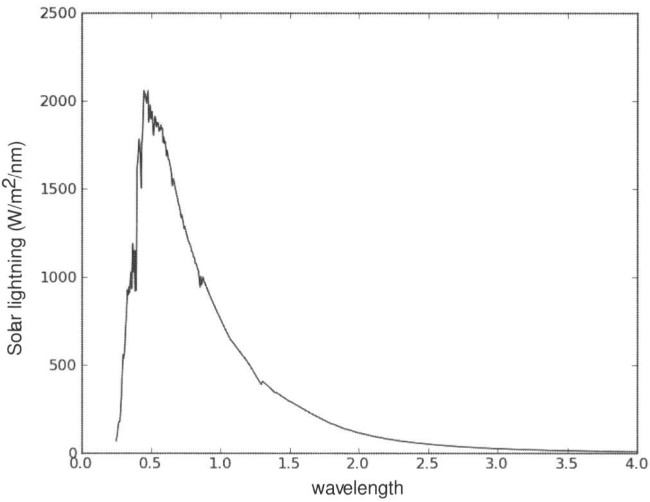

Figure 3.2. *Spectral solar irradiance versus wavelength*

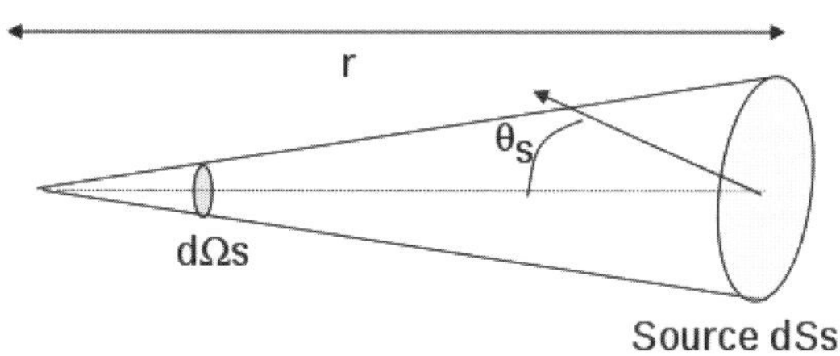

Figure 3.3. *Definition of the solid angle $d\Omega_s = \frac{dS_s \cos\theta_s}{r^2}$ (where θ_s is the angle between the normal of the source and the normal of the detector). The solid angle unit is the steradian (sr)*

The quantity $d^2G = ds\, d\Omega_s \cos\,\theta_s$ is very widely used by opticians and is called the elementary "geometric etendue" or the elementary throughput [MEY 01] (we will return back to the properties of the geometric etendue in the next section).

Let us consider an instrument with a surface S, a field of view with a solid angle Ω_s and a geometric etendue G. The spectral flux obtained at the input of this instrument may be written as:

$$\phi(\lambda) = \int_S \int_{\Omega_s} L(\lambda) \cos\,\theta_s ds\, d\Omega_s = \int \int_G L(\lambda) d^2G \qquad [3.9]$$

As for all the other quantities, we define the equivalent spectral radiance as follows:

$$L_{eq\,[\lambda_1,\lambda_2]} = \frac{\int_{\lambda_1}^{\lambda_2} S(\lambda)L(\lambda)\,d\lambda}{\int_{\lambda_1}^{\lambda_2} S(\lambda)\,d\lambda} = \frac{\int_{\lambda_1}^{\lambda_2} S(\lambda)L(\lambda)\,d\lambda}{S_{[\lambda_1,\lambda_2]}} = \frac{d^2\phi_{eq\,[\lambda_1,\lambda_2]}}{ds.d\Omega_s.\cos\theta_s} = \frac{d^2\phi_{eq\,[\lambda_1,\lambda_2]}}{d^2G} \quad [3.10]$$

When the geometric etendue is small, which happens quite often in remote sensing, the equation can be written as:

$$L_{eq\,[\lambda_1,\lambda_2]} = \frac{\phi_{eq\,[\lambda_1,\lambda_2]}}{G} \quad [3.11]$$

Radiance is very commonly used in remote sensing, since it combines several advantages:

– The radiance no longer depends on the instrumental characteristics.

– The energy measured by a sensor is proportional to the equivalent radiance. Indeed, for an instrument with a geometric etendue G, a spectral sensitivity $S(\lambda)$ in the band $[\lambda_1, \lambda_2]$ and with an integration time t_i, the measurement X obtained by the sensor of the instrument is expressed as follows:

$$X = \int_{\lambda_1}^{\lambda_2} S(\lambda)\varepsilon(\lambda)\,d\lambda = S_{[\lambda_1,\lambda_2]}\varepsilon_{eq\,[\lambda_1,\lambda_2]} \quad [3.12]$$

and by using formulas [3.5] and [3.11], we therefore obtain:

$$X = S_{[\lambda_1,\lambda_2]}\phi_{eq\,[\lambda_1,\lambda_2]}t_i = S_{[\lambda_1,\lambda_2]}L_{eq\,[\lambda_1,\lambda_2]}Gt_i \quad [3.13]$$

In equation [3.13], all the elements are characteristics of the instrument, except the equivalent spectral radiance, which does not depend on the characteristics of the instrument. The measurements obtained are, therefore, proportional to the observed radiance of the pixel, with a factor SGt that solely depends on the instrument. Instrument designers will, therefore, optimize the instrument so that SGt is on two points:

– the largest possible, to increase the sensitivity and signal-to-noise ratio;

– the most uniform possible from one pixel to another, so as to avoid variations in the image that do not actually belong to the observed scene.

3.1.5. *Temperature and emissivity*

The theory of electromagnetism defines a perfect, emitting body, i.e. the black body, which emits a radiance that obeys the "Planck law" defined below. In this case, we often use the "emittance" quantity, but it is in fact a radiance expressed in W/m^2/sr/μm:

$$E_{black\ body}(\lambda, T) = \frac{2hc^2}{\lambda^5} \frac{1}{e^{\frac{hc}{\lambda kT}} - 1} \qquad [3.14]$$

In the above equation, h is the Planck constant: $6.626\,068 \times 10^{-34}$, m^2/kg, k is the Boltzmann constant: $1.380\,650\,3 \times 10^{-23}$ m^{2}kg/s^2/K and c is the speed of light: 299,792,458 m/s.

For a given temperature, this law represents the maximum emittance that a light-emitting body can achieve. Black bodies do not exist in nature, however it is possible to create cavities that closely approximate black bodies. In nature, the emission of a material is always lower than that of the black body. We can express it in the following way:

$$E(\lambda, T) = \varepsilon(\lambda)\, E_{black\ body}(\lambda, T) \qquad [3.15]$$

where $\varepsilon(\lambda)$ is the spectral emissivity, comprised between 0 and 1.

For a body that emits light but also receives it and reflects it, the radiance is the sum of the radiance reflected and emitted:

$$L(\lambda, T) = L_{reflected}(\lambda) + \varepsilon(\lambda) E_{black\ body}(\lambda, T) \qquad [3.16]$$

3.1.6. *Reflectance and albedo*

For wavelengths lower than 3 μm, the radiance of the terrestrial surfaces essentially comes from the reflection of the solar irradiance Es on the Earth (with the exception of several light sources that are very intense: projectors oriented toward the satellite, lava flows in the short wave infrared (SWIR), etc.). To elude the solar irradiance that, as we have seen in section 3.1.3, depends on the Sun–Earth distance (which varies with the season), the introduction of a new quantity is required. The reflectance is defined by:

$$\rho(\lambda) = \frac{\pi L(\lambda)}{Es(\lambda, t) \cos(\theta_s)} \qquad [3.17]$$

Finally, we define (and (phew!) this is the last quantity we need to define) the equivalent reflectance in much the same way:

$$\rho_{eq\,[\lambda_1,\lambda_2]} = \frac{\pi L_{eq\,[\lambda_1,\lambda_2]}}{Es_{eq\,[\lambda_1,\lambda_2]}(t)\cos(\theta_s)} \qquad [3.18]$$

Formulation [3.17] can be surprising at first, as we could expect a much simpler formula, the reflectance being equal to the radiance divided by the irradiance. The adopted formula for defining the reflectance comes from geometric considerations. The received flux on a surface depends on its orientation in relation to the light source. If the surface is perpendicular to the incoming light direction, the flux is at its maximum, whereas if it is parallel, it becomes null. This explains the presence of the $\cos(\theta_s)$ factor. The standardization factor π causes the reflectance to be equal to 1 if the surface reflects the totality of the incidental irradiance toward the sky. To convince ourselves, let us compute the integral of the radiance reflected in all the directions by a surface whose reflectance is equal to 1 (bearing in mind that in order to integrate on the sphere, we must use the factor $\sin\theta_s d\theta_s d\varphi$ and not only $d\theta_s d\varphi$).

$$\begin{aligned}\int_{\theta=0}^{\frac{\pi}{2}}\int_{\phi=0}^{2\pi} L\sin\theta_s d\theta_s d\varphi &= \int_{\theta=0}^{\frac{\pi}{2}}\int_{\phi=0}^{2\pi} \frac{Es\cos\theta_s}{\pi}\sin\theta_s d\theta_s d\varphi \\ &= \frac{E_s}{\pi}\int_{\theta=0}^{\frac{\pi}{2}}\int_{\phi=0}^{2\pi} \cos\theta_s \sin\theta_s d\theta_s d\varphi = E_s \qquad [3.19]\end{aligned}$$

since:

$$\begin{aligned}\int_{\theta=0}^{\frac{\pi}{2}}\int_{\phi=0}^{2\pi} \cos\theta_s \sin\theta_s d\theta_s d\varphi &= 2\pi\int_{\theta=0}^{\frac{\pi}{2}} \cos\theta_s \sin\theta_s d\theta_s \\ &= 2\pi\int_{\theta=0}^{\frac{\pi}{2}} \sin\theta_s d(\sin\theta_s) = 2\pi\frac{\sin^2\frac{\pi}{2}}{2} = \pi\end{aligned}$$

The reflectance is a characteristic of the surface and does not depend on the instrument that observes it, or on the received irradiance. It depends, however, on the illumination and observation angles, and also on the wavelength. Sometimes additional quantities may be useful, such as the spectral albedo (see [3.20]), which is the average of the reflectance in all the directions, or the albedo, which is the average of the spectral albedo on the entire spectrum. Brave readers who would like to know more about new photometric quantities could turn to the definitions of the different types of directional reflectance in [NIC 77, SCH 06].

$$a(\lambda) = \frac{1}{\pi} \int_{\theta=0}^{\pi} \int_{\phi=0}^{2\pi} \rho(\lambda) d\theta_s d\varphi \qquad [3.20]$$

From hereon, we will no longer have to define new photometric units, but we will study some of their properties in sections 3.4 and 3.5.

3.1.7. *Example of the use of photometric quantities*

Let us suppose that we have a reflectance spectrum $\rho(\lambda)$ that is sampled nanometrically. How should we compute the reflectance observed in a spectral band that is defined by its sensitivity $S(\lambda)$? The following erroneous formulation is very common:

$$\rho_{eq\,[\lambda_1,\lambda_2]} = \frac{\int_{\lambda_1}^{\lambda_2} S(\lambda)\rho(\lambda)d\lambda}{\int_{\lambda_1}^{\lambda_2} S(\lambda)d\lambda} = \frac{\int_{\lambda_1}^{\lambda_2} S(\lambda)\rho(\lambda)d\lambda}{S_{[\lambda_1,\lambda_2]}}$$

This formulation, in fact, neglects the fact that the instrument is sensitive to energy, proportional to a radiance and not to a reflectance. We must, therefore, first calculate the equivalent radiance for the band, using definition [3.17] of the reflectance :

$$\begin{aligned} L_{eq\,[\lambda_1,\lambda_2]} &= \frac{\int_{\lambda_1}^{\lambda_2} S(\lambda)L(\lambda)d\lambda}{\int_{\lambda_1}^{\lambda_2} S(\lambda)d\lambda} = \frac{\int_{\lambda_1}^{\lambda_2} S(\lambda)L(\lambda)d\lambda}{S_{[\lambda_1,\lambda_2]}} \\ &= \frac{\cos\theta_s}{\pi} \frac{\int_{\lambda_1}^{\lambda_2} S(\lambda)\rho(\lambda)E_s(\lambda)d\lambda}{S_{[\lambda_1,\lambda_2]}} \end{aligned}$$

and, in the same way, calculate the equivalent solar irradiance:

$$Es_{eq\,[\lambda_1,\lambda_2]} = \frac{\int_{\lambda_1}^{\lambda_2} S(\lambda)Es(\lambda)d\lambda}{\int_{\lambda_1}^{\lambda_2} S(\lambda)d\lambda} = \frac{\int_{\lambda_1}^{\lambda_2} S(\lambda)Es(\lambda)d\lambda}{S_{[\lambda_1,\lambda_2]}}$$

Therefore, the equivalent reflectance is written as:

$$\rho_{eq\,[\lambda_1,\lambda_2]} = \frac{\pi L_{eq\,[\lambda_1,\lambda_2]}}{Es_{eq\,[\lambda_1,\lambda_2]} \cdot \cos\theta_s} = \frac{\int_{\lambda_1}^{\lambda_2} S(\lambda)Es(\lambda)\rho(\lambda)d\lambda}{\int_{\lambda_1}^{\lambda_2} S(\lambda)Es(\lambda)d\lambda} \qquad [3.21]$$

3.2. Geometric etendue, sensitivity of an instrument

The geometric etendue has the property of remaining constant when we exchange the light source and the sensor (see Figure 3.4), it can be easily demonstrated as:

$$\begin{aligned} d^2G_r = ds_r d\Omega_s \cos\Theta_r &= ds_r \frac{ds_s \cos\Theta_s}{r^2} \cos\Theta_r = ds_s \frac{ds_r \cos\Theta_r}{r^2} \cos\Theta_s \\ &= ds_s d\Omega_r \cos\Theta_s \end{aligned} \qquad [3.22]$$

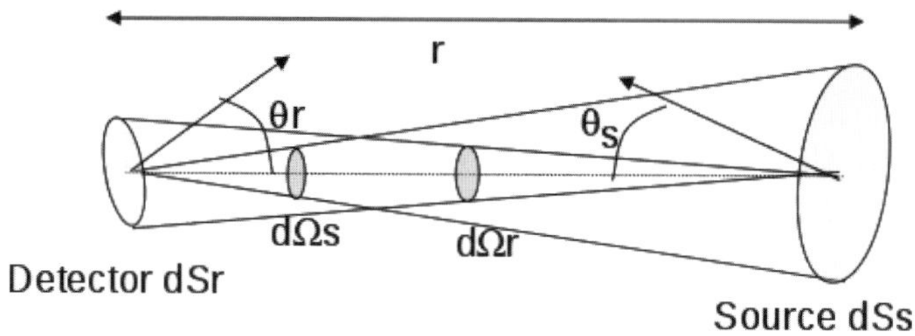

Figure 3.4. *Descriptive scheme for defining the geometric etendue*

We will use this property to determine the sensitivity of an instrument and compare several types of space instruments, already described in Chapter 1:

– The mechanic scanners: a mobile mirror moves the pixel from left to right or from right to left to make a line. The different lines are acquired successively due to the satellite motion. The instruments of the satellites LANDSAT 4, 5 and 7, as well as the instruments AVHRR and MODIS (on Terra and Acqua satellites) are scanner instruments.

– The "electronic scanners" (called *pushbroom*, see section 1.4.4): they work as a photocopier: a linear array of detectors acquire an image line simultaneously. The acquisition of the different lines of the image is ensured by the motion of the satellite. The SPOT, Pléiades, QuickBird, Landsat 8 instruments, as well as the future Venµs and Sentinel-2, are all pushbroom instruments.

3.2.1. *Axis sensor*

Let us study the case shown in Figure 3.5. The instrument is made up of a detector with a surface S, and a circular optic whose diameter is D and whose surface is therefore $ds_s = \pi\frac{D^2}{4}$. This optic serves to collect a large quantity of light and to focus it on the detector. We may view this system in two ways:

– considering that the terrestrial surface is the source, and that the sensor that collects the light is the optics (using this system is not convenient, because the altitude of the satellite above the surface may vary);

– since all the light received by the optics is focused on the detector, it is equally possible to consider a second system where the source is the lens and the light is received by the sensor.

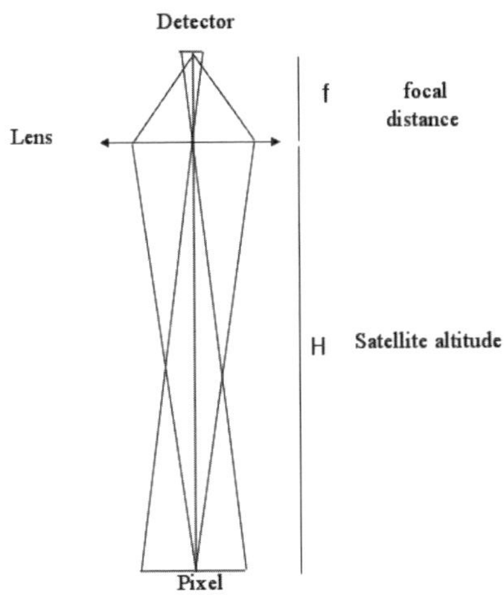

Figure 3.5. *Mono sensor instrument on the optic axis*

The calculation of the geometric etendue of this second system is straightforward: the sensor and the source are perpendicular to the optic axis, the cosines in formula [3.22] are equal to 1. The geometric etendue is thus written as:

$$d^2G_1 = ds_r\frac{ds_s}{r^2} = S\frac{\pi\frac{D^2}{4}}{f^2} = \frac{\pi}{4}S\frac{D^2}{f^2} \qquad [3.23]$$

The magnitude D/f is often called the "aperture number" and marked with N.

3.2.2. *Scanners*

Let us study the case shown in Figure 3.6(a). It represents a scanner instrument. The calculation of the geometric etendue of the pixel seen from the optics will be quite complicated, and it is much simpler to calculate, as previously, the geometric etendue of the lens seen from the sensor. The top part of Figure 3.6, between the sensor and the lens, is in fact identical to the one in Figure 3.5, at about one rotation. We find ourselves in the same case as the sensor on the axis: the geometric etendue seen by the sensor of a scanner remains constant regardless of the pointing angle.

3.2.3. *Pushbrooms*

Figure 3.6(b) corresponds to an instrument with a linear array of detector installed in the focal plane. The different elementary detectors are, therefore, regularly spaced on the focal plane. Let d be the size of an elementary detector (usually around 10 μm). As opposed to the case of the scanner, the sensor is no longer on the optical axis. The geometric etendue is calculated starting from the equation below, which is derived from equation [3.22].

$$d^2G_r = ds_r \frac{ds_s \cos \Theta_s}{r^2} \cos \Theta_r$$

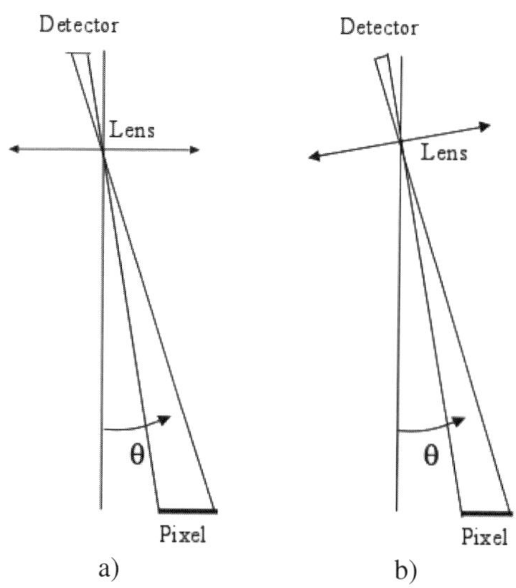

Figure 3.6. *a) Push-broom instrument, b) scanner instrument*

We have:

– on the one hand, $\cos \Theta_s = \cos \Theta_r = \cos \Theta$ because the surfaces of the pixel and of the lens (at the level of the optic center O) are parallel;

– on the other hand, as in the previous case, $ds_r = d^2 = S$ and $ds_s = \pi \frac{D^2}{4}$;

– finally, the distance r is larger than the focal and its value is $r = \frac{f}{\cos \theta}$.

We deduce from this that the value of the geometric etendue is:

$$dG_2 = \frac{\pi}{4} S \frac{D^2}{f^2} \cos^4 \Theta = dG_1 \cos^4 \Theta \qquad [3.24]$$

Therefore, we note that the geometric etendue, and therefore the sensitivity, will decrease strongly depending on the angle Θ of the field of view. For Θ = 30°, the geometric etendue is reduced by 44%. The VEGETATION instrument, present on the satellites SPOT4 and SPOT5, has a field of 55° on both sides of the nadir. In this case, the geometric etendue could be reduced by 90%. Fortunately, opticians have designed optic combinations, called telecentric combinations, such that the surface of the entering pupil ds_s varies like $\frac{1}{\cos^4 \Theta}$, which compensates the variation of the geometric etendue. However, these optics are complex and need a large number of lenses, out of which at least one is not spherical; they are therefore very expensive.

3.3. Atmospheric effects

Even though this book is dedicated to the observation of continental surfaces, it is necessary to talk about the atmosphere, since it is situated between the satellite and the surface and it can alter the observations in different ways:

– The clouds can hide the surface completely or cover it partially. They also project shadows on the surface. Because of their spatial non-uniformity, it is nearly impossible to correct their effects, even when they only cover the surface slightly. The preprocessing of data generally aims just at detecting them, which is not always easy.

– Besides clouds, the atmosphere perturbs the observations of the surface with two different phenomena: absorption and scattering. We will detail these two phenomena in the following.

3.3.1. *Absorption*

The gas molecules in the atmosphere can absorb the radiation at certain wavelengths. When a photon runs into a molecule, it can, if it has the right wavelength, put this molecule in vibration or rotation, which results in the absorption of the photon. The position and the significance of the absorption rays depend on the kind of molecule (see Figure 3.7):

– the ozone O_3 strongly absorbs in the ultraviolet, but also slightly in the green;

– the di-oxygen O_2 has a very narrow absorption band around 760 nm;

– the water vapor H_2O has several absorption bands spread all over the spectrum. For example, it is the cause of the wide absorption bands at wavelengths greater than 1000 nm;

– the carbon dioxide CO_2 presents strong absorption bands in the thermal infrared.

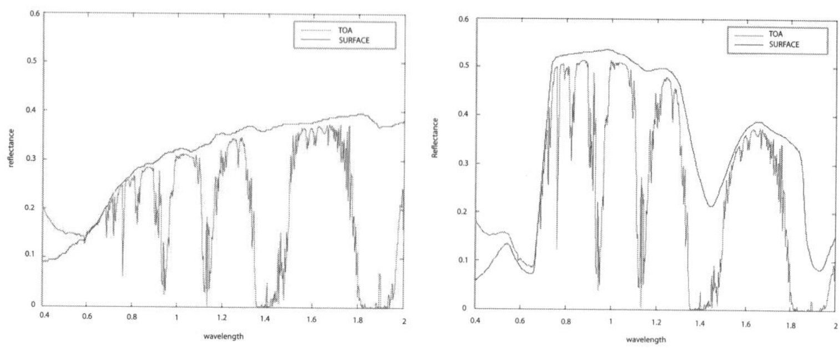

Figure 3.7. *Sand spectra (left) and vegetation spectra (right) at the surface (smooth line) and at the top of the atmosphere (rough line), emphasizing the absorption effects by the atmosphere (TOA = top of atmosphere)*

3.3.2. *Scattering*

The molecules and the particles suspended in the atmosphere (aerosols) can deflect the light: this phenomenon is called "scattering".

The scattering by the air molecules, which are much smaller than the visible wavelengths, was modeled by Rayleigh in 1871. For the spherical aerosol particles whose size is close to the magnitude of the wavelength, the scattering theory was described by Mie [BOR 70].

Scattering is not isotropic and depends on the scattering angle shown in Figure 3.8. It also depends on the quantity of matter it runs through and on the wavelength. For molecules, the magnitude of scattering varies with λ^{-4}, while for aerosols, the variation is of the order λ^{-2} to $\lambda^{-0.5}$. Because of these variations, the bands situated in the blue range are much more sensitive to scattering than the bands situated in the near infrared (NIR: 700–1,000 nm) or in the short wave infrared (SWIR: 1–3 μm).

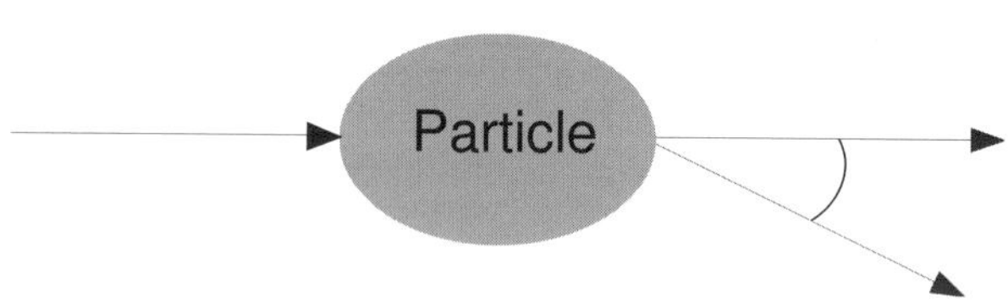

Figure 3.8. *Scattering angle*

3.3.3. *Radiative transfer in the atmosphere*

We will develop here a very simple modeling of the effects of the atmosphere. A much more rigorous and detailed modeling can be found in [LEN 93].

When an atmospheric layer is being run through, a part of the reflectance (at altitude z) propagated in the direction perpendicular to the atmosphere layer is absorbed, and another part is scattered. Furthermore, the exiting reflectance (in $z + dz$) receives an additional input, because some light coming from another direction is scattered or (in the thermal infrared) some light is emitted by the layer.

This can be written as follows, not accounting for the thermal emission:

$$\frac{d\rho(z)}{dz} = \frac{\rho(Z + dz) - \rho(Z)}{dz} = -(\sigma_d + \sigma_a)\rho(z) + \rho_d \qquad [3.25]$$

where σ_d is the extinction coefficient for scattering and σ_a is the extinction coefficient for absorption. The additional reflectance ρ_d is due to the scattering of the upward reflectance by the atmospheric layer.

Equation [3.25] is named the "radiative transfer equation". Integrating this equation in the entire atmosphere and for all the propagation directions is an art of its own (see Chapter 11 as well as [LEN 93]). The integration is not generally feasible analytically and must be done numerically, using a "radiative transfer code".

The approximation used by [TAN 90] in the 5S radiative transfer model is based on the modeling of the four types of path followed by photons in the atmosphere. These four paths are described in Figure 3.10. With this approximation, the reflectance at the top of the atmosphere $\rho(Z_T)$ can be written as a function of the surface reflectance $\rho(Z_0)$:

$$\rho(Z_t) = T_a^{\downarrow}.T_a^{\uparrow}\left(\rho(Z_0)\frac{(T_{dir}^{\downarrow} + T_{dif}^{\downarrow}).(T_{dir}^{\uparrow} + T_{dif}^{\uparrow})}{1 - a\rho_{env}(Z_0)} + \rho_{atm}\right)$$

We will analyze this formulation for the four types of trajectories shown in Figure 3.10. The different notations will be described gradually in the following sections. This formulation needs several hypotheses: the Earth and the atmospheric layers are flat and parallel, the surface reflectance is "Lambertian", it does not depend on the solar or viewing direction, and the reflection on the atmosphere is assumed isotropic. We must keep in mind that this is only an approximation, and that the errors can reach several percent in the least favorable cases.

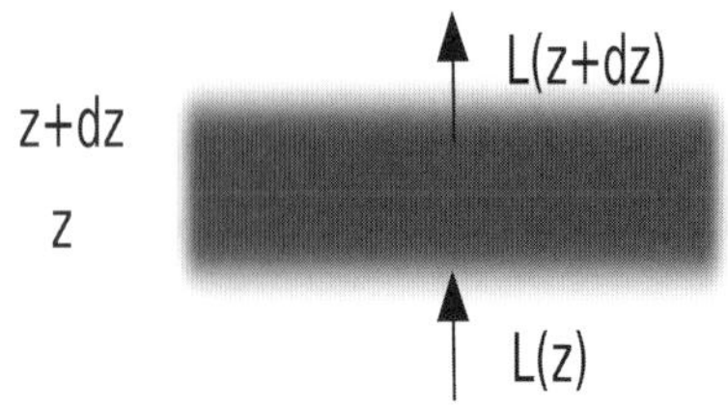

Figure 3.9. *Radiative transfer through an atmospheric layer*

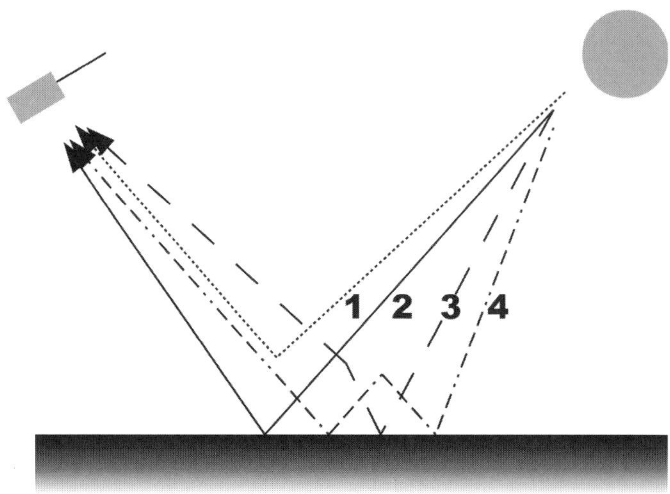

Figure 3.10. *The light follows different types of path to reach the satellite. These trajectories are discussed in section 3.3.3*

3.3.3.1. *Gas absorption*

In this modeling, we assume that the absorbing layer of the atmosphere is situated above the scattering layers, which is true for the absorption of ozone, but less true for the absorption by the oxygen or water vapor. The corresponding approximation is the following:

$$\rho(Z_t) = T_a^{\downarrow}.T_a^{\uparrow}\rho^*(Z_t)$$

where $T_a^{\downarrow}T$ and $a^{\uparrow}$ are, respectively, the upward and downward gas transmissions due to the air molecules and $\rho^*(Z_s)$ is the reflectance at the top of the atmosphere, in the absence of absorption. The upward gas transmission is the rate of photons reflected by the surface that are not absorbed by atmospheric gas molecules. The transmission magnitude is lower than or equal to 1, and is very dependent on the wavelength and on the concentrations of the different molecules. For instance, the atmospheric content in water vapor is highly variable.

In practice, radiative transfer codes such as 6S [VER 97] allow us to calculate these two terms. Some more accurate codes, so-called "ray per ray" codes, provide absorption calculations with a higher accuracy.

3.3.3.2. *Paths reflected by the terrestrial surface (paths 2, 3 and 4)*

The following is dedicated to the modeling of paths 2, 3 and 4 shown in Figure 3.10:

$$\rho_{2,3,4}(Z_0) = \rho(Z_0)\frac{(T^{\downarrow}_{dir} + T^{\downarrow}_{dif}).(T^{\uparrow}_{dir} + T^{\uparrow}_{dif})}{1 - a\rho_{env}(Z_0)} \qquad [3.26]$$

where $T^{\uparrow}_{dir}$ and $T^{\downarrow}_{dir}$ are, respectively, the direct, upward and downward transmissions, $T^{\uparrow}_{dif}$ and $T^{\downarrow}_{dif}$ are diffuse transmissions, a is the atmospheric albedo and ρ_{env} is the environment (or neighborhood) reflectance. We will detail all these terms in the following sections.

3.3.3.2.1. Path 2

The term of equation [3.26] would value $\rho(Z_0).T^{\downarrow}_{dir}.T^{\uparrow}_{dir}$, if it only had type 2 paths. The direct upward transmission is the rate of photons reflected by the surface that are not scattered by the atmosphere. Similarly, the direct downward transmission is the rate of photons coming from the sun that reach the surface without being scattered by the atmosphere. It is possible to derive an analytical expression for these two transmission terms, using equation [3.25] for a vertical trajectory in the atmosphere (with $\sigma_a = 0$: no absorption, and $\rho_d = 0$ since we only consider path 2). This is now written as:

$$\frac{d\rho(z)}{\rho(z)dz} = -\sigma_d$$

We can easily integrate it, and we obtain:

$$T^{\uparrow}_{dir} = \exp(-\int_{Z_0}^{Z_S} \sigma_d dz) = \exp(-\tau)$$

The quantity $\tau = \int_{Z_0}^{Z_S}(\sigma_d)dz$ is called the "*optical thickness*", and it is a unitless number. The atmospheric optical thickness is the sum of the optical thickness due to molecules (see Table 3.1) and of the optical thickness of the aerosols. Opticians often use aerosol optical thickness (AOT) to indicate the quantity of aerosols suspended in the atmosphere. Generally, the optical thickness of aerosols, at 550 nm is comprised between 0 and 2, and in most cases, it is lower than 0.5.

Wavelength (nm)	450	550	670	865	1,650
Molecular optical thickness	0.23	0.09	0.04	0.016	0.001

Table 3.1. *Molecular optical thickness (Rayleigh), as a function of wavelength, at sea level (1,013.25 hPa)*

If the path in the atmosphere is not vertical, the atmospheric optical thickness that must be run through increases with a factor $\frac{1}{\cos\theta^{\uparrow}}$ and therefore $T^{\uparrow}_{dir} = \exp(-\frac{\tau}{\cos\theta^{\uparrow}})$. Similarly $T^{\downarrow}_{dir} = \exp(-\frac{\tau}{\cos\theta^{\downarrow}})$, with $\theta^{\downarrow}$ and $\theta^{\uparrow}$ as the solar zenith angle (downward path) and the viewing zenith angle (upward path).

3.3.3.2.2. Path 3

The path of type 3 includes photons that are scattered by the atmosphere during their downward path or upward path, and that are also reflected once by the Earth's surface. If there were only type 3 paths, the term from equation [3.26] would be written as:

$$\rho(Z_0).(T^{\downarrow}_{dir}.T^{\uparrow}_{dif} + T^{\downarrow}_{dif}.T^{\uparrow}_{dir} + T^{\downarrow}_{dif}.T^{\uparrow}_{dif}) \qquad [3.27]$$

where $T^{\uparrow}_{dif}$ and $T^{\downarrow}_{dif}$ are, respectively, the upward and downward diffuse (or scattered) transmissions (the rate of photons scattered at least once by the atmosphere that reach the surface in the downward path, or the satellite in the upward path). There is no simple way of expressing this term (except by considering it null), and we must compute it using a radiative transfer code.

3.3.3.2.3. Path 4

Let us now look at the set of paths of type 2, 3 and 4. These paths can be classified as a function of the number of reflections on the ground and then on the atmosphere. Paths 2 and 3 correspond to a single reflection on the ground. With each new reflection on the atmosphere and the ground, the reflectance is multiplied by a factor much lower than 1, equal to $a\rho_{env}(Z_0)$, where a is the albedo of the atmosphere, and $\rho_{env}(Z_0)$ is the reflectance of the pixels surrounding the area that we are interested in. The reflectance for paths 2, 3 and 4 is therefore written as:

$$\begin{aligned}\rho_{2,3,4}(Z_0) &= \rho(Z_0)(T^{\downarrow}_{dir} + T^{\downarrow}_{dif}).(T^{\uparrow}_{dir} + T^{\uparrow}_{dif}) \\ &\times \left(1 + a\rho_{env}(Z_0) + a^2\rho^2_{env}(Z_0) + ...\right)\end{aligned}$$

Or, by identifying a geometric series:

$$\rho_{2,3,4}(Z_0) = \rho(Z_0)\frac{(T^{\downarrow}_{dir} + T^{\downarrow}_{dif}).(T^{\uparrow}_{dir} + T^{\uparrow}_{dif})}{1 - a\rho_{env}(Z_0)}$$

3.3.3.3. *Atmospheric path (path 1)*

The reflectance of the paths of type 1 in Figure 3.10, ρ_{atm}, is due to the scattering by the atmosphere. It is also the reflectance observed when the surface reflectance is null, in the absence of absorption. To calculate this term, the best way is to resort to calculations using a radiative transfer code. However, to obtain orders of magnitude, we can use the following equations obtained after breaking down the atmospheric reflectance into its components connected to molecules and aerosols $\rho_{atm} = \rho_{molecules} + \rho_{aerosols}$, which is once again an approximation that neglects photons scattered successively by molecules and aerosols.

$$\rho_{molecules} = \frac{\tau_{molecules}.P_{molecules}(\Theta)}{4.\cos\theta^{\downarrow}\cos\theta^{\uparrow}}$$

where:

– τ is the optical thickness of the molecules in the wavelength;

– $\theta^{\downarrow}$ and $\theta^{\uparrow}$ are the solar zenith angles (downward path) and the viewing angles (upward path).

The same goes for the aerosols: $\rho_{aerosols} = \frac{\tau_{aerosols}.P_{aerosols}(\Theta)}{4.\cos\theta^{\downarrow}\cos\theta^{\uparrow}}$

Figure 3.11 shows two examples of aerosol phase functions. The aerosols made up of large particles have the specificity of scattering the light very slightly, and the phase function exhibits a strong peak toward the forward direction (scattering angle equal to 0). Conversely, the small aerosols have a phase function that is close to that of molecular scattering.

3.3.4. *Magnitude orders of the atmospheric effects*

Figure 3.12 shows several simulations of reflectances at the top of the atmosphere, as a function of surface reflectance, for three aerosol abundances. In the blue, the atmospheric reflectance (for a null surface reflectance) is approximately 0.1 without aerosols and nears 0.2 when the optical thickness increases. In the NIR, the atmospheric reflectance is nearly null. We may notice that, for each wavelength, the curves cross at the same "critical" surface reflectance, situated between 0.1 and 0.2. For this critical surface reflectance, when the optical thickness increases, the increase in the atmospheric reflection is exactly compensated by the decrease in the transmittance, which means that the reflectance at the top of the atmosphere no longer depends on the quantity of aerosols.

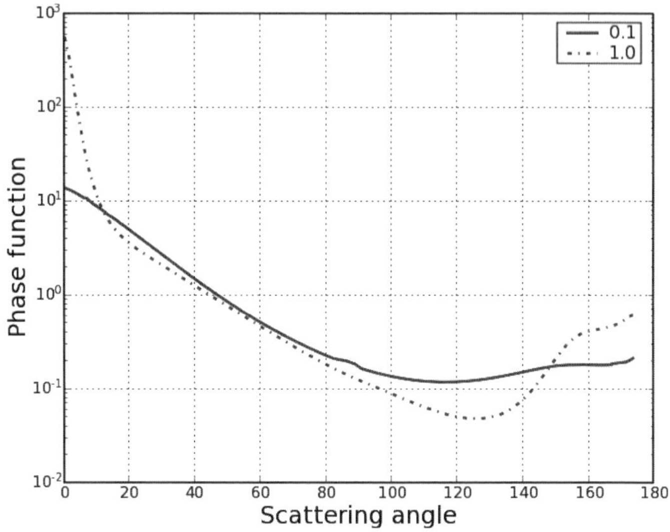

Figure 3.11. *Example of aerosol phase functions, for the particles radius 0.1 μm (solid, to 2 μm (dashed)*

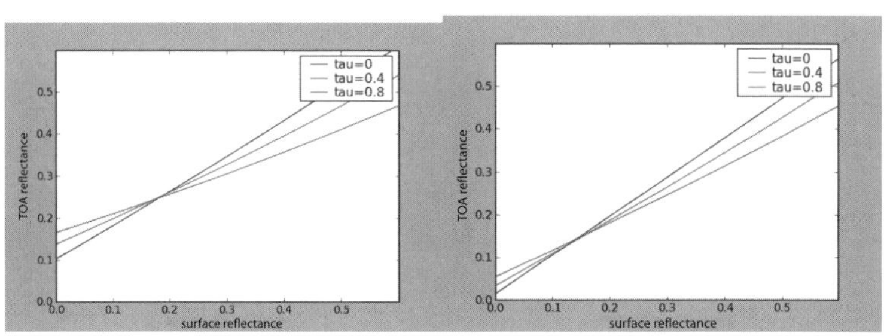

a) b)

Figure 3.12. *TOA reflectance curves as a function of surface reflectance and of optical thickness (τ) at 550 nm for continental aerosols: for a spectral band a) in the blue (440 nm) b) in the PIR (870 nm), for a solar zenith angle of 45° and a viewing zenith angle of 30°. The critical reflectance for which the curves cross is close to 0.3 in the blue, and 0.15 in the PIR*

3.4. Spectral properties of the surfaces

The variability of the reflectance spectra of natural or artificial surfaces is infinite, and there are several examples that can be found in the available spectral databases that are available to the public (for example the ASTER database[1]).

This section aims primarily at comparing the interest of the different spectral bands that are usually available on the satellites for the observation and the discrimination of different terrestrial surfaces. We will provide here some details about the spectra of bare ground, vegetation, water and snow. Figure 3.13 shows several examples of ground, vegetation or snow spectra. The three spectra of green leaves, or even of lawn, display characteristics that are very similar with few, subtle differences. The dry leaf is very easily distinguishable, with much higher reflectances in the visible range that show the absence of chlorophyllous activity (chlorophyll absorbs the light in the visible). For healthy plants, even though the absorption made by the chlorophyll is very strong throughout the entire visible range, it is less strong in the green (550 nm), which explains the green color of these plants. The absorption increases again in the red, and the reflection reaches a new low toward 670 nm. The chlorophyll abruptly stops absorbing the light beyond 720 nm, which induces a strong surge of reflectances that reach a plateau in the NIR. This area of strong reflectance surge is named *red-edge*. It is worth knowing that the sensitivity of human eye pigments stops just before the red-edge, maybe so as to avoid being dazzled by vegetation.

Beyond 1μm, the more significant absorption is associated with the thickness of the leaf, and above all with the light absorption by the water contained in the leaf. Unfortunately, it is difficult to use this property to detect the leaf water content, because the absorption by the water in the leaf is hidden by the absorption by atmospheric water vapor, as we can see in Figure 3.7.

There are several, subtler characteristics of the leaf spectra that are connected to the chlorophyllous activity: the chlorophyll re-emits part of the radiation that is absorbed at a higher wavelength through a phenomenon of fluorescence. Fluorescence light is, however, a very faint signal, very difficult to measure, but it may be detected in narrow spectral bands where the irradiance is low, for instance in Sun Fraunhofer lines. The European Space Agency is studying a satellite project called Flex that seeks to measure the fluorescence. The PROSPECT model [JAC 90] helps simulate the absorption spectra and transmission spectra of the leaves depending on their characteristics.

Finally, we must note that we only showed leaf spectra, but the reflectance of the whole vegetation cover also depends on the density and the geometry of the cover, on

1 ASTER = Advanced Spaceborne Thermal Emission and Reflection Radiometer; http://asterweb.jpl.nasa.gov/).

the leaf orientation and on the ground reflectance. There are several models that try to describe the reflectance of vegetation cover: the simple model SAIL is one of the most commonly used [JAC 09].

Several soil spectra are also shown in Figure 3.13. We note that the wetter the soil, and the more humus it contains, the darker it is. The reflectance often increases quite slowly from blue to red, before reaching a plateau in the NIR area. However, there are several white soils that have a larger reflectance in the blue range. This is also the case for artificial materials such as asphalt and concrete, and for this reason, cities often have a gray color on satellite images in the visible spectrum.

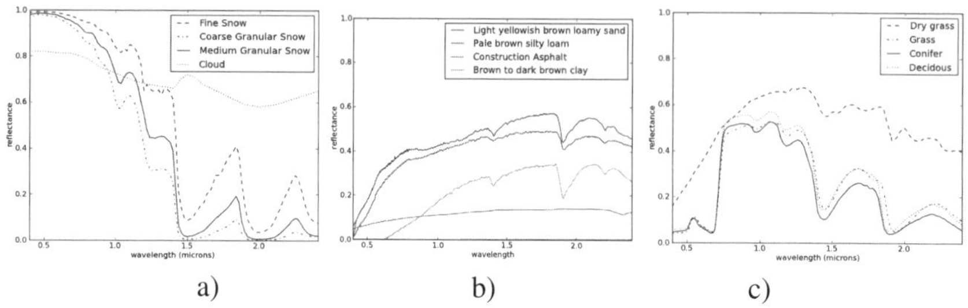

Figure 3.13. *Reflection spectra measured in the laboratory a) for snow b) for different types of soil as well as c) for different types of leaves (source: ASTER, NASA databases)*

Snow is a very white surface for the human eye, but in fact its reflectance decreases quite abruptly in the NIR and especially in the SWIR. This feature allows us to discriminate the snow from the clouds: the clouds have higher reflectances in the SWIR, as we can see in Figure 3.13.

This is visible in Figure 3.15 that compares a color composite using natural colors (blue, green and red), on which it is difficult to differentiate snow from clouds, to a color composite (blue, near infrared (NIR) and short wave infrared (SWIR)), where the snow appears in blue and can be easily separated from the clouds that have remained white.

3.5. Directional properties of the surfaces

The reflectance of a surface depends on the illumination direction and the observation direction. A Lambertian surface is a theoretical surface whose reflectance is constant, regardless of the solar angles and the viewing angles. Several natural surfaces are quasi-Lambertian: this is the case for a flat surface of fresh snow, or for mat sand. Most surfaces, however, show a strong anisotropy, and in this case, we speak of “directional effects”. There are two phenomena that concur to increasing the directional effects:

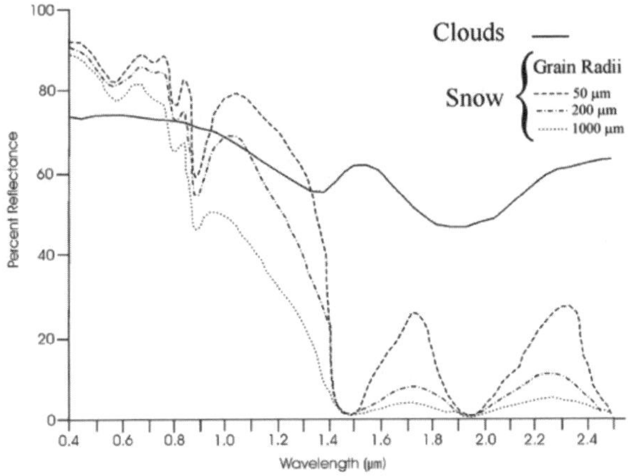

Figure 3.14. *Typical spectra of clouds and snow (for different grain sizes)*

a) b)

Figure 3.15. *Comparison of two colored compositions obtained from a LANDSAT image. a) The colored composition is (blue, green and red). b) red, PIR and MIR . For a color version of this figure, see www.iste.co.uk/tupin/RSImagery.zip*

– *The specular reflections*: the surface behaves partly like a mirror and therefore reflects the light around the specular direction. The specular direction is such that the normal at the surface is the bisector of the incident direction and the specular direction (see Figure 3.16). At a high resolution, reflectances greater than 10 can be observed in the specular direction above roof skylights or car windscreens. Such specular reflections often cause a saturation of the sensors that are not designed to observe such strong radiances. Light reflection on a liquid surface is also mainly specular. If the water surface is rough, the light gets reflected into a cone that is centered on the specular direction. In the case of the reflection on the ocean, at a low resolution, the observed specular reflectance is generally comprised between 0.1 and 0.6 (see Figure 3.17). At high resolutions, higher reflectances can be observed occasionally.

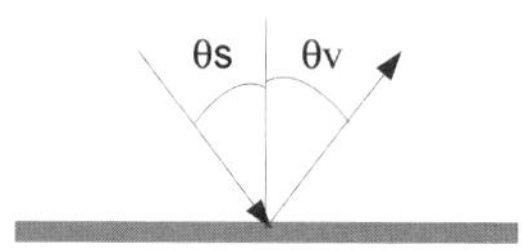

Figure 3.16. *The specular direction is symmetrical to the incident direction with regard to the normal of the surface*

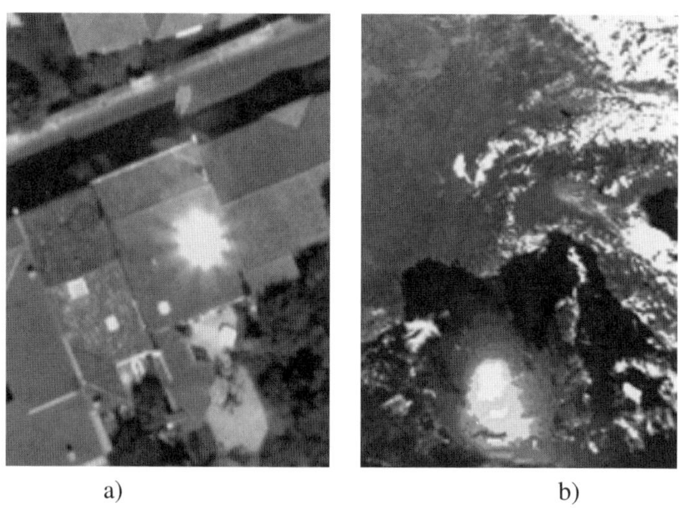

Figure 3.17. *Two examples of specular reflections: a) on a roof window, and b) on the sea surface*

– *The shadow or volume effects*: the surface being neither flat nor homogeneous (a tree cover, rocks), the light enters the volume, and some parts of the solid are shadowed. This effect creates very strong variations of the reflectance near the backscattering direction illustrated in Figure 3.18. This effect, called a *hot spot*, is visible in Figure 3.19(b). In this figure, it may be noticed that reflectance variations due to directional effects on vegetation can exceed a factor 2. This phenomenon is truly a nuisance when we use a time series of observations of the same pixel under variable viewing angles, or at variable hours during the day. If the directional effects are not taken into account (and considering them is not easy), the series is affected by a considerable noise. The *hot spot* phenomenon is surely visible to the naked eye, as we can see in Figure 3.20.

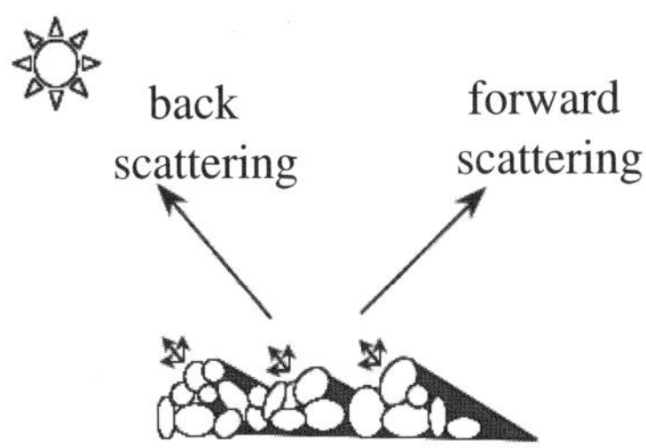

Figure 3.18. *The backscattering direction corresponds to the incident direction. In this direction, only the lit surfaces are visible, while as soon as we deviate from this direction, the shadows appear and cause a strong decrease in the reflectances*

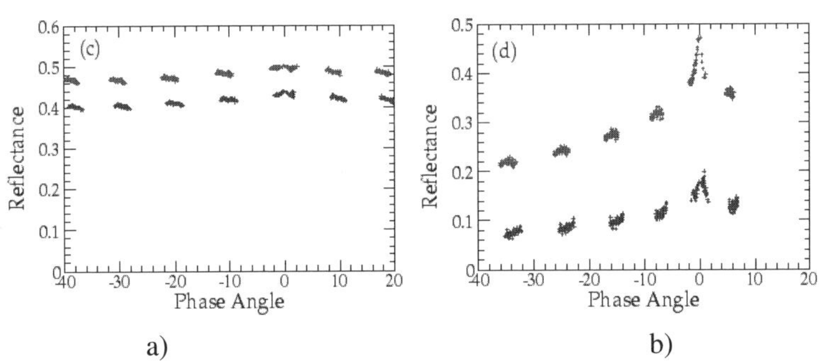

Figure 3.19. *Variation of the reflectances measured by the POLDER instrument as a function of the angular deviation from the backscattering direction. a) For a desert, and b) for cultivated areas, for which the backscattering effect is very pronounced (the hot spot situation) (plot by F.M. Bréon). For each plot, the top curve correponds to the NIR band, and the bottom one to the red band*

3.6. Practical aspects: products, atmospheric corrections, directional corrections

We saw in section 3.3 (Figure 3.12) that because of atmospheric effects, the reflectances observed at the top of the atmosphere above a vegetation parcel can double when the optical thickness of aerosols goes from 0 to 0.3. We have also seen that there are radiative transfer codes whose accuracy is generally enough for correcting these effects. In fact, the difficulty of the correction lies in the knowledge of the atmospheric composition: absorbing molecules, aerosols, etc.

Figure 3.20. *Which hot-air balloon was this picture taken from? (photo A. Deramecourt)*

3.6.1. *Absorption correction*

Generally, radiative transfer codes process absorption and scattering separately, and do not allow us to consider the fact that the absorbing molecules and the scattering particles are mixed together. The method that is currently used for implementing the atmospheric corrections in bands where the absorption is weak consists of correcting the gas absorption, then the effects of the atmospheric scattering. Implementing this correction in this order means that we must suppose that the absorbing molecules are situated above scattering particles of the atmosphere. This hypothesis is verified for ozone, but not for oxygen and water vapor. However, in most cases, the satellite spectral bands have been defined to avoid strong absorption bands and therefore so that the error margin in this approximation stays insignificant.

3.6.2. *Scattering correction*

To be able to correct for the scattering effects, we must know the scattering particles of the atmosphere. The abundance and distribution of the gas in the atmosphere are sufficiently well known and depend mainly on the surface atmospheric pressure, and therefore on the altitude. So, if we have accurate information about the aerosols, the correction of atmospheric effects can be done quite accurately. Unfortunately, there are not operational, global and accurate information sources yet on the abundance and type of aerosols continuously available. There are, however, measurement devices, photometers, which give this type of information at a local level. We can, therefore, equip sites with this kind of device in order to perform atmospheric correction. The Aeronet network collects data from 500 sites and makes it available on its site[2]. However, the number of these instruments is definitively not sufficient for allowing us to correct the effects of the aerosols everywhere, and there is, for example, only half a dozen of such instruments in France. Spaceborne instruments (such as MODIS and VIIRS) currently only provide partial and incomplete information on aerosols, solely at the overpass time and only for cloud-free pixels. The combination of atmospheric transport models, source modeling, on-site measurements and satellite observations is promising, but not yet operational.

For the time being, it is therefore necessary to rely on earth observation data themselves to estimate the optical properties of aerosols. The principles of several methods are explained in the following sections. As the optical thickness of aerosols varies slowly in space, it is not necessary to process the aerosol estimates at full resolution, a resolution comprised between 100 and 500 m being generally enough for estimating the optical thickness.

3.6.2.1. *The dark pixel method*

The method consists of finding the darkest pixel in the image (a lake, a forest). This pixel has a reflectance ρ_{TOA}. We attribute to this pixel a quasi-null surface reflectance (0.01). If we know the aerosol model to be used, we can find the optical thickness that allows us to observe the reflectance ρ_{TOA}. The uncertainty about the dark pixel surface reflectance introduces, however, a strong uncertainty into the optical thickness. Moreover, the pixels in the shadow of clouds must be avoided, since their reflectance at the top of the atmosphere can be lower than that of the atmospheric reflectance, as they are not fully illuminated.

3.6.2.2. *The spectral method*

This method is based on the hypothesis of relations between surface reflectances. This method works particularly well for dense vegetation (dark dense vegetation

---

2 Aeronet network: http://aeronet.gsfc.nasa.gov.

(DDV), which have very stable spectra. The best-known version is the one used by NASA for the MODIS project [REM 05]. It connects the surface reflections in the blue and in the red with those in the short infrared wave. These two assumptions provide two equations that allow us to estimate the two unknowns: the type and optical thickness of aerosols. This method works very well in temperate and Northern areas, but not in arid areas, where it is difficult to find dense vegetation.

The first versions of the DDV method used the following equations:

$$\rho_{Blue} = 0.5\,\rho_{SWIR} \qquad \text{and} \qquad \rho_{Red} = 0.25\,\rho_{SWIR} \qquad [3.28]$$

The following versions have complicated these equations a little bit, without modifying the principle.

There are other recent works [HAG 08] that have shown that the use of an equation $\rho_{Red} = 0.5\rho_{Bue}$ allows for more precise determination of the optical thickness, for less-dense vegetation covers. On the contrary, it does not allow us to determine the model of the aerosols.

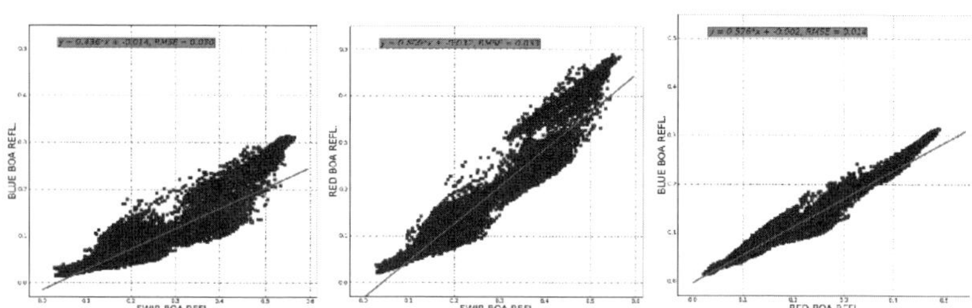

Figure 3.21. *Comparison of surface reflectances on different types of scenes, in different bands of the satellite LANDSAT (from left to right, blue and SWIR, red and SWIR, blue and red). The decrease of the scatter on the right curve is obvious*

3.6.2.3. *The multitemporal method*

This method needs dense image time series obtained with constant observation angles, for example the ones provided by the satellites LANDSAT, FORMOSAT-2, or the future VENμS and Sentinel-2. Between two successive acquisitions, in most cases, we may assume that the surface reflectance has not changed. Under this assumption, the top of the atmosphere reflectance changes can be attributed to the changes in the AOT which can be used for determining the optical thickness [HAG 08]. This method works as well on the vegetation as it does on the bare grounds, but can fail when the assumption of stable reflectance is not true, for instance:

– when the vegetation cover varies quickly (in spring or autumn);

– in case of agricultural intervention (harvest);

– on bare ground, after rain.

Part of these problematic cases can be detected by a thresholding on the variations in the SWIR (this band being much less sensitive to the presence of aerosols, a variation in top of atmosphere reflectance is often due to a variation of surface reflectance).

Figure 3.22. *Two images acquired one day apart, on the same site and from the same angle. A significant aerosol content is obvious on the image on the left*

3.6.3. *Examples of atmospheric correction results*

Using time series of FORMOSAT-2 data, we were able to compare the performances of the AOT retrieval obtained with the multispectral or multitemporal methods. This comparison shows the superiority of the multispectral method over the multitemporal method on sites covered with vegetation, but provides the inverse result on arid sites. It is, however, possible to combine both methods in order to improve the results in both arid and vegetation cases.

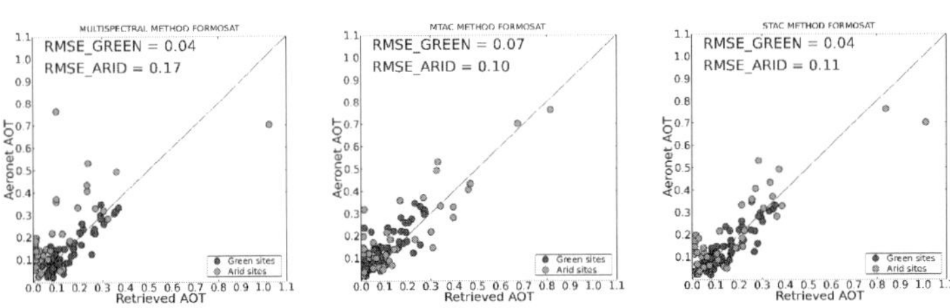

Figure 3.23. *From left to right, validation of the thicknesses obtained by the multispectral and multitemporal methods, as well as a combination of both. The combination of both methods gives slightly superior results in all cases*

Chapter 4

The Physics of Radar Measurement

This chapter deals with the physics principles involved in radar measurements, radar imaging and radar scatterometry. The chapter begins with the laws and formalisms involved in the propagation (section 4.1.1) and the polarization (section 4.1.2) of electromagnetic waves. The radar equation and the respective definitions (radar cross-section (RCS) and backscattering coefficient) are presented in section 4.2. In section 4.3, we then deal with the main principles of electromagnetic wave interaction with natural environments: bare ground, vegetation and sea surface. The quality of the radiometric estimation is discussed in section 4.4. The last section (section 4.5) is devoted to radar polarimetry.

4.1. Propagation and polarization of electromagnetic waves

4.1.1. *Propagation of electromagnetic waves*

4.1.1.1. *Maxwell equations*

Maxwell equations govern the propagation of electromagnetic waves. Defined in their definite form by James Clark Maxwell in 1873, they formalized and extended

---

Chapter written by Jean-Claude SOUYRIS.

Faraday's intuitions (1831), before having been experimentally validated by Hertz (1888):

$$\begin{cases} \nabla \times \vec{e}(\vec{r},t) + \frac{\partial}{\partial t}\vec{b}(\vec{r},t) = 0 \\ \nabla \times \vec{h}(\vec{r},t) - \frac{\partial}{\partial t}\vec{d}(\vec{r},t) = \vec{j}(\vec{r},t) \\ \nabla \vec{b}(\vec{r},t) = 0 \\ \nabla \vec{d}(\vec{r},t) = \rho(\vec{r},t) \end{cases} \qquad [4.1]$$

Equations [4.1] connect the vectors $\vec{e}(\vec{r},t)$, $\vec{h}(\vec{r},t)$, $\vec{d}(\vec{r},t)$, $\vec{b}(\vec{r},t)$, which are, respectively, the electric and magnetic fields, the electric displacement and the magnetic induction. They are the unknowns of any propagation problem. More specifically, we are concerned with the electric and magnetic fields, as well as electric and magnetic inductions. The terms appearing on the right-hand side of the equations, $\rho(\vec{r},t)$ and $\vec{j}(\vec{r},t)$, known as source terms, represent the electric charge density and the electric current density (these are the given data of the problem).

The introduced vectors depend on four parameters (three space parameters x, y and z, and a time parameter t), and are then subjected to a set of mathematical operators: the curl operator ($\nabla\times$), the divergence operator (∇) and the partial time derivative ($\partial/\partial t$). The operators $\nabla\times$ and ∇ act on the space coordinates, while the operator $\partial/\partial t$ acts on the time coordinate. The first two equations, called Faraday's law and Ampere's generalized law, reveal the way in which the space variations of the electric and magnetic fields generate time variations of magnetic (and electric, respectively) induction. The last two equations are Gauss' laws for electric displacement and magnetic induction. They are deduced from the first ones, and from the energy conservation principles during the spatiotemporal ballet of the electromagnetic fields. In this respect, they do not bring any additional information on the structure of the propagation. Thus, Maxwell equations limit the number of independent scalar equations to 6 (2×3).

4.1.1.2. *Constitutive equations*

The unknown vectors $\vec{e}$, $\vec{h}$, $\vec{d}$, $\vec{b}$ are each formed of three spatiotemporal components; thus, the problem posed by equations [4.1] has 12 unknowns. It is therefore necessary to attach a complementary information, suitable for the propagation medium. This is the role of the constitutive equations, which express $\vec{d}$ and $\vec{b}$ depending on $\vec{e}$, $\vec{h}$, and on the permittivity and permeability characteristics of the studied medium:

$$\begin{cases} \vec{d}(\vec{r},t) = \bar{\bar{\varepsilon}}\,\vec{e}(\vec{r},t) + \bar{\bar{\xi}}\,\vec{h}(\vec{r},t) \\ \vec{b}(\vec{r},t) = \bar{\bar{\varsigma}}\,\vec{e}(\vec{r},t) + \bar{\bar{\mu}}\,\vec{h}(\vec{r},t) \end{cases} \qquad [4.2]$$

these equations refer to the "reaction" of the medium to the electromagnetic illumination: the fields $\vec{e}$ and $\vec{h}$ of the incident wave create the inductions $\vec{d}$ and $\vec{b}$, which characterize the spatiotemporal geometry of the propagation within the considered medium. We will not approach the general sense that must be given to the tensors $\bar{\bar{\varepsilon}}$, $\bar{\bar{\xi}}$, $\bar{\bar{\varsigma}}$ and $\bar{\bar{\mu}}$ (see, for example, [KON 90]). When the electromagnetic wave propagates in the vacuum, the constitutive equations [4.2] take the simplified form:

$$\begin{cases} \vec{d}(\vec{r},t) = \varepsilon_0\,\vec{e}(\vec{r},t) \\ \vec{b}(\vec{r},t) = \mu_0\,\vec{h}(\vec{r},t) \end{cases} \qquad [4.3]$$

The vacuum is an isotropic medium (the vectors $\vec{d}$ and $\vec{e}$ on the one hand, and $\vec{b}$ and $\vec{h}$ on the other hand are collinear), characterized by the universal constants $\varepsilon_0 \approx 8.85{\times}10^{-12}\text{Farad m}^{-1}$ and $\mu_0 \approx 4\pi 10^{-7}\text{Henry m}^{-1}$. Furthermore, in the vacuum, which is a non-charged medium:

$$\vec{j}(\vec{r},t) = \vec{0} \qquad \text{and} \qquad \rho(\vec{r},t) = 0 \qquad [4.4]$$

The hypothesis of a propagation medium, assimilated to the vacuum, implies that we should neglect the interactions of the incident waves with some of the constitutive layers of the atmosphere (i.e. troposphere and ionosphere). The validity of these hypotheses can be questioned for some bandwidths (typically, the emergence of interactions with the ionosphere for the frequencies < 1 GHz, and the interactions with a troposphere beyond 10 GHz) (see section 4.2.1).

4.1.1.3. *The limit conditions*

The Maxwell equations are finally complemented with boundary conditions, thus specifying a given problem. They express the need to impose initial conditions to ensure the uniqueness of the solution of differential equations. The presence of a perfectly conductive wall will, for example, impose locally the orthogonal aspect of the electric field to its surface.

4.1.2. *Polarization of the electromagnetic waves*

4.1.2.1. *Structure of the electromagnetic waves in the vacuum*

Starting from Maxwell equations [4.1] and the constitutive equations for the vacuum [4.2], we establish the following propagation equation:

$$\nabla^2\vec{e}(\vec{r},t) - \varepsilon_0\mu_0\frac{\partial^2}{\partial t^2}\vec{e}(\vec{r},t) = 0 \qquad [4.5]$$

where ∇^2 is the Laplace operator. Expressed in a reference basis (x, y, z), it is thus written:

$$\nabla^2 = \frac{\partial^2}{\partial x^2} + \frac{\partial^2}{\partial y^2} + \frac{\partial^2}{\partial z^2} \qquad [4.6]$$

In the absence of boundary conditions (an infinite vacuum), a solution of this differential equation is given by [KON 90]:

$$\begin{aligned} \vec{e}(\vec{r}, t) &= \Re\left\{ \vec{E}_0 \exp\left[j\,(\omega t - k_x x - k_y y - k_z z)\right] \right\} \\ &= \Re\left\{ \vec{E}_0 \exp\left[j\,(\omega t - \vec{k}\,\vec{r})\right] \right\} \end{aligned} \qquad [4.7]$$

where $\Re$ represents the real part; x, y, z and k_x, k_y, k_z are the respective coordinates of the position vector $\vec{r}$ and the propagation vector $\vec{k}$, whose components verify:

$$k_x^2 + k_y^2 + k_z^2 = \omega^2 \varepsilon_0 \mu_0 \qquad [4.8]$$

Relation [4.8] is known as the equation of dispersion. The pulsation of the wave, ω, is related to the frequency $f(\omega = 2\pi f)$ and $\vec{E}_0$ is a constant vector, with complex components, contained in a plane that is perpendicular to $\vec{k}$. The structure of $\vec{e}(\vec{r}, t)$ characterizes a monochromatic wave mechanism (both in time and space, simultaneously), with a unique frequency, that propagates along the direction $\hat{k} = \vec{k} \Big/ \left\| \vec{k} \right\|$, k being the "wave number" verifying $k = \frac{2\pi}{\lambda}$.

Equiphase and equiamplitude surfaces are the planes that are orthogonal to the direction of energy propagation, represented by the vector $\hat{k}$: the wave is called a plane wave. Finally, by confronting it once again with the Maxwell equations, the magnetic field $\vec{h}(\vec{r}, t)$ is given by:

$$\vec{e}(\vec{r}, t) = \xi \vec{h}(\vec{r}, t) \wedge \hat{k} \qquad [4.9]$$

where $\xi = \sqrt{\mu_0 / \varepsilon_0} \approx 120\pi$ is the impedance of the wave in the vacuum.

Equation [4.9] defines the structure of the plane wave propagating in the vacuum: the vectors triplet $\left[\, \vec{e}(\vec{r}, t),\ \vec{h}(\vec{r}, t),\ \hat{k} \,\right]$ makes up a right-handed orthogonal basis. The wave has a transverse electromagnetic structure (TEM), where the fields $\vec{e}(\vec{r}, t)$ and $\vec{h}(\vec{r}, t)$ are in the plane perpendicular to the propagation direction (Figure 4.1). The electromagnetic energy transfer thus operates in a direction perpendicular to the electric or magnetic vibrations. As they propagate, the electromagnetic fields undergo a helical rotation centered around $\hat{z}$. Projected onto the transversal plane, this helix is reduced to an ellipse called a "polarization ellipse".

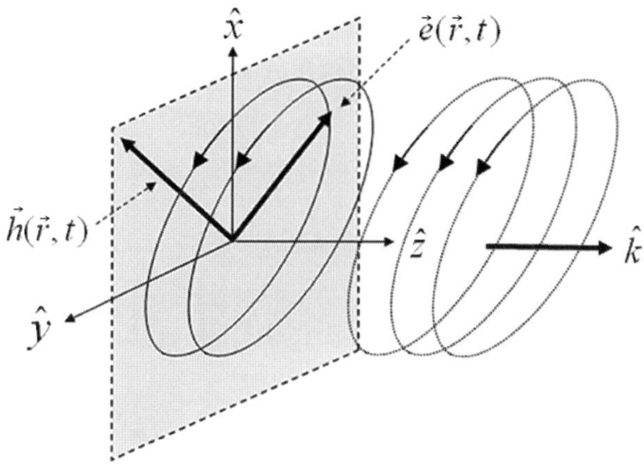

Figure 4.1. *Wave propagation in free space*

4.1.2.2. *Polarization ellipse*

The wave's polarization is defined by the projection, onto a plane perpendicular to the propagation direction, of the curve traced by the leading edge of the electric field $\vec{e}(\vec{r},t)$. The transversal structure of $\vec{e}(\vec{r},t)$, explained according to equation [4.7], yields:

$$\vec{e}(x,y,z,t) = e_x\hat{x} + e_y\hat{y} = \Re\left\{ \vec{E}_0 \exp\left[j(\omega t - kz) \right] \right\} \qquad [4.10]$$

where the vector $\vec{E}_0$ (called the Jones vector) is thus expressed:

$$\vec{E}_0 = E_{0x}\hat{x} + E_{0y}\,\hat{y} = a_x e^{j\delta_x} + a_y e^{j\delta_y}\,\hat{y} \qquad [4.11]$$

This representation of the electric field independently of its spatiotemporal variations is also called a "wave polarization". Starting with equations [4.10] and [4.11], it becomes:

$$\begin{cases} e_x = a_x \cos\left(\omega t - kz + \delta_x\right) \\ e_y = a_y \cos\left(\omega t - kz + \delta_y\right) \end{cases} \qquad [4.12]$$

The variations of the components e_x and e_y only depend on the spatiotemporal variable $(\omega t - kz)$. By considering the square of the difference between e_x and e_y, it becomes:

$$(e_x/a_x)^2 + (e_y/a_y)^2 - 2(e_x/a_x)(e_y/a_y)\cos(\delta_y - \delta_x) = \sin^2(\delta_y - \delta_x) \quad [4.13]$$

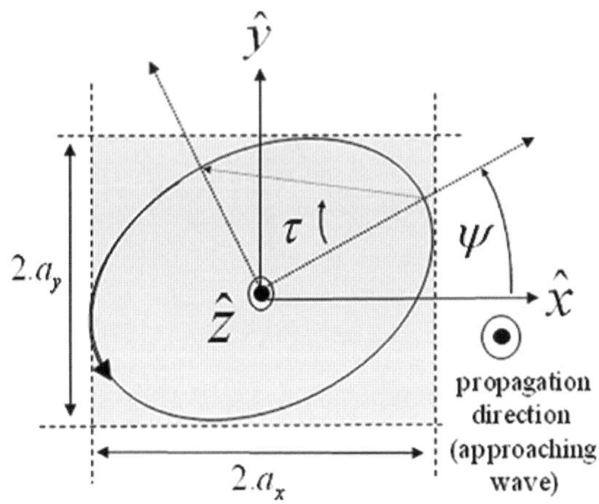

Figure 4.2. *Configuration of the polarization ellipse in the polarization plane*

We obtain the equation of an ellipse, characterized as a polarization ellipse. We choose, arbitrarily, a transversal reference plane, from which we will observe the movement of the electric field, for example the plane z = 0. In this reference plane, the polarization ellipse is characterized by two angular parameters: ψ and τ (Figure 4.2). ψ characterizes the orientation of the major axis of the ellipse with respect to the axis $\hat{x}$, and τ its ellipticity (through its tangent). The sign of the parameter τ indicates the direction in which $\bar{e}(x, y, 0, t)$ rotates around the polarization ellipse. Finally, the magnitude of the semi-major axis of the ellipse informs us about the amount of energy carried by the wave.

By manipulating the trigonometric formulas, we derive the values (ψ, τ) from the values $(a_y/a_x, \delta_y - \delta_x)$ [BOR 70]: the orientation of the major axis of the polarization ellipse, as well as its ellipticity, can be deliberately set if two orthogonal elementary excitations, whose amplitude and relative phase difference ratio can be controlled, are given. Thus, every wave with elliptic polarization can be considered as a linear combination of the two waves with orthogonal, linear polarizations (along the directions $\hat{x}$ and $\hat{y}$), assigned, respectively, by the complex coefficients $a_x e^{j\delta_x}$ and $a_y e^{j\delta_y}$. The complex character of the weighting coefficients expresses the notion of coherent summing defined below.

The Jones vector (equation [4.11]) is also expressed as a function of ψ and τ [LEE 08]:

$$\begin{aligned}\vec{E}_0 &= A\begin{pmatrix}\cos\psi & -\sin\psi\\ \sin\psi & \cos\psi\end{pmatrix}\begin{pmatrix}\cos\tau & j\sin\tau\\ j\sin\tau & \cos\tau\end{pmatrix}\begin{pmatrix}e^{-j\alpha} & 0\\ 0 & e^{j\alpha}\end{pmatrix}\hat{x}\\ &= A\left[\bar{\bar{U}}(\psi)\right]\left[\bar{\bar{U}}(\tau)\right]\left[\bar{\bar{U}}(\alpha)\right]\hat{x}\end{aligned} \qquad [4.14]$$

where A is the amplitude of $\vec{E}_0$ and α is the absolute phase term. To within the complex amplitude $Ae^{j\alpha}$, expression [4.14] reveals the coexistence of the two in quadrature phase vibrations (term j), respectively, with amplitudes proportional to $\cos\tau$ and $\sin\tau$, in a basis that is rotated by an angle ψ with respect to the reference base $(\hat{x}, \hat{y})$.

4.1.2.2.1. The linear polarizations

The linear polarizations have a constant orientation of the propagated electric field. It assumes either a phase difference $(\delta_y - \delta_x)$, which is a multiple of π (equation [4.12]), or a null value $\tau = 0$ [4.14]. For a radarist, a linear polarization is called horizontal when the electric field remains parallel to the surface, and vertical when the electric field belongs to the incident plane (Figures 4.3(a) and (b)) [1].

4.1.2.2.2. The circular polarizations

The circular polarizations require (equation [4.12]) that the following two conditions are satisfied: $a_y = a_x$ and the phase difference $(\delta_y - \delta_x)$ is equal to an odd multiple of $\pi/2$ (the two elementary orthogonal excitations have the same amplitude and are in quadrature $\tau = \pm\pi/4$, see equation [4.10]).

4.1.2.2.3. Orthogonal polarizations

Two polarizations are said to be orthogonal if their respective Jones vectors $\vec{E}_0$ and $\vec{E}_0^{\perp}$ are orthogonal in terms of the Hermitian scalar product $({}^T\vec{E}_0\vec{E}_0^{\perp *} = 0)$. Starting from equation [4.9], the vector $\vec{E}_0^{\perp}$ can be expressed in two ways:

$$\vec{E}_0^{\perp} = A\left[\bar{\bar{U}}(\psi)\right]\left[\bar{\bar{U}}(\tau)\right]\left[\bar{\bar{U}}(\alpha)\right]\hat{y} = A\left[\bar{\bar{U}}(\psi + \pi/2)\right]\left[\bar{\bar{U}}(-\tau)\right]\left[\bar{\bar{U}}(\alpha)\right]\hat{x} \qquad [4.15]$$

1 The European satellites ERS-1/2 operated with vertical polarization and the Canadian satellite RADARSAT-1 operated with horizontal polarization. For these first-generation instruments, with a unique polarization, the choice of polarization depends on the main mission objectives. For RADARSAT, designed for tracing maritime routes in arctic areas (by discriminating water and ice), the horizontal polarization, which minimizes the sea clutter, was chosen. For the ERS missions, originally designed for maritime studies, vertical polarization has been selected. Since 2007, new missions with various polarizations (TerraSAR, Cosmo-Skymed, etc.) have helped avoid the difficult choice between polarizations.

the transposition operator being noted with T.

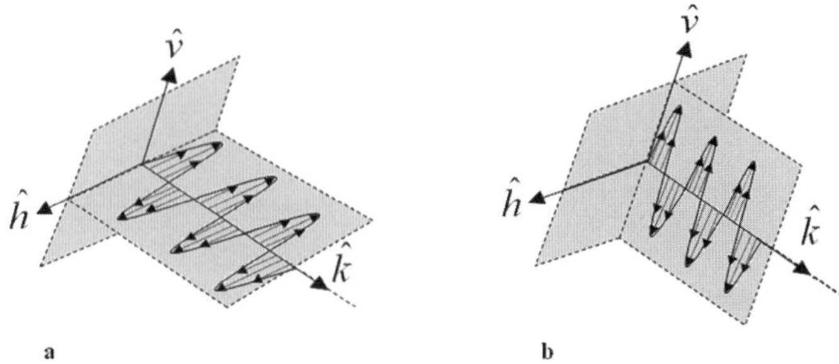

Figure 4.3. *Spatial variations of the electric field in the case of a horizontal linear polarization (RADARSAT type) a) and vertical linear polarization(ERS type) b)*

In the polarizations space, two orthogonal polarizations always form a basis.

The pairs of orthogonal polarizations consist of linear polarizations oriented in orthogonal directions ($\psi = \psi_0$ for one, $\psi = \psi_0 + \pi/2$ for the other), circular polarizations in opposite rotation directions ($\tau = \pi/4$ for one, $\tau = -\pi/4$ for the other) and finally, more generally, elliptic polarizations with the same ellipticity ratio, with orthogonal major axis and with opposite rotation directions.

4.1.2.2.4. The waves propagating in opposite directions

Which relation connects the Jones vectors $\vec{E}_0(\hat{k})$ and $\vec{E}_0(-\hat{k})$ of the two waves of the same polarization structure that propagate in opposite directions? By definition, the propagation in opposite directions means a change of the propagation term $\exp\left[j(\omega t - kz)\right]$ of equation [4.10] in $\exp\left[-j(\omega t - kz)\right]$. Consequently, we also need to conjugate the Jones vector $\vec{E}_0(\hat{k})$ for preserving the same spatiotemporal structure $\vec{e}(x, y, z, t)$. Thus:

$$\vec{E}_0(-\hat{k}) = \vec{E}_0^*(\hat{k}) \qquad [4.16]$$

4.1.2.2.5. The direction of the polarization

The sign of τ sets the direction in which the edge of the electric field rotates on the polarization ellipse so that the "right" polarizations differ from the "left" polarizations. This "left to right" denomination must be handled carefully. Depending on the publications, the authors use contradicting definitions for the direction of the polarization. The IEEE standard definition refers to right polarization when the field $\vec{e}(x, y, 0, t)$ turns in a clockwise direction, for an observer looking in the direction of the propagation (this is the case in Figure 4.2, for a wave that propagates along $\hat{z}$).

Figure 4.4 shows the polarization ellipse (as well as its track direction over time) in the transversal plane $z = 0$, for values from given a_x and a_y, and a variable dephasing $\delta = \delta_y - \delta_x$.

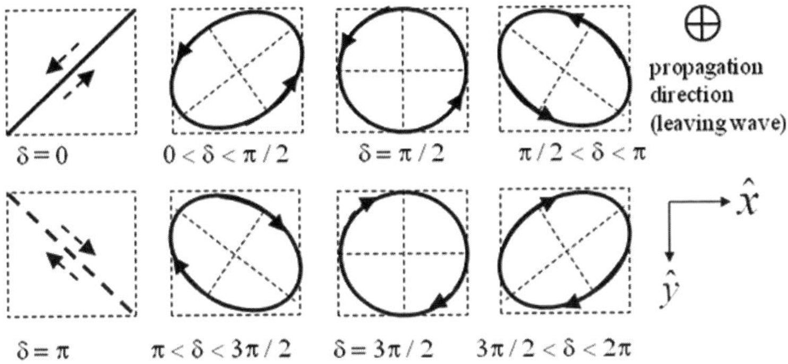

Figure 4.4. *The polarization ellipses in the plane that is transversal to the propagation, for given values ax and ay (with the exception of the particular cases $\delta = \pi/2$ and $\delta = 3\pi/2$ where they are chosen to be equal), and a variable phase difference $\delta = \delta x - \delta y$. [BOR 70]*

4.1.3. *Partially polarized waves*

Until now, the wave polarization (ψ and τ, or dually $a_x e^{j\delta_x}$ and $a_y e^{j\delta_y}$) was assumed to be independent of time and space. We now consider the general case of the partially polarized waves, which inform this hypothesis.

Equation [4.10] shows time variations in the monochromatic electric field of sinusoidal type, throughout the term $\exp\,[j2\pi ft]$. The terms $a_x e^{j\delta_x}$ and $a_y e^{j\delta_y}$ (such as ψ and τ) are fixed magnitudes, specific to the wave polarization. In a more general case, these terms can be time dependent, yielding:

$$\begin{aligned} \vec{e}(x, y, z, t) &= \Re\left\{\left(a_x(t)e^{j\delta_x(t)}\,\hat{x} + a_y(t)e^{j\delta_y(t)}\,\hat{y}\right)\exp\left[j(\omega t - kz)\right]\right\} \\ &= \Re\left\{\vec{E}_0(t)\exp\left[j(\omega t - kz)\right]\right\} \end{aligned} \qquad [4.17]$$

This time dependence may be due to the scattering of a fluctuating point target, throughout time or, in a dual way, due to the measurement samples collected from a large static target. The time fluctuation of $\vec{E}_0(t)$ is significant if the correlation time of the target (duration below which we can consider the target as being static) is lower than the duration of the measurement. By putting $\delta(t) = \delta_y(t) - \delta_x(t)$, the time variations of $a_x(t)$, $a_y(t)$ and $\delta(t)$ produce, via a Doppler effect, a spectral spread Δf of the monochromatic wave. By assuming $\Delta f \ll f$, and by writing the measurement duration as Δt, the time dependency of the polarization parameters can be neglected as long as $\Delta t \ll 1/\Delta f$. Under this condition, the wave keeps its fully polarized character. Otherwise, the wave is said to be partially polarized.

4.1.3.1. *The coherence matrix*

An electric field that propagates into the free space becomes "understandable" when it is received by an antenna pointing in the direction of the propagation and characterized by its receiving polarization $\hat{E}_0^{rec}$. This polarization corresponds to the emitted polarization if this antenna was used as a transmitter (the reciprocity theorem). The coupling of the reception antenna and the incident field produces, at the output, a detected voltage $V(t)$, with an average power proportional to:

$$P_{mean} = \langle V(t)V^*(t) \rangle = \left\langle \left| {}^T\hat{E}_0^{rec}\vec{E}_0(t) \right|^2 \right\rangle \quad [4.18]$$

The operator $\langle . \rangle$ corresponds to an average throughout the time. Formula [4.18] presents two remarks:

– $\hat{E}_0^{rec}$ is a dimensionless vector, that is independent of time, with an unitary norm;

– ${}^T\hat{E}_0^{rec}\vec{E}_0(t)$ is a Hermitian scalar product, whose second vector is conjugated in two stages. The first conjugation is inherent to the mathematical formulation of the Hermitian product, while the second conjugation is of physical origin: it expresses the coupling of the two waves propagating in opposite directions (see section 4.1.2.2.4).

The average detected power can be written as:

$$P_{mean} = {}^T\hat{E}_0^{rec} \left[\bar{\bar{J}} \right] \hat{E}_0^{rec*} \quad [4.19]$$

where $\left[\bar{\bar{J}}\right]$, called the wave coherence matrix, is the 2×2 Hermitian matrix (it is equal to its conjugated transposed matrix):

$$\left[\bar{\bar{J}}\right] = \begin{bmatrix} J_{xx} & J_{xy} \\ J_{yx} & J_{yy} \end{bmatrix} = \begin{bmatrix} \langle E_{0x}E_{0x}^* \rangle & \langle E_{0x}E_{0y}^* \rangle \\ \langle E_{0y}E_{0x}^* \rangle & \langle E_{0y}E_{0y}^* \rangle \end{bmatrix} \quad [4.20]$$

The degree of coherence characterizes the correlation between the electric components E_{0x} and E_{0y} along the directions $\hat{x}$ and $\hat{y}$:

$$\mu_{xy} = \frac{J_{xy}}{\sqrt{J_{xx}}\sqrt{J_{yy}}} = \frac{\langle E_{0x}E_{0y}^* \rangle}{\sqrt{\langle E_{0x}E_{0x}^* \rangle}\sqrt{\langle E_{0y}E_{0y}^* \rangle}} \quad [4.21]$$

Because of the Schwarz inequality, the module μ_{xy} is comprised between 0 and 1. The extreme cases found are:

1) *The fully polarized waves:* the polarization parameters do not depend on time $|\mu_{xy}| = 1$; det ($\left[\bar{\bar{J}}\right]$) = 0.

2) *The fully unpolarized waves:* radiation whose intensity of the components of the electric field are all equal on average (there is no privileged vibration direction). The relative phase $\delta(t)$ is randomly distributed between 0 and 2π (this is the case of natural light); $J_{xy} = J_{yx} = 0$; $J_{xx} = J_{yy}$; $|\mu_{xy}| = 0$. The coherence matrix of a fully unpolarized matrix is thus reduced to the identity matrix. According to relation [4.19], the average power received by the antenna is independent of the choice of its reception polarization. This detected power is assimilated to a polarization noise.

Between the two extreme cases ($0 < |\mu_{xy}| < 1$), the waves carry a part of the energy in polarized form (fixed polarization ellipse), and the other part under a fully unpolarized form. The potential polarizations' diversity in reception will help draw the best part (maximum coupling) of the polarized component of the wave, without, however, escaping the "polarization noise" associated with its unpolarized part. The aim of the next section is to quantify the part of each of these components in the composition of the wave.

4.1.3.2. *The wave decomposition theorem*

The previous section has intuitively led to representing the partially polarized wave by the sum of a fully polarized wave and a fully unpolarized wave. But what do we precisely understand by "wave sum" and how do we achieve it? The natural tendency would be to add their respective Jones vectors (i.e. their polarization states). This electric field addition (thus called the *coherent sum*) is only possible if there is an invariance of the phase differences between the two waves, which requires that they are both perfectly polarized. This is not the case here. Therefore, the only meaningful addition is a "wave addition" applied on energy terms. Such an addition is qualified as an *incoherent sum*; it operates on the coherence matrices.

If we admit the uniqueness of the decomposition, we immediately show that $\left[\bar{\bar{J}}\right]$ is written as:

$$\left[\bar{\bar{J}}\right] = \left[\bar{\bar{J}}^{upol}\right] + \left[\bar{\bar{J}}^{pol}\right] = \begin{bmatrix} A & 0 \\ 0 & A \end{bmatrix} + \begin{bmatrix} B & D \\ D^* & C \end{bmatrix} \qquad [4.22]$$

where A, B and C are real quantities verifying:

$$B\,C - |D|^2 = 0 \qquad [4.23]$$

The matrix $\left[\bar{\bar{J}}^{upol}\right]$ is characteristic of an unpolarized wave, whereas relation [4.23] expresses the polarized aspect. By substitution, we can write:

$$\left\{\begin{array}{l} A = \frac{1}{2}\left(J_{xx}+J_{yy}\right) - \frac{1}{2}\sqrt{\left(J_{xx}+J_{yy}\right)^{2} - 4\left|\bar{\bar{J}}\right|} \\ B = \frac{1}{2}\left(J_{xx}-J_{yy}\right) + \frac{1}{2}\sqrt{\left(J_{xx}+J_{yy}\right)^{2} - 4\left|\bar{\bar{J}}\right|} \\ C = \frac{1}{2}\left(J_{yy}-J_{xx}\right) + \frac{1}{2}\sqrt{\left(J_{xx}+J_{yy}\right)^{2} - 4\left|\bar{\bar{J}}\right|} \\ \quad D = J_{xy} \end{array}\right. \qquad [4.24]$$

$$\text{with :} \qquad \left|\bar{\bar{J}}\right| = J_{xx}J_{yy} - J_{xy}J_{yx} \geq 0 \qquad [4.25]$$

4.1.3.3. *The degree of polarization*

The *degree of polarization P* quantifies the respective energetic weights of the polarized and unpolarized components in the composition of the wave. P belongs to $[0;1]$. We get (equations [4.20]–[4.24]):

$$P = \sqrt{1 - 4\left|\bar{\bar{J}}\right| / \left(J_{xx}+J_{yy}\right)^{2}} \qquad [4.26]$$

The expression of P only affects the matrix $\left[\bar{\bar{J}}\right]$ by its determinant factor and its trace. These two quantities are invariant under a rotation in the polarization plane of the wave, so that P does not depend on the choice of the basis $[\hat{x},\hat{y}]$ where the fields are expressed. In this respect, P is qualified as a polarimetric invariant (which is not the case with μ_{xy}).

4.1.4. *The group of Pauli matrices and the Stokes parameters*

The polarization state of a wave propagating in free space is given by its coherence matrix (see section 4.1.3.1). We propose here to express it in the Pauli matrix basis ($[\bar{\bar{\sigma}}_0]$,$[\bar{\bar{\sigma}}_1]$,$[\bar{\bar{\sigma}}_2]$,$[\bar{\bar{\sigma}}_3]$):

$$[\bar{\bar{\sigma}}_0] = \frac{1}{\sqrt{2}}\begin{bmatrix} 1 & 0 \\ 0 & 1 \end{bmatrix} \qquad [\bar{\bar{\sigma}}_1] = \frac{1}{\sqrt{2}}\begin{bmatrix} 1 & 0 \\ 0 & -1 \end{bmatrix} \qquad [4.27]$$

$$[\bar{\bar{\sigma}}_2] = \frac{1}{\sqrt{2}}\begin{bmatrix} 0 & 1 \\ 1 & 0 \end{bmatrix} \qquad [\bar{\bar{\sigma}}_3] = \frac{1}{\sqrt{2}}\begin{bmatrix} 0 & -1 \\ 1 & 0 \end{bmatrix} \qquad [4.28]$$

$\left[\bar{\bar{J}}\right]$ is written as:

$$\left[\bar{\bar{J}}\right] = \frac{1}{\sqrt{2}}\left\{s_0\left[\bar{\bar{\sigma}}_0\right] + s_1\left[\bar{\bar{\sigma}}_1\right] + s_2\left[\bar{\bar{\sigma}}_2\right] + js_3\left[\bar{\bar{\sigma}}_3\right]\right\} = \frac{1}{2}\begin{bmatrix} s_0+s_1 & s_2-js_3 \\ s_2+js_3 & s_0-s_1 \end{bmatrix}$$

The parameters s_i were introduced by Stokes at the end of the 19th Century, within the studies dedicated to partially polarized light. They provide the four components of the Stokes vector $\tilde{S}$:

$$\tilde{S} = \begin{pmatrix} s_0 \\ s_1 \\ s_2 \\ s_3 \end{pmatrix} = \begin{pmatrix} J_{xx}+J_{yy} \\ J_{xx}-J_{yy} \\ J_{xy}+J_{yx} \\ j(J_{xy}-J_{yx}) \end{pmatrix} = \begin{pmatrix} \langle E_{0x}E_{0x}^*\rangle + \langle E_{0y}E_{0y}^*\rangle \\ \langle E_{0x}E_{0x}^*\rangle - \langle E_{0y}E_{0y}^*\rangle \\ 2.\Re\langle E_{0x}E_{0y}^*\rangle \\ -2.\Im\langle E_{0x}E_{0y}^*\rangle \end{pmatrix} \qquad [4.29]$$

the fact that the determinant of the matrix $\left[\bar{\bar{J}}\right]$ is positive implies (equation [4.25]) is rewritten as:

$$s_0^2 \geq s_1^2 + s_2^2 + s_3^2 \qquad [4.30]$$

and the degree of polarization P as:

$$P = \sqrt{s_1^2 + s_2^2 + s_3^2}/s_0 \qquad [4.31]$$

Particular cases:

1) $P = 1$, fully polarized wave: $s_0^2 = s_1^2 + s_2^2 + s_3^2$.

2) $P = 0$, fully unpolarized wave: $s_1^2 + s_2^2 + s_3^2 = 0$.

The wave decomposition theorem (see section 4.1.3.2) is thus rewritten as:

$$\tilde{S} = \begin{pmatrix} s_0 \\ s_1 \\ s_2 \\ s_3 \end{pmatrix} = \tilde{S}^{upol} + \tilde{S}^{pol} = \begin{pmatrix} s_0 - \sqrt{s_1^2+s_2^2+s_3^2} \\ 0 \\ 0 \\ 0 \end{pmatrix} + \begin{pmatrix} \sqrt{s_1^2+s_2^2+s_3^2} \\ s_1 \\ s_2 \\ s_3 \end{pmatrix}$$

Through this construction, the polarizing degree $\tilde{S}^{upol}$ is null (fully unpolarized wave), that of $\tilde{S}^{pol}$ is equal to 1.

It offers a unified representation of the waves, regardless of their polarization state. A Stokes vector wave $\tilde{S}$, with a normalized power density ($s_0 = 1$), is associated with a point contained in a unitary radius sphere, with a reference coordinate system $[0, \hat{x}, \hat{y}, \hat{z}]$ (Figure 4.5). The quantities $s_1,\ s_2,\ s_3$ are expressed by means of two angles ψ and τ such that:

$$\begin{cases} s_1 = \sqrt{s_1^2 + s_2^2 + s_3^2}.\cos(2\tau)\ \cos(2\psi) \\ s_2 = \sqrt{s_1^2 + s_2^2 + s_3^2}.\cos(2\tau)\ \sin(2\psi) \\ s_3 = \sqrt{s_1^2 + s_2^2 + s_3^2}.\sin(2\tau) \end{cases} \qquad [4.32]$$

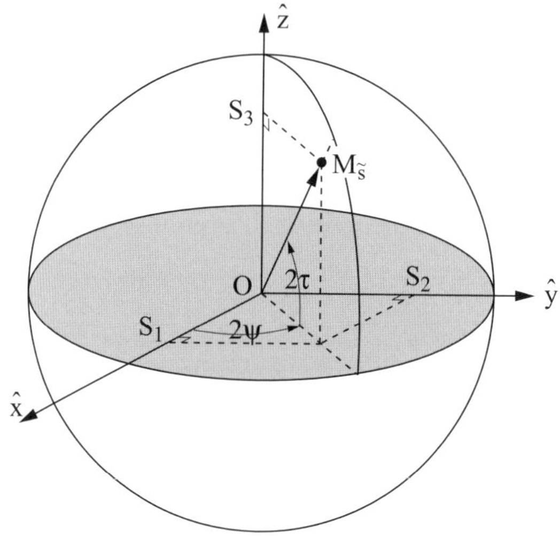

Figure 4.5. *The Poincaré sphere*

The vector $\tilde{S}$ is associated with a point inside the Poincaré sphere, with azimuth and elevation angles of 2ψ and 2τ, respectively, separated from the sphere center by a distance that is equal to its degree of polarization $\sqrt{s_1^2 + s_2^2 + s_3^2} \leq 1$.

For a fully polarized wave ($s_0^2 = s_1^2 + s_2^2 + s_3^2$), the parameters ψ and τ resulting from equation [4.32] are those mentioned in section 4.2. Its representative point is located on the surface of the Poincaré surface. For a partially polarized wave, ψ and τ characterize the polarization ellipse of its polarized part: a representative point $M_{\tilde{S}}$ of the wave "sinks" into the sphere when its degree of polarization decreases. The O center of the sphere corresponds to a unpolarized wave. The linear polarizations ($\tau = 0$) are located in the equatorial plane, the right elliptical polarizations in the southern hemisphere ($\tau < 0$) and the left elliptical polarizations in the northern hemisphere ($\tau > 0$). The north and south poles represent, respectively, the left and right circular polarizations of fully polarized waves.

4.2. Radar signatures

4.2.1. *RCS of a point target*

The *RCS* of a point target characterizes the strength of its radar echo. The wave/target interaction is expressed by the radar equation, which is a sequence of three events:

– The radar emits a pulse with peak power P_e, amplified by the antenna gain G in the direction of the target. Located at a distance R from the radar, it has a power density (expressed in $\mathrm{W/m^2}$) equal to:

$$dP^{inc} = P_e G/(4\pi R^2) \qquad [4.33]$$

It all happens as if the antenna emits an isotropic radiation in the radar–target axis[2]. The quantity $P_e G$ is called the effective isotropic radiated power (EIRP).

– Illuminated by this emitted pulse, the target acts as a radiating element, with an apparent isotropic attitude, returning a power density toward the radar:

$$dP^s = dP^{inc}\sigma/(4\pi R^2) \qquad [4.34]$$

which suggests that dP^{inc} has been intercepted and reflected by a surface σ (the RCS, expressed in m^2)). If the target radiates isotropically and independently of the observation distance (e.g. conductive sphere of a large radius R_s with regard to the wavelength), the RCS is equal to the surface seen from the radar (here πR_s^2, see Figure 4.6(b)). In the other cases, a target producing a radiation mainly in the backscattering direction has a RCS that is higher to the target surface (Figure 4.6(a)), as opposed to a target that "evacuates" the energy in a third direction (Figure 4.6(c)).

Figure 4.7 shows typical RCS values of point targets as well as their radiation patterns in the main axes. As an example, the trihedral, based on three orthogonal triangular plates, presents a maximum RCS (illumination in the axis of the trihedral) equal to:

$$\sigma^{trihedral} = 4\pi a^4/(3\lambda^2) \qquad [4.35]$$

where a is the side of a plate. If $a = 1$ m, we have $\sigma^{trihedral} = 1,289\ \mathrm{m}^2$ with 5.3 GHz (a sphere with an equivalent RCS has a diameter of 40 m!): this strong signature, connected to a very open radiating pattern (we can consider this target as isotropic with a sidelobe of 40°), is widely used in radar calibration.

2 Actually, the radiation of the antenna is not at all isotropic. However, the formulation of incorrect hypotheses in other directions than the radar line of sight has no consequences on the final result.

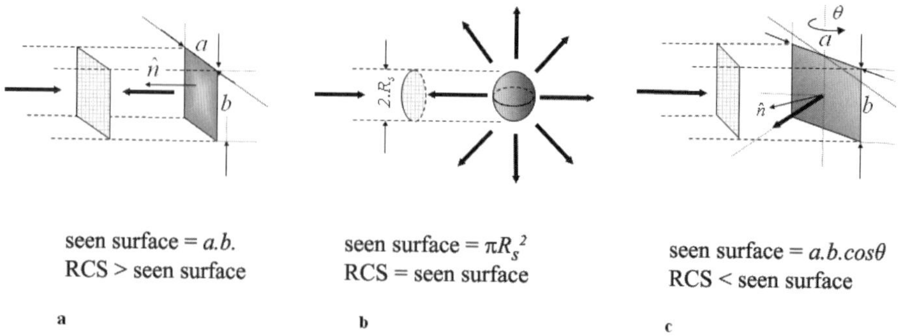

Figure 4.6. *Comparison between RCS and the surfaces seen in canonical situations*

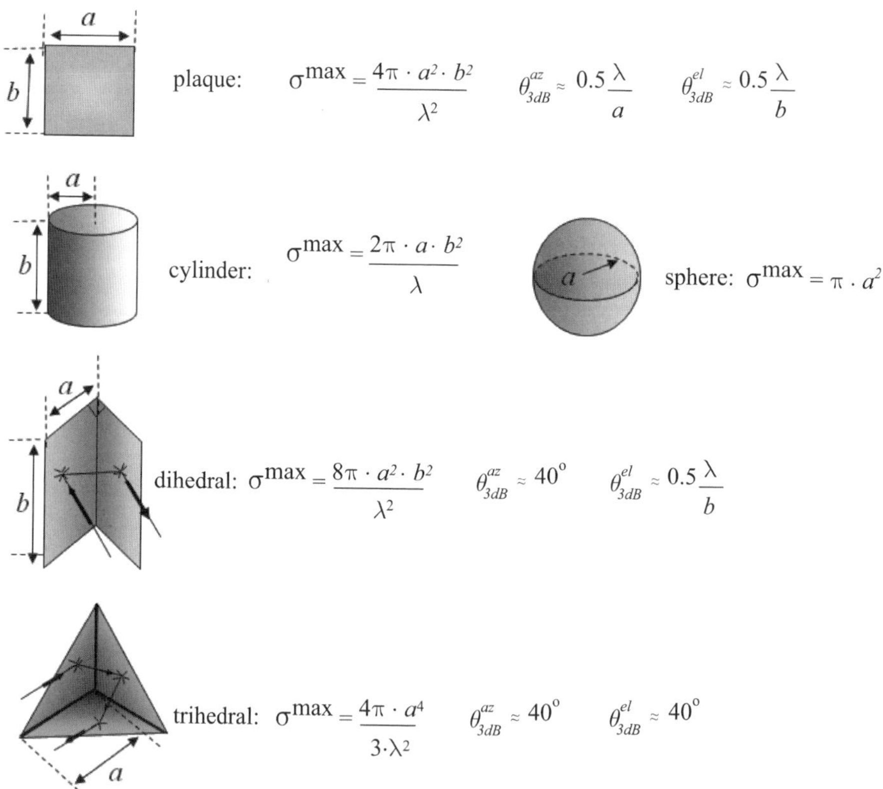

Figure 4.7. *Examples of classic targets RCS [RUC 70]. The indicated values correspond to the configurations generating the maximum signatures. θ^{az}_{3dB} and θ^{el}_{3dB} represent 3 dB aperture for radiation patterns, respectively, in the azimuth direction and in the elevation direction*

– Finally, the power P^{rec} collected by the reception antenna with an effective area A_{eff} is given by:

$$P^{rec} = dP^s \, A_{eff} \qquad [4.36]$$

equations [4.33], [4.34] and [4.36], along with the relation that connects the effective area of the antenna with its gain ($G = 4\pi A_{eff}/\lambda^2$), allow us to establish the radar equation (assuming the same antenna gain both in emission and in reception):

$$P^{rec} = P_e \frac{\lambda^2}{(4\pi)^3} \frac{G^2}{R^4} \sigma \qquad [4.37]$$

The power received is all the more important as the antenna is directional (a high G value). This link budget is otherwise affected by a term in λ^2/R^4. Equation [4.37] predicts the signal levels expected at the input of the chain of radar reception. For a parabolic antenna of $1\,\mathrm{m}^2$, emitting a $1\,\mathrm{kW}$ peak power at C band ($\lambda \approx 5.7\,\mathrm{cm}$) in the direction of a 1 m radius conducting sphere, located at $1\,\mathrm{km}$, the received echo peak power will be, at best, 10^{-8} W.

Until now, all have been done assuming that no loss would affect the radar/target link. Actually, the power received is lower than the power predicted. We can identify losses of various origins, in order of decreasing magnitude:

– *Losses in the emission and the reception devices (antenna subsystem, preamplifier, radio-frequencies losses, encoding losses)*: because of these losses, we distinguish between the radiated power P_e (see the radar equation) and the power produced by the high-power amplifiers, the efficiency being of the order of 50%.

– *Atmospheric propagation losses*: the propagation models predict for the radar bands (from P to X) small attenuations in the atmosphere, except small wavelengths (band X) encountering rain. Figure 4.8 shows an example of tropospheric attenuation revealed on an image from the German Synthetic Aperture Radar (X-SAR) mission acquired in Mali in 1994. All of these losses (atmospheric gas, clouds, rain and ionospheric scintillation) equally depend on the geographic location.

– *Losses linked to polarization effects*: these are of two types:

- The possible rotation of the polarization of the wave that is being propagated (a Faraday effect when crossing the ionosphere). This effect can become critical for frequencies lower than 1 GHz.

- The depolarization effects connected to the wave/surface interactions.

The set of all of these losses is combined in a $Loss > 1$ factor, which integrates the propagation effects as much as the receptor's losses. This loss factor has a magnitude in the order of 2–3 dB for each one-way trip of the radar wave.

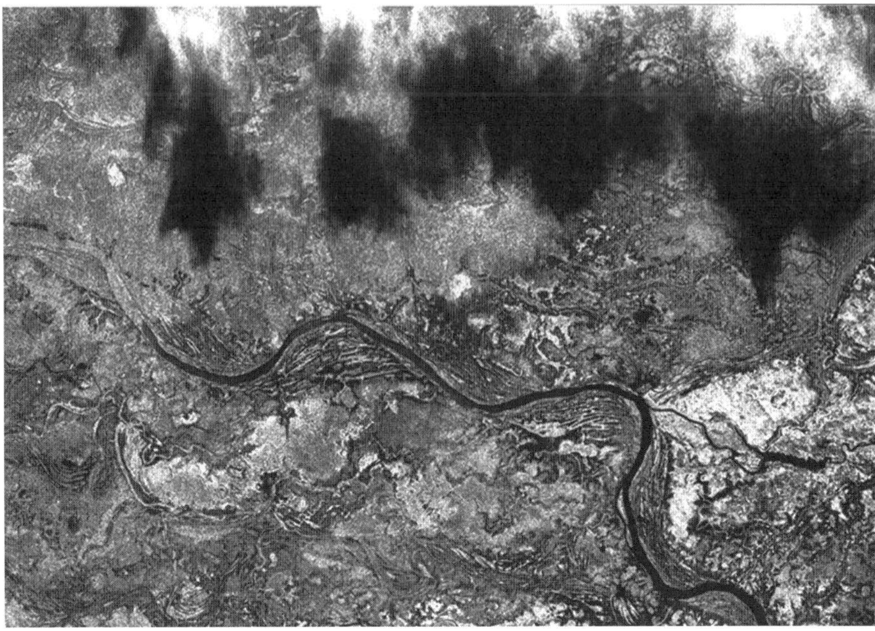

Figure 4.8. *Radar image band X (X-SAR) acquired on the interior delta of the Niger in Mali in 1994 (mission SIRC/X-SAR). The high part of the image reveals the presence of clouds that reflect the radar signal (thus creating the effects of electromagnetic shade). This phenomenon is rarely observed and requires very dense clouds, as we can find in West Africa during the rainy season. The radar radiation comes from above. Copyright JPL-DLR/SIRC-XSAR*

4.2.2. *Radar signature for extended targets – the backscatter coefficient* σ^o

How can we adapt the notion of RCS to extended targets? If the extended targets are stationary, they backscatter a quantity of energy that is proportional to their ground surface S. To remove this dependency, we introduce a normalized signature:

$$\sigma^0(dB) = 10\log_{10}(d\sigma/dS) \approx 10\log_{10}(\sigma/S) \qquad [4.38]$$

where σ^0 is the backscattering radar coefficient, expressed in $\mathrm{m}^2/\mathrm{m}^2$ (the exponent 0 informs about the normalized character of the RCS)[3]. Most natural surfaces have negative normalized backscatter coefficients σ^0 (expressed in $\mathrm{dBm}^2/\mathrm{m}^2$ on a logarithmic scale): $1\ \mathrm{m}^2$ of their surface has a RCS lower than $1\ \mathrm{m}^2$.

3 Because of the significant dynamics that were observed in the measurements, the logarithmic scale is generally used.

4.3. The basics of radar measurement physics: interaction between waves and natural surfaces

4.3.1. *Introduction*

Illuminated by an electromagnetic radiation, natural surfaces may generate surface and/or volume effects. Surface scattering depends mainly on surface roughness while volume scattering chiefly depends on the structure and the organization of the inner scatterers (as, for example, within vegetation). In both cases, the dielectric properties, which reflect the water content of the observed media, play a key role in backscattering. The radar echo is finally modulated by the local ground slope, which causes strong radiometric variations. In absolute terms, electromagnetic and interaction models are required for untangling the complex issues of all these contributions, but their elaboration quickly runs into the difficulty – if not the impossibility – of simulating nature with exact mathematical models. As an alternative, the semi-empirical approaches, which are influenced both by theoretical behavior – the first-order surface scattering, and volume radiative transfer – and by experimental observations, offer an alternative to the use of exact models. We briefly describe here these fundamental mechanisms. By using several examples, we will show the main noticeable effects on the radar images: roughness, water content, volume mechanisms, penetration of waves and slope effects. We dedicate the last part of the section to radar scatterometry for the study of waves and winds at the surface of the oceans.

4.3.2. *Bare soil scattering*

4.3.2.1. *Description of the surface*

We are dealing with a dielectric interface that separates two homogeneous semi-infinite media (air/ground, air/sea, etc.) (Figure 4.9). The dielectric constant ε^r of the lower layer expresses the resistance of the medium to the wave penetration. For a ground layer, it depends on its free water content, and, to a lesser degree, on its texture. The roughness is characterized here by a random process $z(x, y)$ reflecting the local variations of the profile in relation to a mean reference plane. $z(x, y)$ is usually assumed to be a zero mean stationary Gaussian variable.

In the vertical direction, height *root mean square* (r.m.s.) s characterizes the standard deviation of the rough surface with respect to the reference plane[4]:

$$s = \left\langle\, z^2(x, y) \,\right\rangle^{\frac{1}{2}} \qquad [4.39]$$

4 For agricultural surfaces, the typical values of s range between 1 and 2.5 cm for sown or harrowed parcels, and between 2 and 7 cm for plowed parcels.

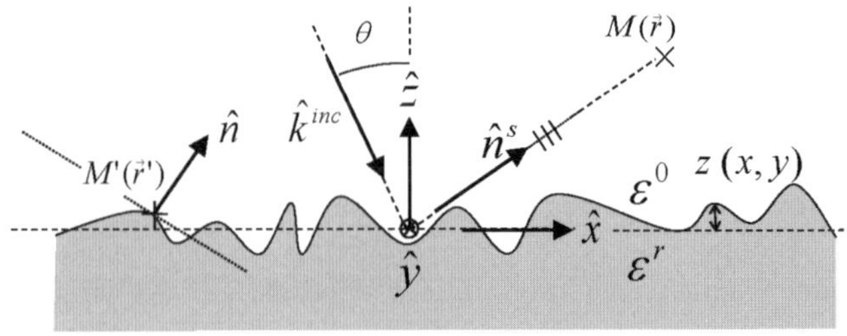

Figure 4.9. *Electromagnetic scattering on a rough surface*

In the horizontal plane, the surface autocorrelation function informs, in a given direction, about the rate of the roughness variations:

$$K(\tau) = \langle\, z(x, y)\, z(x - \tau, y)\,\rangle \,/\, \langle\, z^2(x, y)\,\rangle \qquad [4.40]$$

for the direction $\hat{x}$. The function $K(\tau)$ is an autosimilarity measure after a translation of the length τ. The correlation length l is the translation beyond which:

$$K(l) \leq 1/e \qquad [4.41]$$

with $\ln(e) = 1$. Rapid roughness variations (rough terrain) yield an autocorrelation function that is quickly decreasing, i.e. having a short correlation length. The nature of the autocorrelation function (generally assumed to be Gaussian or exponential) has a significant impact on the morphology of the profile (see Figure 4.10).

REMARKS.–

– 1) Mathematically speaking, $K(\tau)$ is the spatial representation of the surface roughness spectrum (it is its Fourier transform).

– 2) Using formula [4.40], we implicitly admit the stationary trend of profiles, so that the autocorrelation function does not depend on x. More realistic roughness statistics (such as fractal statistics) cannot allow this type of simplification.

– 3) The stationary hypotheses, and those of a single roughness scale, do not necessarily fit with reality. An experimental soil profile acquired on an agricultural surface (Figure 4.11) shows that the soil/air interface consists of an aggregation of clumps of different shapes and sizes, with various gaps and included volumes: an excessive simplification of reality often leads to a poor modeling (comparing Figures 4.9 and 4.11).

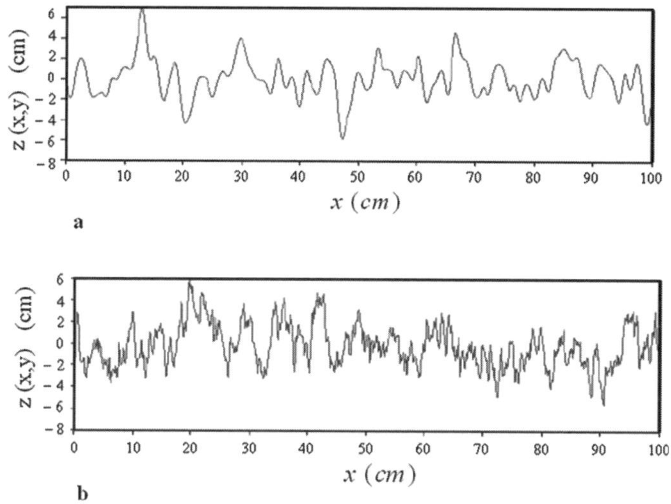

Figure 4.10. *Ground profiles with r.m.s. height s = 2 cm and with correlation length l = 12 cm. The autocorrelation has a Gaussian shape a) and an exponential shape b)*

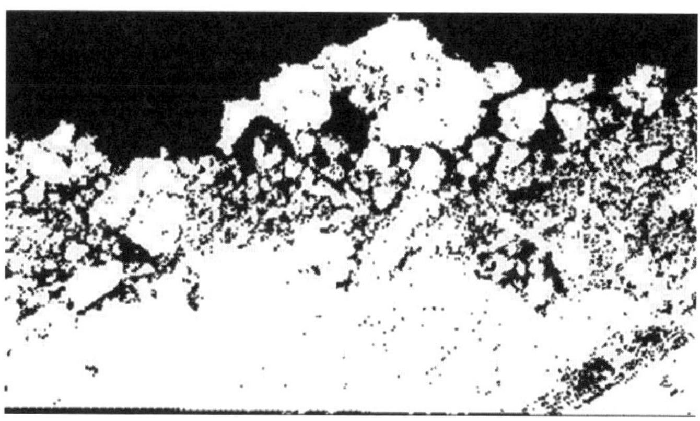

Figure 4.11. *Experimental ground profile acquired on agricultural land. Courtesy of INRA*

4.3.2.2. *The Rayleigh criterion*

The roughness effect on electromagnetic scattering depends on the transmitted wavelength: indeed, the same surface appears smooth in L band ($\lambda = 25\,\mathrm{cm}$), and rough in X band ($\lambda = 3\,\mathrm{cm}$). The Rayleigh criterion stipulates that the roughness effect is negligible as soon as the phase difference $\Delta\phi$ generated by the altimetric variations of the surface (Figure 4.12) remains less than $\pi/2$. Considering the round-trip path of the wave, if h is the maximum altimetric variation, it becomes:

$$h \leq \frac{\lambda}{8\cos\theta} \qquad [4.42]$$

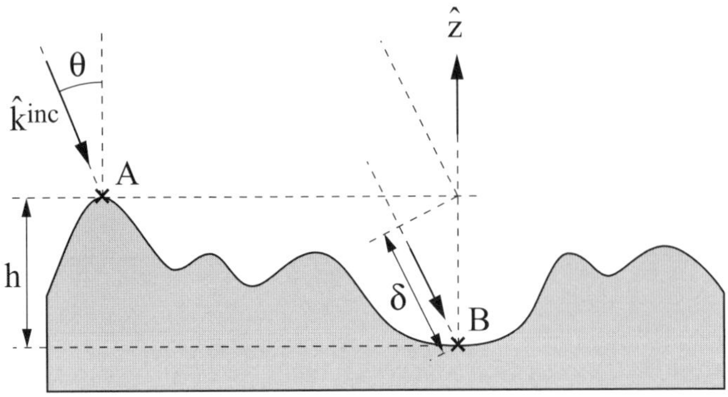

Figure 4.12. *Rayleigh criterion on a rough surface*

For $\theta = 45°$, the bands L ($\lambda = 24$ cm), C ($\lambda = 5.7$ cm) and X ($\lambda = 3$ cm) lead to maximum altimetric variations of 4.2, 1 and 0.5 cm, respectively: almost all natural surfaces behave as rough surfaces on the X and C bands.

In Chapter 7, we will see the consequences of this roughness on the image: this is what we call the *speckle*. We will then examine its role in the statistic properties of the pixels and therefore its significance in signal processing. However, we will examine here the interaction phenomena of the electromagnetic wave with natural medium.

4.3.2.3. *Calculation the electromagnetic scattering*

This calculation is governed by the Stratton–Chu equations, which derive from the Maxwell equations when boundary conditions shown in Figure 4.9 are assumed. The field $\vec{E}^s(\vec{r})$ scattered at a point $\vec{r}$ located far from the surface is given by [ULA 86]:

$$\vec{E}^s(\vec{r}) = \alpha\hat{n}^s \times \int \left(\left(\hat{n} \times \vec{E}^s(\vec{r}')\right) - \eta\hat{r} \times \left(\hat{n} \times \vec{H}^s(\vec{r}')\right)\right) \exp(-j\vec{k}\vec{r}')dS' \qquad [4.43]$$

with $\alpha = -jk\exp(-jkR)/(4\pi R)$. The vector $\hat{n}^s$ characterizes the propagation direction of the scattered wave, and $\hat{n}$ the normal direction on the exterior surface at a point M' (situated in $\vec{r}'$) on the surface, the integration operator acting on the entire surface. η is the impedance of the medium (expressed in Ohms). Finally, $\hat{n} \times \vec{E}^s(\vec{r}')$ and $\hat{n} \times \vec{H}^s(\vec{r}')$ are the unknowns of the integral equation, which represent the tangential components of the electric and magnetic fields, respectively, on the interface. The simultaneous presence of electric and magnetic components expresses the dielectric character of the interface (in the case of a perfect conducting surface, $\hat{n} \times \vec{E}^s(\vec{r}') = 0$).

The integral equation [4.43] takes into account any kind of scattering mechanisms, whether they are simple or multiple. It is only resolved by using digital methods (e.g. the moments method). In the field of remote sensing, asymptotic methods are generally preferred.

4.3.2.3.1. Asymptotic methods

The backscattering behavior of rough surfaces is described by making approximations both for the description of the interface and for the respective electromagnetic calculations. The description of the interface is based on a "standard" representation (stationary, without any volume inclusions, at a single roughness scale).

– The *physical optics* approximation (or *Kirchhoff* approximation) [ULA 86] divides the surface in a succession of facets that are connected to the local tangent planes (Figure 4.13(a)). The validity condition on the wave number k is translated as $kl > 6$ and $l^2 > 2,76\, s\lambda$.

– For very rough surfaces ($ks \geq 3$), the approximation of the *geometric optics* reduces the interaction to the contribution of only the orthogonal planes that are tangent to the incident wave (Figure 4.13(b)).

– The model of the *small perturbations* is based on the exact resolution of equation [4.43] to a given order. The order 1 is enough to characterize the backscattering in direct polarization (*HH* or *VV*). However, the assessment of the (*HV*) cross-polarization needs to go to order 2. The method of small perturbations only applies to slightly rough surfaces, which can only slightly alter the structure of the wave (Figure 4.13(c)). The validity condition influences the r.m.s. slope ($\sqrt{2}s/l$ for a Gaussian autocorrelation function) and the product ks, both being ≤ 0.3.

– The *integral equation model (IEM)* [FUN 92] admits a validity domain that encompasses and connects those of the physical optics and the small perturbations[5], this extension having been made possible by including it in the specificity calculation that is proper to backscattering geometry.

4.3.2.4. *Simulations and experimental observations*

Despite their limitations, the asymptotic models reveal the main tendencies of the radar behavior of bare soils. Figure 4.14 shows, for the configuration (band C, 23°, *VV* polarization), the backscattering coefficient σ^0_{VV}, simulated by the IEM model,

5 As an example, it can be applied to the case $ks \leq 3$.

for different vertical roughness states (r.m.s. height) and water content. The autocorrelation function has an exponential form: the length of the correlation is set to a typical value of $l = 8\,\mathrm{cm}$ (l is difficult to estimate and often works as a fitting parameter).

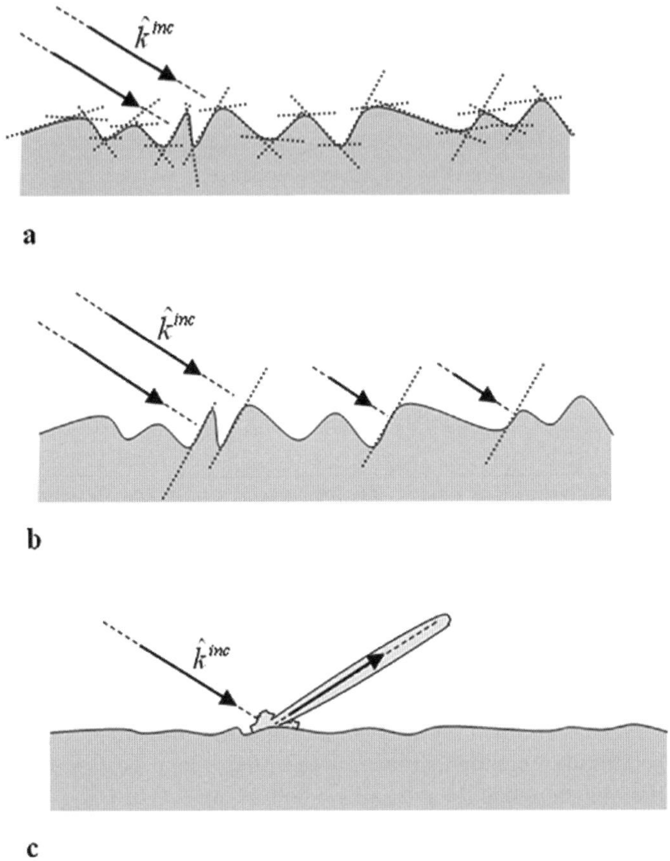

Figure 4.13. *a) Pysical optics approximation (or Kirchhoff approximation), b) geometric optics approximation and (c) small perturbation approximation for a rough surface*

– For a given water content, we notice an *increase in the radar response when roughness increases*: on a radar image, the bright surfaces are rough surfaces, and the dark surfaces are flat surfaces, as proven by the RADARSAT image (Figure 4.14(b)), indicating a clear-cut radiometric segmentation between a flat soil (loam fragments) and a rough soil (limestone).

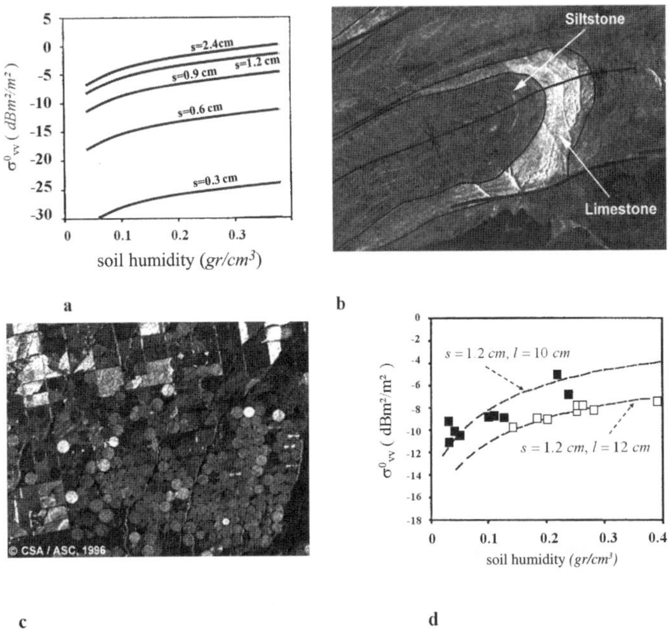

Figure 4.14. *a) Theoretical bare soil backscattering coefficient for the ERS configuration (band C, 23°, VV) depending on the vertical roughness (r.m.s. height) and on the water content. The autocorrelation function has an exponential form, and the correlation length is set at 8 cm [LET 93]. b) Roughness effect on the SAR imaging radar. RADARSAT image (band C, 45°, resolution: 20 m) [RAD 97]. c) Farming region in the state of Washington, USA. Canadian Space Agency© processed and provided courtesy of RSAT International, Inc. d) Experimental backscattering coefficient acquired by ERS (band C, 23°, VV) on two types of bare soil, depending on the local water content (agricultural area of Gharb, Morocco). The modeling (- - - -) is carried out with the values s = 1.2 cm and l = 10 cm for the measures in the first area (■), and s = 1.2 cm and l = 12 cm for those of the second area (□). For exponential autocorrelation function, see [LET 93]*

– For a given roughness (fixed $r.m.s.$ height), we notice an *increase in the radar response when water content increases* (of the order of 7 $\mathrm{dBm}^2/\mathrm{m}^2$): moist soils are "brighter" than the dry soils for comparable roughness values. The RADARSAT image of an agricultural image in the state of Washington (United States) (Figure 4.14(c)) displays a collection of circular patches with variable radiometric content, which are agricultural parcels irrigated to various degrees[6]. The ERS

6 To achieve this conclusion, we must "bet" on the similarity of the roughness states between two plots, as the lack of vegetation.

experimental observations (Figure 4.14(d)) confirm a measuring dynamic of the order 6–7 $\mathrm{dBm}^2/\mathrm{m}^2$ for the water content rates going from 0 to $0.4\ \mathrm{g/cm}^3$.

In the light of these observations, the attempts to turn the radar measure into water content prove to be difficult. The entangled effects of roughness and water content give the *same backscattering level to a dry rough soil as to a more humid, flatter surface*. The separation of the water content and roughness effects on the radar measure needs the diversification of the radar information (for example by using radar polarimetry).

4.3.3. *Sea surface scattering*

4.3.3.1. *General approach*

At centimetric frequencies, the sea is nearly "impermeable" to electromagnetic waves. Its radar signature depends on several scales of roughness (waves of capillarity for the small scales, swell fields for the larger scales, etc.).

For backscattering, the incidence angle range can be divided into three areas: the specular area (weak incidence), the flat zone (average incidence) and the grazing incidence angles (high incidence) (Figure 4.15(a)). For an average sea condition (wave heights of approximately 2 m), the semi-empirical model of [MOR 90] presents on (0.30°) a backscattering coefficient σ^0 (averaged with respect to the polarization effects and wind direction), whose main specular lobe does not depend on the frequency. On the other hand, on the flat zone (where incoherent scattering predominates), σ^0 increases with the frequency (with a $1/\lambda$ dependency). For a given frequency, there are high radiometric variations (of the order of $20\ \mathrm{dBm}^2/\mathrm{m}^2$) between 20° and 50°, corresponding to the transition between the specular area and the flat zone. This type of pattern has been observed experimentally on a RADARSAT image (Figure 4.15(b)) (variations of $10\ \mathrm{dBm}^2/\mathrm{m}^2$ for the swath of a standard-S1 mode between 20° and 27°). We equally notice this on the radar profiles revealed by the TRMM mission.

– *The effect of the wind*: the orientation of the wind with respect to the radar line of sight also affects the response of the sea. For a "downwind" target, (i.e. the direction of the wind makes an angle of 180° with that of the radar observation) and for the intermediate incidences (around 45°), σ^0 decreases from 7 to 8 dBm^2/m^2 on average with respect to the corresponding "upwind" configuration (the direction of the wind and the direction of observation are identical, the sea having a smoother aspect). The σ^0 drop reduces to 3 to 4 dB with the intermediate "crosswind" configuration.

– *Polarization effect*: we predict a higher backscattering level in vertical polarization (*VV*) than in horizontal polarization (*HH*) for the incidences that are higher than 10°. When the incidence increases, the deviation between *HH* and *VV* increases to the disadvantage of *HH*, the predictions locating σ^0_{HH} at approximately 15 dBm^2/m^2 below σ^0_{VV} at a 60° incidence.

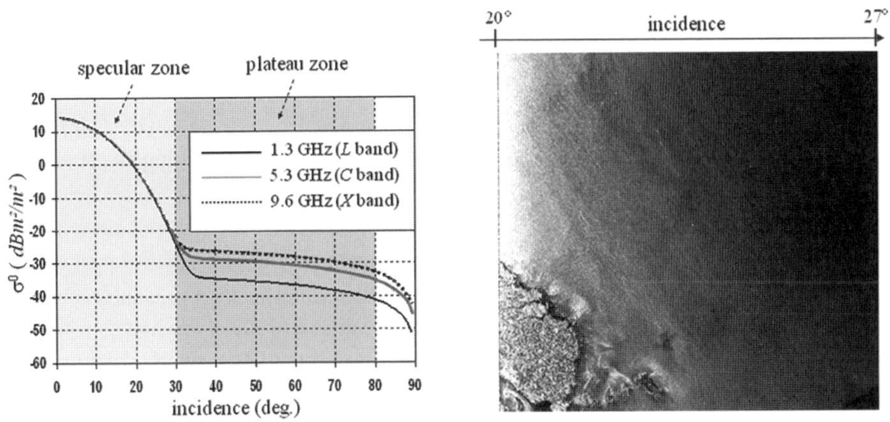

a b

Figure 4.15. *a) Semi-empirical behavior of the backscattering coefficient σ^0 depending on the incidence, for an average sea state. b) The incidence effect observed on a RADARSAT image, standard-S1 mode (20°–27°), acquired on the sea (Ouessant rail, France), 9 March, 1999. Canadian Space Agency©*

Now, we detail the backscattering behaviors for the weak incidences (10°) by using radar scatterometers, specifically for the study of the sea wave fields.

4.3.3.2. *Radar scatterometry on sea surfaces*

4.3.3.2.1. Wave spectrum

As in the case of bare soil (see section 4.3.2.1), the sea surface can be described by its elevation $z\ (x, y)$. Its dynamic character, however, needs to put first the approaches based on a statistic representativeness that is well adapted to reality (while describing, for example, the surface through a wave field associated with an amplitude, a direction and a wavelength). To this end, we consider on the elevation power spectrum density, called directional wave spectrum $F(k_x, k_y)$, expressed in the wave numbers space (i.e. the frequency space):

$$F(k_x, k_y) = F(k, \Phi) = |\, FT(z(x, y))\,|^2 \qquad [4.44]$$

where "FT" represents the Fourier transform.

The wave spectrum is generally split into two independent contributions, the radial spectrum $F(k)$ and the anisotropy $\xi(k, \Phi)$:

$$F(k, \Phi) = F(k)\, \xi(k, \Phi) \qquad \text{with :} \qquad \int_{-\pi}^{+\pi} \xi(k, \Phi)\, d\Phi = 1 \qquad [4.45]$$

There are several models for F and G. They express the different sea states that are usually considered. The most common is the so-called fully developed sea wind model, associated with the Pierson–Moskowitz model [PIE 64]. There are more recent models, such as the JONSWAP model (*Joint North Sea Wave Project*) [HAS 80], the swell model [DUR 85] and the mixed sea models. However, describing them would be outside the scope of this work.

4.3.3.3. *Radar signature behavior around 10°*

The sea condition is generally associated with the simultaneous presence of wind and waves. The local wind creates small-scale roughness (capillarity waves), on which wave fields of different origin settle (these could be propagating swells, bathymetry effects, beach-combers, etc.). Close to 10° incidence, we notice that the back-scattered signal is basically proportional to the slopes of the large-scale waves in the direction of the line of sight and no longer depends on capillary waves, that is on wind (Figure 4.16). Indeed, when the incidence is low ($\ll 10°$), the probability of having normal sea surface perpendicular to the emitted wave decreases when the wind increases, which contributes to the diminution of the backscattering coefficient (we move away from the specular backscattering situation on a plane surface). The phenomenon is inversed at large incidences ($\gg 10°$) since the Bragg scattering ("incoherent" scattering) gets the relay on the specular reflection. There is, therefore, a transition area, around 10° incidence, for which the radar signature becomes independent on the force of the wind. This independence property is used for measuring the wave fields. In light of Figure 4.16, there are therefore two typical choices of incidence for using the scatterometry on the ocean surfaces:

– The high incidence areas (>20°) display a strong sensitivity to wind. Several missions have selected these configurations called "wind-scattering" configurations: ERS-1/2, advanced SCAtterometer (ASCAT), tropical rainfall measuring mission (TRMM), etc. we will not detail them, since there is a rich literature on the topic.

– The incidences between 6° and 10° are relatively insensitive to wind conditions. The Chinese–French mission Chine French Ocean SATellite (CFOSAT) will be the first one to send this concept into space [TIS 11]. Hereon, we will be drawing the main measuring principles from this.

4.3.3.3.1. "Wave" scatterometry, interests and principles of the measurement

Weather prediction was largely improved throughout the assimilation of surface winds ("wind" scatterometry) or altimetric measures (coming from radar altimetry missions TOPEX/POSEIDON, JASON, ERS, etc.). This, however, also requires wave prediction tools that have not yet reached their operational maturity. The global wave fields have an impact on the seasonal variations of ocean storms, on the heat and CO_2 exchanges between the ocean and the atmosphere, and on the surface albedos. They also contribute to the knowledge of coastal phenomena such as sediment transport and coastline changes. The current radiofrequency measurements only partially fulfill the requirement to know the wave fields. The radar altimetry provides the average height of the waves, but not their direction. The SAR imagery only detects the wavelengths that are more than 150–200 m, and that generally move in the direction that is radial to the sensor. Furthermore, this measurement is quite incomplete (because the swaths are reduced and the revisiting time of SAR systems is quite far apart). Consequently, we had to develop an adequate instrumentation.

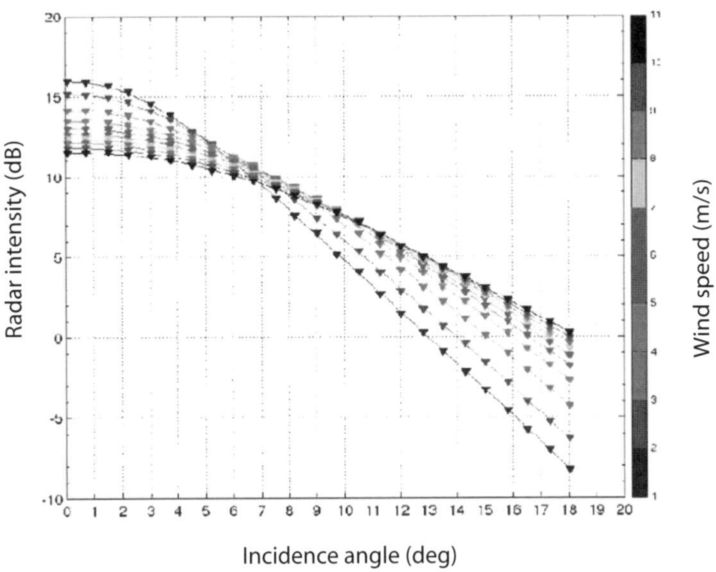

Figure 4.16. *The evolution of the radar signature of the ocean's surface depending on the incidence angle and the speed of the wind (origin: mission TRMM). The incidences between 6° and 8° prove to be relatively insensitive to wind conditions*

The objective of the "wave" scatterometry is the estimation of the directional spectrum of the waves $F(k, \Phi)$, which represents the spectral density of the sea heights in the azimuthal direction Φ (Figure 4.17). The principle of the measure is connected to the effect that the waves have on the radiometric variations of the collected radar signatures. A series of observations of a given area will thus present a variability $d\sigma^0/\sigma^0$ (where σ^0 represents the radar backscattering coefficient) almost exclusively connected to the structure of the waves if we specifically choose the observation incidence angle (between 6° and 10° for the above-mentioned reasons). This variability creates a modulation in the received signal compared to the average signal that would correspond to a plane surface.

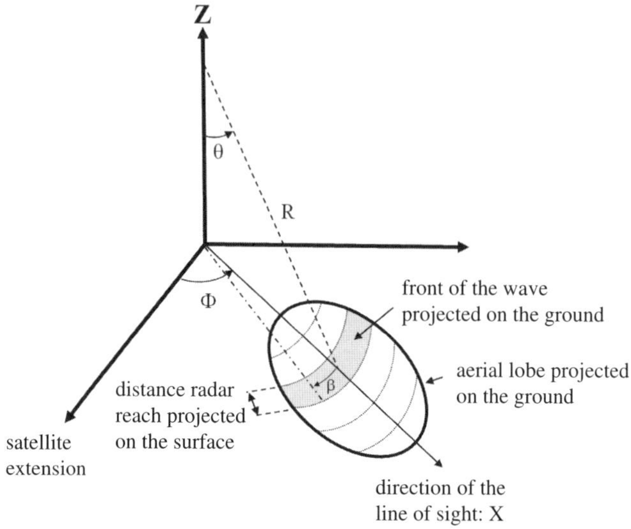

Figure 4.17. *Principle of the azimuth radar scatterometry measurement*

In the azimuth direction, the signals received on a given range sample (the gray area in Figure 4.17) are integrated inside the azimuth antenna lobe. The modulation $m(X, \Phi)$ of the signal in the direction of the line of sight Φ is defined by:

$$m(X, \Phi) = \frac{\int G^2(X, \beta) \frac{d\sigma^0}{\sigma^0}(X, \beta)\, d\beta}{\int G^2(X, \beta)\, d\beta} \qquad [4.46]$$

where $G(X, \beta)$ represents the antenna gain associated with the direction β.

Not being dependent on wind effects, $m(X, \Phi)$ only depends on the distribution of slopes in the line of sight direction. It is therefore possible to demonstrate that the spectral density of $m(X, \Phi)$, noted $P_m(k, \Phi)$, is proportional to the directional wave spectrum $F(k, \Phi)$ in the direction Φ:

$$P_m(k, \Phi) = |TF(m(X, \Phi))|^2 = f(\theta)\, k^2\, F(k, \Phi) \qquad [4.47]$$

where $f(\theta)$ is the function that depends only on θ. Equation [4.47] represents the decisive result of the "wave" scatterometry: it links the spectral density of $m(X, \Phi)$ to the unknown being researched: the wave spectrum $F(k, \Phi)$. An observation system with a turning antenna will cover the totality of the azimuth directions Φ during the motion of the satellite.

4.3.4. *Volume scattering*

The interaction between an electromagnetic wave and a cloud of scattering particles is described by the radiative transfer equation [CHA 60]:

$$\frac{d\,\tilde{S}(\vec{r},\hat{u})}{du} = -\left[\bar{\bar{\kappa}}_e\right]\,\tilde{S}(\vec{r},\hat{u}) + \int_{4\pi}\left[\bar{\bar{R}}\,(\vec{r},\hat{u},\hat{u}')\right]\,\tilde{S}\,(\vec{r},\hat{u}')\,d\Omega' \qquad [4.48]$$

We see here the Stokes vector $\tilde{S}(\vec{r},\hat{u})$ (see section 4.1.4) of a wave propagating in the direction $\hat{u}$, which is simultaneously damped by the matrix $\left[\bar{\bar{\kappa}}_e\right]$ – the medium extinction matrix – (path 1), but reinforced by the indirect radiation scattered by each particle in the propagation direction (path $2+3$) (Figure 4.18). In $\vec{r}$, the phase matrix $\bar{\bar{R}}\,(\vec{r},\hat{u},\hat{u}')$ governs the deflection of the energy beams from $\hat{u}'$ to $\hat{u}$. The joint effects of damping and reinforcing of $\tilde{S}(\vec{r},\hat{u})$ cause backscattering behaviors that are difficult to predict, and depend on several parameters: density, size, shape, scatterer organization, associated dielectric properties, associated surface mechanisms (soils), etc.

Moreover, the radiative transfer does not allow making a coherent addition (both in amplitude and phase) of the contributions of different kinds, reviewed and illustrated in the case of a forest canopy (Figure 4.19(a)). With a few fine tunings, the modeling requires us to assimilate the medium to a stack of layers, made up of particles settled in a host medium (Figure 4.19(b)).

We then limit ourselves to characterizing the "average" behavior of the wave within each layer, without the possibility of preserving the phase information. This imperfection is corrected by the "coherent" approaches, tending toward an exact calculation of the interaction (in amplitude and phase), which is difficult to apply outside the low-frequency range ($\approx$ 100 MHz), which limits the number of scatterers that need to be taken into consideration.

Stands of pinaster pines, at different stages of development (i.e. different rates of biomass) (Figure 4.20), illustrate the complexity of the mechanisms that result through the association between surface and volume effects. In L band ($f = 1.25$ GHz), polarization *VV* (Figure 4.20(a)), the decreasing radar response of the sparse forest (ranging from 0 to 33 $\mathrm{t/h}$) reveals a progressive damping of the intrinsic response of the soil (at a level 10 $\mathrm{dBm^2/m^2}$), because of the increasing extinction of the vegetation. Beyond a critical density (33 $\mathrm{t/h}$), the multiscattering mechanisms dominate, taking the response 12 $\mathrm{dB}m^2/\mathrm{m}^2$ to a saturation of 10 $\mathrm{dBm^2/m^2}$. The L band signal *HV* (emission in H, reception in V) (Figure 4.20(c)) indicates a large sensitivity of the response to the biomass level. This is enhanced in two respects: (1) *HV* is mainly generated by the multiscatterings mechanisms; (2) the contribution of the soil, dominated by the surface effects, hardly produces any crossed polarization at all.

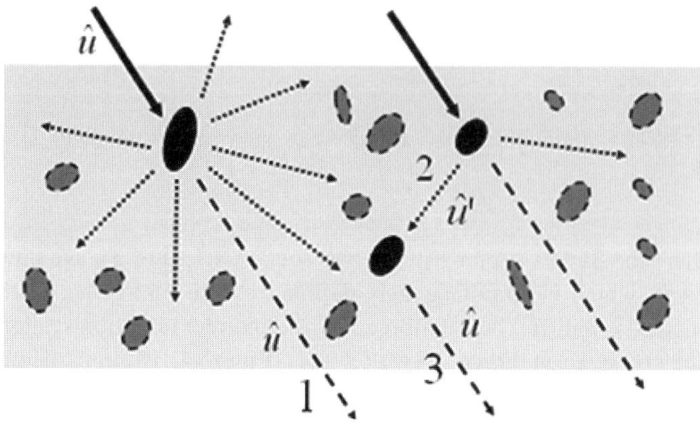

Figure 4.18. *The theory of the radiative transfer on a cloud of particles*

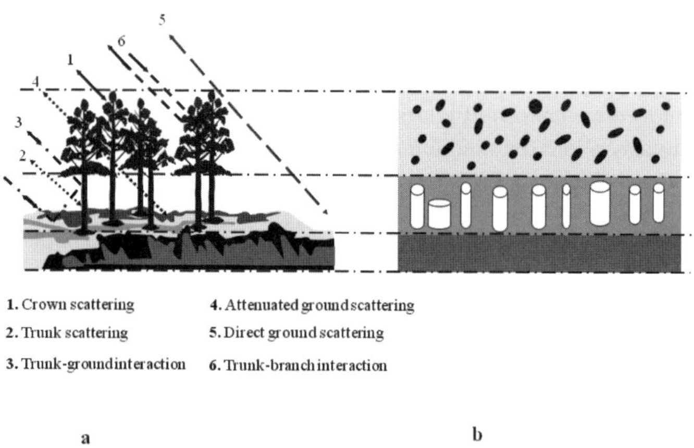

Figure 4.19. *a) Wave/tree cover interaction: main interaction mechanisms. b) Tree cover modeling in accordance with the radiative transfer approach*

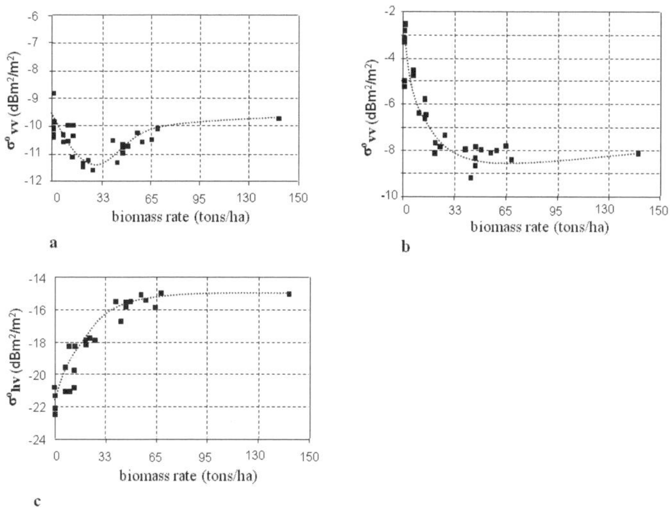

Figure 4.20. *Behavior of forest stands of various ages. SIR-C data, Landes forest, France (origin: CESBIO). The dotted curves come from the interpolation of experimental data. a) Band L, polarization VV, 26°. b) Band C, polarization VV, 26°. c) Band L, polarization HV, 26°*

Finally, in band C, *VV* polarization (Figure 4.20(b)), the intrinsic contribution of the soil becomes preponderant (the roughness effects increase when the frequency increases as well, with an intrinsic level of the order 3 dBm^2/m^2), up to wiping out the transition phase observed in L band: we are quite far from the common idea that the radar response systematically increases when the biomass increase.

4.3.5. *The penetration properties of electromagnetic waves*

To what extent can the electromagnetic waves penetrate and propagate in the studied structures? To answer this question, let us go back to the expression of an electric field $\vec{e}(z,t)$ that propagates in the $\hat{z}$ direction of a homogeneous medium with a dielectric constant ε_r (see section 4.1.1.2):

$$\vec{e}(z,t) = \Re\left\{\vec{E}_0 \exp\left[j(\omega t - k\,z)\right]\right\} \qquad [4.49]$$

The wave number k in the propagation medium is given by $k = k_0\sqrt{\varepsilon_r}$ where $k_0 = 2\pi/\lambda$ is its equivalent in the vacuum. Generally speaking, the dielectric constant ε_r is a complex quantity, whose imaginary part expresses a possible attenuation of the wave that is being propagated. The extraction of its square root leads to a sign ambiguity, which is resolved by physical considerations. By putting $k = \alpha - j\,\beta$, it becomes:

$$\vec{e}(z,t) = \exp\left[-\beta z\right] \Re\left\{\vec{E}_0 \exp\left[j\,\left(\omega t - \alpha z\right)\right]\right\} \qquad [4.50]$$

where β is a positive real number (if it is not, the wave would amplify while it propagates).

By definition, the penetration depth is the depth δp beyond which the amplitude of $\vec{e}(z,t)$ is attenuated with a factor larger than $1/\sqrt{e}$. From equation [4.50], it follows that:

$$\delta p = \frac{1}{2\beta} = -\frac{1}{2k_0\,\Im(\sqrt{\varepsilon_r})} \qquad [4.51]$$

where $\Im$ represents the imaginary part. In the case of slightly dispersive media[7], we have: $|\Im(\varepsilon_r)| < 0.3\,\Re(\varepsilon_r)$; therefore:

$$\delta p \approx -\lambda\frac{\sqrt{\Re(\varepsilon_r)}}{2\pi\Im(\varepsilon_r)} \qquad [4.52]$$

These results present the following remarks:

1) The penetration increases with wavelength: low frequencies (L, P) penetrate deeper than the higher frequencies (C, X), hence the interest that the scientific community has shown toward very low frequencies SARs (several tens to several hundred megahertz) for probing (vegetation, planetary undergrounds, comet-like structures, etc.).

2) A total dissipation lack ($\Im\,(\varepsilon_r) = 0$) leads to an infinite penetration depth. At the opposite end, a metallic surface (or of a water with a high salinity content) induces a very strong dissipation ($|\Im\,(\varepsilon_r)| = \infty$), which prohibits any penetration. The penetration of the radar wave allows us to observe (Figure 4.21(a)) the former bed of the Nile, covered in sand, on a combination of Shuttle Imaging Radar of third (C) generation (SIR-C) radar images acquired on bands C and L. The corresponding infrared optic image (Figure 4.21(b)) cannot reveal this underground trace.

In the case of bare soils, with the help of the estimation of dielectric constants given by [HAL 85], equation [4.5] shows a decrease in δp since the water content rate increases. The X band does not penetrate the soils with more than 15% water

7 This is the case for the majority of natural media, except water.

content (Figure 4.21), whereas the L band penetrates several centimeters deep up to 40% water content. In the C band, the electromagnetic wave penetrates an average water content soil to a depth of approximately 5 cm.

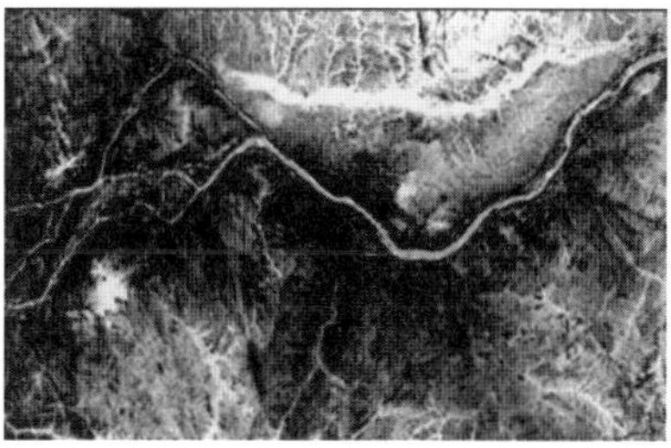

a

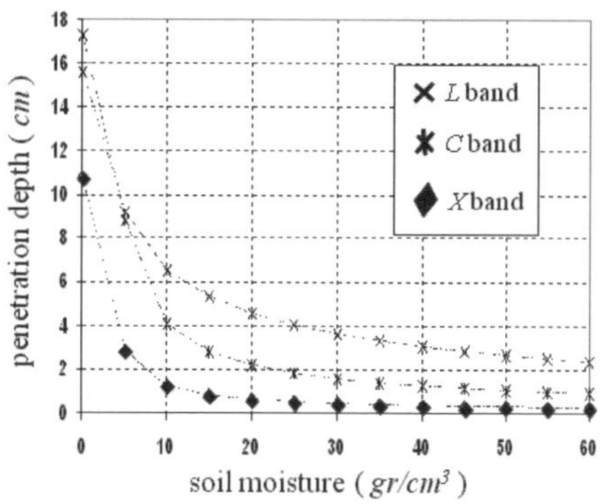

c

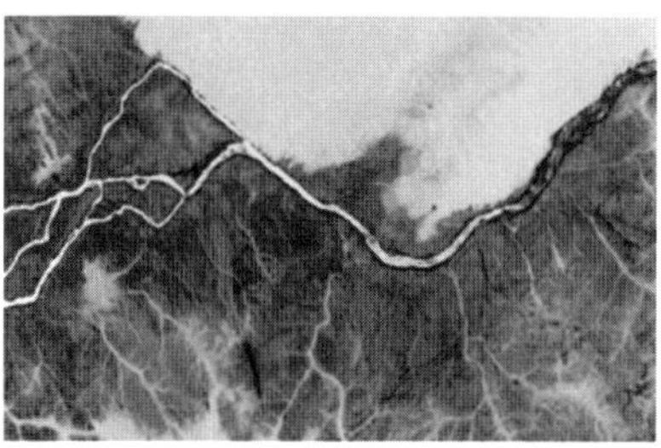

b

Figure 4.21. *Penetration depth for waves in bare soils as a function of their water content. For a color version of this figure, see www.iste.co.uk/tupin/RSImagery.zip*

In the case of forested areas, typical penetration values inside the forest layer are 1 m in X band, 6 m in C band and up to 20 m in L band.

4.3.6. *The effects of slope on radiometry*

Finally, the local slopes of the ground produce significant radiometric variations. Their impact may be observed on a "macroscopic" scale in contrast to the scale of the effects mentioned previously (roughness, water content, etc.). By assimilating the landscape to a succession of flat facets, the orientation of each of them affects, locally, the level of the signal response: a facet is a passive antenna that re-emits a signal whose power increases as the facet orientation is close to the radar line of sight. The comparison of Figures 4.22(a) (actual ERS image) and 4.22(b) (slope effect simulations) gives an opinion about the slope effect on the global radiometry of an image.

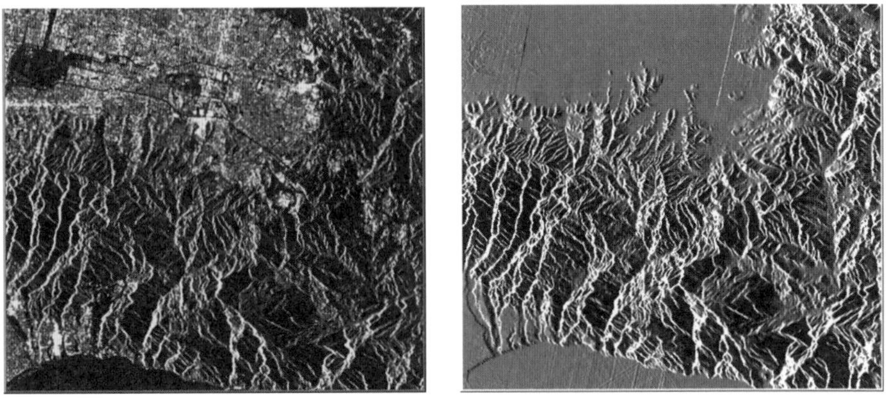

a b

Figure 4.22. *a) Actual ERS image in mountainous area. b) Simulated ERS image*

4.4. Calibration of radar images

The purpose of calibration is to associate an absolute radiometric content with each image pixel, related to the physics of the measurement. A large number of applications (cartography, interferometry, photo-interpretation, etc.) do not, however, make calibration a prerequisite, contrary to those that are specifically based on the inversion of the radar measurement (estimation of water content, roughness, biomass rate, etc.). The technical details of the calibration can be read in [MAS 08].

The requirements for the quality of the calibration come from the scientific application we are concerned with, via the estimation precision expected on the respective parameters. The calibration coefficients generally depend on a series of factors that are subjected to time and/or space fluctuations. The time fluctuations act not only throughout an orbital revolution (cyclical temperature variations affecting the electronic circuits), but also along the lifetime of the satellite. regarding the space segment, platform attitude variations, orbital drifts and drifting of the radiation pattern are so many sources of instability. In this context, the specifications of the calibration require the use of several scales of time and space.

Ulaby [ULA 98] underlines the progress carried out in the field of calibration since the 1970s, a high quality calibration thus meaning a precision on the relative measurement of the order ± 2 dB, and an absolute precision of the order ± 3 dB. To take the measure of accomplished progress, let us emphasize the exceptional radiometric stability of ERS-1/2 (absolute precision of the order ± 0.25 dB).

Calibration becomes more critical for the multimodal radar imaging devices, because of the uncertainties arising from incidence angle and resolution agility.

Table 4.1 gives the calibration requirements for a series of scientific objectives relating to the continental biosphere. The majority are extremely constraining, both for absolute level (of the order ± 1 dB), and relative level (from ±0.5 to ±1 dB for the short term, and of the order ±1 dB for the long term).

Geophysical thematic	Required precision	Absolute calibration	Relative calibration (short term) (long term)
Classification of the ice into three classes (winter)	Correct classification rate >80%	± 2.0 dB	± 0.5 dB ± 2.0 dB
Ice motion	Detection probability >95 %		... ± 0.5 dB
Snow cover water equivalent	For layer <20 cm and snow mantle depth > 20 cm	± 1.0 dB	± 1.0 dB ± 1.0 dB
Ground moisture (five levels)	20 % of variations per level (< ±0.3 g/cm^3)	± 1 dB	< ±0.5 dB ± 1 dB
Surface roughness	...	± 1 dB ...	...
Cartography/vegetation monitoring	Biomass density to within 25%	± 1 dB	< ±0.5 dB ± 0.5 dB
Lava ages	Precise classification of three classes of age	± 3 dB	± 1 dB ± 1 dB

Table 4.1. *Calibration requirements regarding several scientific objectives connected to continental biosphere studies [FRE 92]*

4.5. Radar polarimetry

4.5.1. *Introduction*

The diversity of polarizations has been uncommon in spaceborne SAR. The civil systems from the 20th Century were designed to transmit an electromagnetic wave with a single polarization, whether it is ERS-1&2 (linear, vertical polarization, band C), or RADARSAT-1 (linear, horizontal polarization, band C) and JERS (linear, horizontal polarization, band L). The later space missions all include experimental polarimetric modes, to various extents. ASAR/ENVISAT (launched in March 2002) offered, for the first time, polarization agility. This innovating step was followed by full polarimetry payloads (RADARSAT-2, TerraSAR-X and ALOS), operating, respectively, in C band, X band or L band, and in the resolution range from 3 to 10 m. The objective of this section has been to briefly present the key concepts for understanding the polarimetric measurement, and to present the potential applications by using several examples. More details can be found in [MAS 08, LEE 08].

4.5.2. *Operating principles of the polarimetric radar*

On a full polarimetric system, two waves with orthogonal polarizations are quasi-simultaneously emitted to receive the echoes on two orthogonal polarizations having the same phase reference. In our example, the antenna first emits a linear polarization $\hat{E}_0^{inc1}$ that is directed according to $\hat{x}$. The target scattering produces a scattered field $\vec{E}_0^{s1}$ of an unknown orientation[8]; because of the diversity of reception polarizations, we can capture the two components of this radiation $\hat{z}\,\vec{E}_0^{s1}$ and $\hat{t}\,\vec{E}_0^{s1}$. Having switched the emission polarization $\hat{E}_0^{inc1}$ toward its orthogonal polarization $\hat{E}_0^{inc2}$ directed along $\hat{y}$, we gather the quantities $\hat{z}\,\vec{E}_0^{s2}$ and $\hat{t}\,\vec{E}_0^{s2}$. The set of measures is stored in the following matrix form:

$$\left[\bar{\bar{S}}\right] = \begin{bmatrix} S_{xz} & S_{yz} \\ S_{xt} & S_{yt} \end{bmatrix} = \begin{bmatrix} \hat{z}\vec{E}_0^{s1} & \hat{z}\vec{E}_0^{s2} \\ \hat{t}\vec{E}_0^{s1} & \hat{t}\vec{E}_0^{s2} \end{bmatrix} \quad [4.53]$$

where $\left[\bar{\bar{S}}\right]$ is called the "target scattering matrix". The indices i and j of the complex coefficients S_{ij} refer, respectively, to the transmission modes (x or y), and the reception modes (z or t).

For a monostatic radar ($\hat{z} = \hat{x},\ \hat{t} = \hat{y},\ \hat{k}^s = -\hat{k}^{inc}$), we show that $S_{yx} = S_{xy}$. $\left[\bar{\bar{S}}\right]$ has been renamed the "backscattering matrix". In the following, we will stick to this configuration. The notation $\hat{h}$ is changed to $\hat{x}$ (horizontal direction), and $\hat{v}$ to $\hat{y}$ (vertical direction).

At this step, we know the behavior of a target that has been successively subjected to two orthogonal rectilinear illuminations: $\hat{h}$ and $\hat{v}$. Every wave with an elliptical polarization is a linear combination of these two canonical modes:

$$\hat{E}_0^{inc} = a_h e^{j\delta_h}\,\hat{h} + a_v e^{j\delta_v}\,\hat{v} \quad [4.54]$$

---

8 For reasons of simplicity, we suppose here that the target has no depolarizing effect. It is worth noting the difference between the notions of "polarization change" (i.e. modification of the orientation of the electric field), and the notion of depolarization (i.e. creation of an incoherent component in the radiation, see section 7.2.2).

$a_h,\ a_v,\ \delta_h,\ \delta_v$ being the real coefficients that are specific to the polarization ellipse (ellipticity rate and orientation of the large axis). The linearity of the operator $\left[\bar{\bar{S}}\right]$ then informs about what would have been the response $\vec{E}_0^s$ of the target to this elliptical polarization, whose existence is still virtual:

$$\vec{E}_0^s = a_h e^{j\delta_h}\ \vec{E}_0^{s1} + a_v e^{j\delta_v}\ \vec{E}_0^{s2} \qquad [4.55]$$

This "back and forth" is known as the "polarization synthesis" (Chapter 7, section 7.2.2).

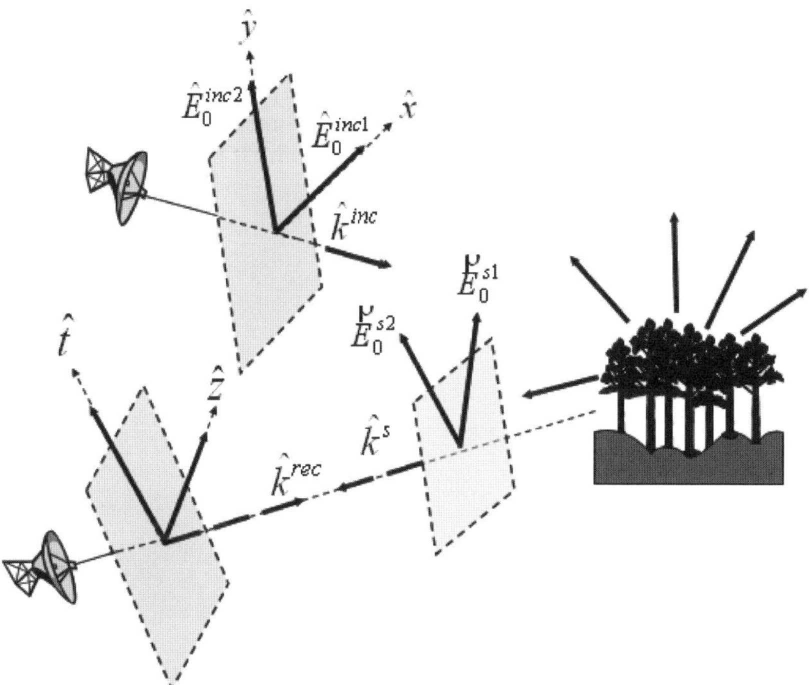

Figure 4.23. *The principle of polarimetric radar functioning*

To various degrees, the polarimetry appears systematically in the SAR architectures, both current ones and future ones. The physics of the measure associated with polarimetry can be systematically described in two stages: first, an isolated pixel manifests itself by its ability to modify the orientation of the waves that illuminate it (without reducing the degree of polarization, i.e. the backscattered wave is fully polarized). A behavior average is then naturally elaborated, when several pixels are considered for the local estimation of the polarimetric response. The averages operated lead to a partial loss of the polarization state, which fatally decreases the efficiency of polarimetry. Reducing these harmful effects means reducing the number of scatterers that contribute, locally, to the construction of the radar response. The low frequencies (L or P), much like high resolutions (metricrange), lead to the sought reduction. Let us finally note that polarimetry, through the technique of polarization synthesis, represents a plus-value in the field of interferometry. The fusion between polarimetry and interferometry currently plays a key part in the interest that the radar imaging scientific community has shown in the use of polarimetry [CLO 98]. As we will see in more detail in Chapter 7, all these polarization effects are put to good use in image processing techniques.

Let us finally note that polarimetry, through the technique of polarization synthesis, represents a plus-value in the field of interferometry. The fusion between polarimetry and interferometry currently plays a key part in the interest that the radar imaging scientific community has shown in the use of polarimetry [CLO 98]. To conclude, the polarimetry is not a high-resolution palliative, but it is a complementary tool. It does not compete with interferometry, but strengthens it.

As we will see in more detail in Chapter 7, these polarization effects are put to good use in image processing techniques.

PART 2

Physics and Data Processing

Chapter 5

Image Processing Techniques for Remote Sensing

5.1. Introduction

This chapter introduces the methods for signal and image processing used in a large number of applications in remote sensing. First and foremost, we present the models and the so-called "low-level" processing that are closer to pixel level: the statistic models (section 5.2), and then the preprocessing, sampling, deconvolution and denoising (section 5.3). Section 5.4 describes the segmentation methods and section 5.5 details the information extraction methods (i.e. punctual, linear and extended objects). Section 5.6 is dedicated to the classification methods and dimensionality reduction techniques. The chapter concludes with a presentation of fusion methods.

The techniques presented here can be applied to monochannel images (a single spectral band in optical imagery or a single polarization in radar imagery), but, when necessary, we will show how they can be extended to multichannel cases (vectorial data) whether it is several spectral bands or different radar polarizations. Figure 5.1 shows an extract of an optical and a radar image from the same region in Toulouse where we will illustrate certain processing stages[1].

Chapter written by Florence TUPIN, Jordi INGLADA and Grégoire MERCIER.

1 The results used for illustrating this chapter have been obtained using the Orfeo Toolbox (OTB) `http://www.orfeo-toolbox.org/`. OTB is a C++ library that provides tools for remote sensing image processing, from geometric and radiometric corrections to object recognition, going through filtering, segmentation and image classification.

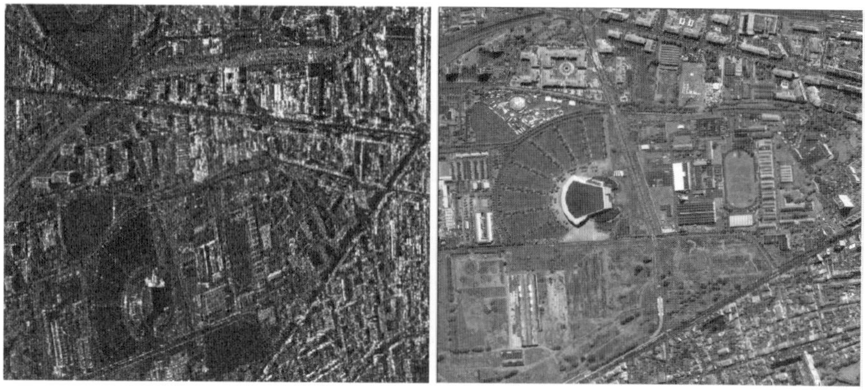

a) b)

Figure 5.1. *The same region imaged by an optical sensor and a radar sensor: a) the radar data TerraSAR ©DLR 2008; b) the Pleiades data ©CNES 2013, Distribution Astrium Services/Spot Image SA*

5.2. Image statistics

This section presents the statistical models that are necessary for certain segmentation, detection or classification techniques that will be presented in the following sections. We begin by presenting the statistical models used in optical imagery, and then we will detail the particularities of the models used in radar imaging. In the images, these statistics are provided by the histogram that expresses the frequency with which each gray level appears.

5.2.1. *Statistics of optical images*

The noise that can be found in optical images is generally quite low. It comes from several sources whose main contributions are the electronic noise of the components in the detection chain and the compression processing. The compression noise is quite difficult to characterize and can be quite different from one sensor to another. We will not consider it here. The electronic noise can be modeled by an additive white Gaussian noise that is added to the signals following the convolution with the sensor

---

OTB is a free software developed and maintained by CNES. It is used for research and training purposes, as well as for the development of operational processing chains, including satellite ground segments.

transfer function. The distribution of the image I for a surface with reflectance R is then given by:

$$p(I) = \frac{1}{\sqrt{2\pi}\sigma} \exp\left(-\frac{(I-R)^2}{2\sigma^2}\right) \qquad [5.1]$$

where σ is supposed to be constant throughout the entire image for a given sensor (Figure 5.2).

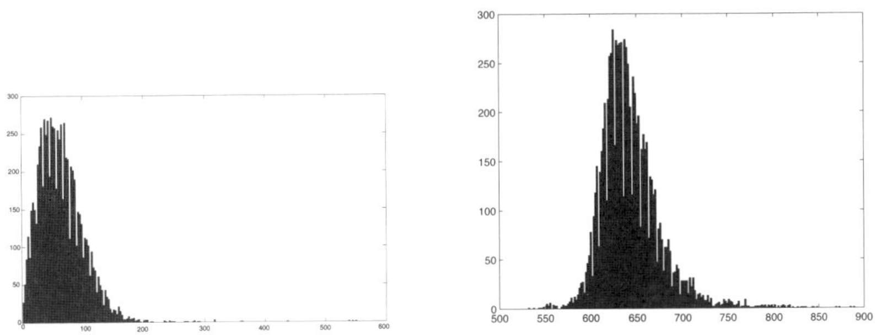

a) b)

Figure 5.2. *Histogram of the same physically homogeneous region for a) the radar sensor (close to a Rayleigh distribution), and b) for the optical sensor (close to a Gaussian distribution)*

5.2.2. *Radar data statistics*

Radar images have very specific statistics, which are due to the speckle phenomenon; we will be detailing this in Chapter 6. We will summarize it here for the amplitude data (noted A), that is for the module of the backscattered electromagnetic field, and for intensity data (noted I), which is the module squared. In the case of the data that we call single-look (no averaging takes place), the distributions follow a Rayleigh law in amplitude and an exponential law in intensity, given by the following formulas:

$$p(A) = \frac{2A}{R} \exp\left(-\frac{A^2}{R}\right) \quad \text{and} \quad p(I) = \frac{1}{R} \exp\left(-\frac{I}{R}\right) \qquad [5.2]$$

and in the case of multilook amplitude, the distributions follow a Nakagami law (where Γ is the gamma function):

$$p_L(A) = \frac{2L^L}{\Gamma(L)} \frac{A^{(2L-1)}}{R^L} \exp\left(-\frac{A^2}{R}\right) \qquad [5.3]$$

where in the three formulas, R is the reflectivity, that is the physical parameter of the region considered, which is proportional to the average of the intensity and to the backscattering coefficient σ^0 mentioned in Chapter 4.

These distributions have the following specificities: their standard deviations increase with the average (we can model them using a multiplicative noise), their modes do not correspond to their averages and the low-value pixels are predominant over the high-value pixels (Figure 5.2). Without going into the specific details of radar image processing, we must, however, remember that the noise model is very different from that of optical images, since instead of being an additive noise, the noise is multiplicative. Generally, image processing methods must be adapted, or else their behavior varies depending on the average radiometry of the regions. The readers can refer to [MAI 01] for more details on this subject.

5.3. Preprocessing

5.3.1. *Sampling and deconvolution*

We have seen in Chapter 2 how products of different levels are released by space agencies from the outputs of Charge-Coupled Device (CCD) sensors and radar antennas. Among the preliminary processing that turns "raw" signals into images, there are two fundamental stages. The *resampling* stage, which turns measured samples into "better" spatially distributed samples, and the stage of *deconvolution*, which seeks to correct potential faults in the acquisition device. These two stages are strongly connected to the sensor that is being considered, but we will present several general processing techniques used in them. For more detail, the readers can refer to [LIE 08].

Resampling consists of passing from a set of values that have a certain spatial distribution imposed by the sensor to a set of values disposed on a regular and usually isotropic grid. A key stage here is interpolation, which allows us to combine values measured in certain points to compute a new value that was not measured directly. The initial sampling can be regular (for example the CCD matrix) but it does not correspond to a regular ground grid, because of geometric distortions related to the acquisition system and relief effects (see the rectification levels in Chapter 2). It can also be irregular when certain data are missing. The interpolation techniques performing the resampling often involve the implicit hypothesis that the signal is smooth and they calculate the missing values starting from the weighted averages of the measured values. It is possible to improve the results obtained using these techniques by introducing constrains of fidelity to the data, such as a low number of discontinuities (in terms of edges). However, there are optimization problems that arise for calculating the solution that fulfills these constraints and respects the acquired data.

The deconvolution stage corrects the defaults that may come up in the response of the sensor. This is generally modeled by the impulse response (or *point spread function* (PSF) defined in Chapter 1), which is the response of the sensor to a punctual source. The image acquired can then be seen as a convolution of the scene by the PSF of the sensor, which generally entails a blurring effect on the image. There are several approaches that exist for deconvoluting an image such as the Wiener filter [MAI 01] or variational approaches that, again, will impose constraints on the solution sought [LIE 08]. If the image noise is also taken into account, then we may speak of *restauration*.

5.3.2. *Denoising*

In this section, we only consider the problem of denoising. For all the presented approaches, we will suppose that there is a distribution model that links the data observed and the parameters sought (for example between I and R by equation [5.1], or between A and R by equation [5.2]). We can distinguish the following three broad families of approaches for image denoising:

– The regularization approach (whether they are variational or Bayesian using Markovian models) where we introduce *a priori* information on the spatial regularity of the solution sought.

– The approaches based on parsimonious transformations (wavelets, dictionaries) which, once applied, make the noise easier to eliminate from the useful signal.

– The non-local approaches that search for the redundant elements in the image in order to denoise it.

Without going into further detail, in the following, we will present the main stages for each of these methods.

For the regularization methods, the problem is formulated as follows. Let g be the image acquired and f the underlying scene (noiseless, unknown); we look for a solution $\hat{f}$ that minimizes the energy (or a functional). This energy can be divided into two terms:

$$E(f) = E_{data}(f, g) + E_{regul}(f) \qquad [5.4]$$

$E_{data}(f, g)$ connecting the observations g to the scene f based on the physics of the sensor, and $E_{regul}(f)$ imposing a certain regularity on the solution. For example, in the case of a Gaussian additive noise, we obtain:

$$E_{data}(f, g) = \int_{(x,y)\in\Omega} (f(x, y) - g(x, y))^2 dxdy$$

There are several choices for $E_{regul}(f)$, which are often expressed using the gradient of f. One very popular choice is the minimization of the total variation (TV), $E_{regul}(f) = \lambda \int_{\Omega} |\nabla f|$, which corresponds to the Rudin–Osher–Fatemi model [RUD 92] and assumes that the images are approximately piecewise constant. After these modeling stages of the problem (the choice of $E_{data}(f, g)$ and $E_{regul}(f)$), the problem of the optimization of this energy arises. Depending on whether the field considered is continuous or discrete, depending on the form of the energies chosen, i.e. convex or not, the minimization of the energy is based on gradient descent approaches [CHA 04b] or on combinatorial optimization approaches, via a minimum graph cuts [ISH 03, DAR 06].

Wavelet-based approaches or parsimonious transformations can be divided into the following stages. We begin by changing the space of the signal representation, by calculating its projections into the new representation space where the parsimony hypothesis must be verified. The coefficients in this space are thresholded (i.e. hard or soft thresholding) by using hypotheses on the expected distribution of the coefficients. After analyzing the coefficients thresholded, the signal is projected back into the initial space, the noise component having been suppressed [MAI 01].

The third family, non-local approaches, is relatively recent [BUA 07]. It is based on the notion of a *patch* (or small image block). The main principle is that an image can be divided into very redundant patches and filtered with the help of these redundant versions (for example the central value of a patch will be denoised by exploiting all of the redundant patches and doing a weighted average of their central values). Let us consider the classic case of a Gaussian filter that averages the pixels that are spatially close in order to suppress the noise. Since the pixels that are close to each other in the image can belong to different populations, we may prefer to choose the pixels that are radiometrically close to each other in similar configurations, rather than selecting the pixels that are spatially close.

The problem then becomes selecting these similar pixels. The notion of the *patch* comes up. Two pixels will be considered to be radiometrically close if the patches that surround them are radiometrically close as well (Figure 5.3). The significance of this principle is that no connectivity constraint is imposed, and that similar patches, which are far from the pixel considered, can also be considered. This framework was developed initially for the Gaussian noise: the denoising is done by averaging pixels and the similarity criterion between patches is the Euclidian distance. We can adapt this comparison framework and the patch combinations to other types of noise by relying on a probabilistic framework [DEL 09].

In remote sensing, the denoising approaches can be used as preprocessing so that the data can easily be processed in the stages that follow. This is particularly true for radar data because of the speckle noise. The use of a denoising method must be assessed with respect to the final application, at the output of the complete processing chain.

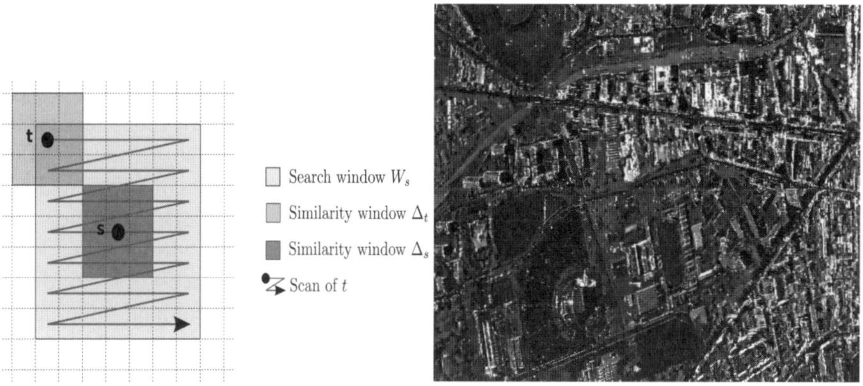

a) b)

Figure 5.3. *The idea of non-local averages is to denoise a pixel s by using the weighted values of the pixels t. a) The weight of a pixel t is calculated by comparing a small neighborhood of pixels around t with an identical one around s. The pixels are considered in a research window W_s [DEL 09]. b) Denoised image by NL means filtering*

5.4. Image segmentation

Image segmentation is the procedure that allows us to obtain a partition of the image (a pixel matrix) into separate, homogeneous regions (groups of interconnected pixels), on the basis of a common given criterion. Segmentation is rarely the last stage of an information extraction chain in remote sensing imagery, but is rather an intermediate simplification stage of the content of the image.

5.4.1. *Panorama of segmentation methods*

We may distinguish between two broad families of segmentation approaches: region-based approaches, which regroup the pixels according to the similarity criteria, and contour-based approaches, which detect the discontinuities, that is the abrupt transitions in the images. These are the dual approaches, which have advantages and drawbacks depending on the type of data that are to be processed and the applications considered.

In this section, we will review a certain number of region-based segmentation methods. We will not consider here the issue of textures, as there are specific methods for them [MAI 03]. There are several textural indices that we will introduce in Chapter 6 (section 6.3.4).

A first family of segmentation methods is given by the methods that are in fact, classification methods. Usually, the result of the classification is considered as a segmentation. These approaches can be supervised or not, and they are essentially based on the histogram and their modes (the local maxima) and they also rely on the Bayesian theory. In a supervised context, the distribution of the different classes sought is learnt. More elaborate algorithms allow us to learn automatically, usually in an iterative manner. When they are applied to each pixel independently, they give good results if the data have little noise. If that is not the case, we must use a global *a priori*, for example with a Markovian model, in order to guarantee the regularity of the solution obtained [GEM 84]. One such example of an automatic approach is detailed within the *mean-shift* algorithm (section 5.4.4). We will also return to the general Bayesian classification methods in section 5.6.1.

A second family of approaches is based, more specifically, on the notion of partition and on region growing, or the splitting and merging of regions. Generally, we use a predicate that defines the homogeneity of a region (i.e. low standard deviation, difference between the maximum and the minimum values, smaller than a certain threshold, etc.) in order to split or merge together two regions. Unfortunately, generally, there is no algorithm that guarantees that we will find the optimal partition, e.g. the minimal number of regions.

A third family of approaches poses the problem as the search, starting from a given image, for a set of boundaries and a regular approximation of the image within those boundaries. We must then globally optimize a functional that involves three terms: a term of data fidelity, a regularity term of the approximating function and a term that controls the number of boundaries (also see section 5.3.2). This formalism was proposed by Mumford and Shah [MUM 89] in a continuous framework, by Geman and Geman [GEM 84] in a discreet and probabilistic framework, and by Leclerc [LEC 82] within information theory (search for *minimum description length (MDL)*).

5.4.2. *MDL methods*

These approaches are based on the stochastic theory of information and try to find the shortest description for an image (in number of bits) using a description language. An image is considered as made up of homogeneous areas with constant values, within which there are fluctuations. The description of an image is thus given by the coding of its partition, parameters within the regions of the partition and the coding of the fluctuations in each region. Among all the possible descriptions, we then seek the one with the shortest coding [LEC 82].

It is difficult to propose an algorithm that allows us to find the optimum distribution. In [GAL 03], the so-called "active grid" allows us to minimize the coding, starting from an initialization with a very fine grid. It works by iterating the following steps:

– Region fusion on the basis of the current grid.

– Node displacement.

– Suppression of the useless nodes.

No modification is accepted if it increases the stochastic complexity of the representation. This approach gives very good results on remote sensing data, particularly in radar imagery, for which the coding term of the fluctuations allows us to consider appropriate distributions (gamma laws, Wishart laws in polarimetry, etc.). One such example is given in Figure 5.4.

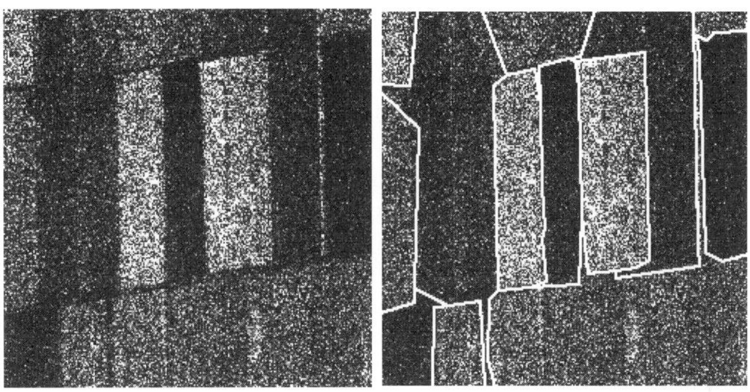

a) b)

Figure 5.4. *a) Original radar image. b) The result of segmentation with an active grid approach [GAL 03]*

5.4.3. *Watershed*

The *watershed* segmentation classifies the pixels of the image into regions by using gradient descent on the characteristics of the image (typically the gray-level value of the pixels), and by researching the elevated points along the boundaries of the regions. It is generally applied to a gradient image, i.e. to an image where the value of each pixel is the modulus of the gradient of the original image. The strong values correspond therefore to the discontinuities, but we will return to this point later on.

Let us imagine the rain that falls on a landscape whose topography represents the values of the pixels of the gradient image: the water streams down along the slopes and is captured in basins (the bottom of the valley). The size of these basins will increase with the quantity of rainfall, up to the point where certain basins will be merged with their neighbors thus forming larger basins.

The regions (catch basins) are formed using the local geometrical structure so as to associate points of the image with the local extremities of a gradient-type measurement. The advantage of this technique with respect to more classical region growing methods is that it is less sensitive to the choice of thresholds. Another advantage is that it does not produce only one segmentation, but rather a hierarchy of segmentations. A unique segmentation can then be obtained by thresholding this hierarchy.

The main idea is to process the image $f(x, y)$ as an elevation function. This image is not the original image that must be segmented, but rather it is the result of preprocessing. Herein lies the flexibility of the watershed method, because this filtering can be a simple gradient of a scalar image, but it can also be a vector gradient on several spectral bands, or even a gradient calculated on a set of features, such as textures. The only hypothesis we make here is that there are elevated values of $f(x, y)$ (or $-f(x, y)$) that indicate the presence of contours in the original image. The gradient descent then associates the regions with the local minima of $f(x, y)$ while using the watersheds of f as shown in Figure 5.5.

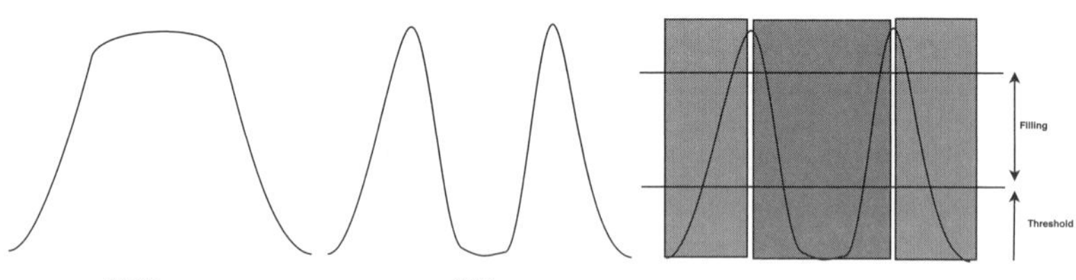

Figure 5.5. *Illustration of the watershed principle: to the left there is the mono-dimensional representation of a line of gray levels; to the center the gradient (its derivative here) corresponding to it; to the right, the slope basins define the watershed line that goes through the local maxima of the gradient; the threshold values and the filling values allow us to select the lines that must be kept (when the minimum filling rate increases, fewer lines are kept)*

Thus, the segments (regions) are made of all the points whose steeper paths lead to a local minimum.

The main drawback of this segmentation technique is that it produces a region for each local minimum, which results in over-segmentation. One way of mitigating this problem is by fixing a threshold and a filling threshold as indicated in Figure 5.5. The algorithm can therefore produce a hierarchy of encapsulated segmentations which depend on those thresholds. Finally, the algorithm can be completed by a criterion of minimum region size.

Figure 5.6. *Examples of the results obtained using a watershed segmentation for different sets of parameters for the image in Figure 5.1. One color is randomly attributed to each region obtained. To the left, for the parameters of 0.05 for the filling level and 0.1 for the threshold; to the right for 0.01 and 0.1 (see Figure 5.5). For a color version of this figure, see www.iste.co.uk/tupin/RSImagery.zip*

5.4.4. *Mean-shift*

The *mean-shift* algorithm is not a segmentation algorithm *per se*. Rather, it is a data clustering technique. When used in a more particular fashion – taking into account the geometric coordinates of the pixels – it gives very interesting segmentation results. The origin of this technique goes back to research work in 1975 [FUK 75] and was rediscovered 20 years later [CHE 95]. In the beginning of the 2000s, Comaniciu and Meer [COM 02] applied it to image segmentation.

The principle is quite simple. We have a data set of dimension N, the pixels of a multi-spectral image with N bands. Each individual in the data set is represented by a point in the N-dimensional space, called the feature space. In order to sum up the information contained in this cloud of points, we decide to only keep a few points for representing the samples of this space (this is the clustering operation).

In order to do this, we will replace the position of each individual in the space by the average position of the individuals lying in a ball with a radius d_c. It is this operation that justifies the name *mean-shift*. We see that this approach is a kind of generalization of the algorithm of k-means [MAC 67]. This procedure can be applied iteratively until convergence.

The algorithm, as described above, can be applied to any data set and has nothing specific to the images. We can however introduce the position of the pixels in the following way: we can add a geometrical distance in the image space constraint to the constraint on the neighborhood in the feature space. Analogously, we use a ball with radius d_s to limit the extent in which we compare the pixels. With this constraint, the algorithm converges towards average values that represent the pixels that are spatially close to each other.

After convergence, we will associate each pixel with the nearest mode, which constitutes the result of the segmentation. Classically, and in order to limit the effects of over-segmentation, we fix a threshold for the minimal size of the regions obtained. The regions of smaller size will be fused with the adjacent region that is closest in the feature space.

Figure 5.7. *Examples of the results obtained by the mean-shift segmentation for different sets of parameters for the image in Figure 5.1. A color is randomly attributed to each region obtained. To the left for the parameters: spatial radius 5, spectral radius 15, minimum region size 100 pixels; to the right, spatial radius 5, spectral radius 30, minimal region size 100 pixels. The increase of the spectral radius allows us to reunite more pixels in the same region L. For a color version of this figure, see www.iste.co.uk/tupin/RSImagery.zip*

5.4.5. *Edge detection*

The contour-based approaches are the dual of the previous approaches, and they rely on the discontinuities of the signal. In the same way that in one dimension the detection of strong transitions is done by detecting the maximum values of the derivative of the signal, in 2D, the gradient presents a strong module on the edges. Their detection is therefore done by localizing the extreme points of the module of the gradient in the direction of the gradient. The computation of the gradient can be done in different ways. It generally combines a derivative filter in one direction (such as the derivative of a Gaussian) and a smoothing filter in the opposite direction in order to maximize the detection probability (robustness to noise) and the localization of the edges [CAN 86]. However, these approaches make an additional noise hypothesis that cannot be verified, for example on radar images. We can show that in the context of multiplicative noise, it is preferable to compute the ratio between local averages in order to have a CFAR (Constant False Alarm Rate) detector [TOU 88a, FJO 97].

The edge detection stage is followed by a closure stage that tries to propagate and connect the contours between them. This stage can be performed through dynamic programming between the contour extremities, or by trying to join the contour elements belonging to the same structure via the Hough transform [MAI 03], or by using *a contrario* methods [DES 00] (see section 5.5.3).

The edges can also be detected with active contour approaches, also called deformable models [KAS 88]. The main idea is to move and deform an initial curve under the influence of the image and regularity constraints that are internal to the curve, until it adjusts to the desired contours. These approaches are widely used in medical imagery, for example to segment certain parts of the body, but they are less widespread in remote sensing, where there is less prior information about the objects present in the images.

5.5. Information extraction

5.5.1. *Point target detection in radar imagery*

The detection of point targets essentially regards radar imagery, where very bright points connected to the specific geometric structures appear frequently. The usual process consists of doing a hypothesis test between the presence and the absence of a target, that is, the two following hypotheses for the pixel considered and its surroundings [LOP 90]:

– there is one single region;

– there are two regions, the target and the background.

If we consider a set of observations I_i in an analysis window V, this can be written as the calculation of the likelihood ratio:

$$\frac{P(I_i, ..., \forall i \in V_1|R_1)P(I_i, ..., \forall i \in V_2|R_2)}{P(I_i, ..., \forall i \in V|R)}$$

with V_1 the region corresponding to the target of reflectivity R_1 (*cf.*eq.5.2) , V_2 the background region of reflectivity R_2, $V = V_1 \cup V_2$ with a reflectivity R. In the case of Gamma 1-look distributions, this ratio is expressed by:

$$\lambda(N_1, N_2, R_1, R_2, R) = \begin{array}{l} -N_1(\ln R_1 + \frac{\hat{I}_1}{R_1}) - N_2(\ln R_2 + \frac{\hat{I}_2}{R_2}) \\ +(N_1 + N_2)(\ln R + \frac{\hat{I}}{R}) \end{array} \quad [5.5]$$

where N_1 and N_2 are the number of pixels of the target and the background regions and $\hat{I}_1$ and $\hat{I}_2$ are the empirical averages calculated on these regions. In the absence of any knowledge on the true reflectivity of the regions, we calculate the generalized likelihood ratio, obtained by replacing the reflectivities by their estimations in the sense of the maximum likelihood, that is, the empirical average of the intensities within each of these two regions. We can show that this ratio depends on $r = \frac{\hat{I}_1}{\hat{I}_2}$, this ratio of local averages plays a key role during the processing of radar data since it also plays a role in the definition of contour detectors. One example of target detection is presented in Figure 5.4.

5.5.2. *Interest point detection and descriptors*

Generally, the images contain particular interest points, such as corners. Several approaches in image processing and computer vision rely on the extraction and characterization of these points for the applications of object registration or change detection. The SIFT (Scale Invariant Feature Transform) algorithm [LOW 04] is particularly popular and several variations and improvements have consequently been proposed for it. It relies on two steps: the extraction of the key points and the computation of a descriptor around the extracted key points.

The extraction phase analyzes the image on several scales. More precisely, a Gaussian image pyramid is made of successive convolutions from the original image with Gaussian kernels of increasing size. The search for extreme 3D points on the pyramid of differences between two successive levels allows us to localize the candidate interest points that are then validated by a corner detector that aims to suppress the points situated along the contours. Finally, with each of the points detected is associated an orientation and a scale. The position of the point extracted and these two parameters are then used to build a descriptor. This descriptor is made by the local histogram of the orientations of the gradient weighted by its modulus. It represents the distribution of the orientations of the contours around the point. These descriptors are then compared to find similar key-points.

The popularity of the SIFTs is due to their robustness because they are invariant to scale, rotation, translation and, partially, lighting changes. They were used in remote sensing for image registration, object recognition or change detection [SIR 09].

5.5.3. *Network detection*

Networks, whether they are road networks or river networks, play a structuring role in remote sensing images. They are therefore exploited in numerous applications, such as image registration, urban area detection, etc. Network detection methods can generally be broken down into two stages. A low-level line detection stage, that allows us to obtain potential pixels, and a higher level stage, that seeks to join these candidates.

As regards the line extraction stage, it depends on the size of the lines (function of the sensor resolution) and the type of noise that is present in the data. When the lines are larger than 5 pixels, we can use contour detection approaches followed by the parallel contour regrouping n each side of the lines. For narrower lines, we use linear, specialized sensors such as the *Duda-road operator* [DUD 72] or methods which rely on ratios for radar imagery [TUP 98].

The connection stage can be done in different approaches. One very popular approach is the Hough transform [MAI 03], which analyzes the geometrical shapes (lines, circles, ellipses, etc.) that go through a set of points, starting from the points themselves. The principle of these approaches is to use a space of parameters, each point of the image voting for a set of parameters. The accumulations in the space of the parameters allow us to detect the geometrical shapes of the image. More recently, a contrario approaches have been developed, which are very resilient to noise [DES 00]. They consist of testing the points' configurations against the hypotheses that this configuration has appeared randomly for a certain noise model. One last family of approaches uses stochastic approaches, for example Markovian approaches on a graph of segments [TUP 98], or stochastic geometry approaches based on marked point processes [DES 11].

5.5.4. *Detection and recognition of extended objects*

5.5.4.1. *Introduction*

The detection of point targets and networks can be approached as a generic problem because the researched objects can be modeled on the basis of simple properties (geometrical and statistical properties, etc.). Extended objects and therefore composite objects are difficult to describe generically in terms of the properties that are accessible on the image. For images with a metric resolution, we can mention buildings, crossroads, bridges, etc. Although relatively easy to recognize visually, these objects are difficult to characterize generally (i.e. all the bridges, all the buildings, etc.) by using only radiometric and geometric features.

One way of avoiding the generic model building problem is to use automatic learning techniques based on examples. Thus, if we have a database of examples of classes of objects that must be detected, as well as the name of the class they belong to, it is possible to use statistic approaches in order to establish the connection between image features that can be computed from the examples, and the class they belong to. The procedure can thus be built into several stages:

– building a labeled example database;

– extracting features from the examples;

– selecting pertinent features;

– learning the model.

In the following sections, we will detail each of these stages.

5.5.4.2. *Building a database of labeled examples*

This is a rather long and boring task that is usually done by a human interpreter and consists of selecting from real images examples of objects of interest. This requires a previous definition of the class nomenclature, that is, of the list of types of objects that we wish to detect.

Depending on the type of algorithm we use, there can be constraints on the way the examples are extracted. Quite often, in remote sensing, we choose to extract image patches of fixed size, centered on the object of interest (Figure 5.8). Furthermore, there are certain approaches that require that all the image patches have the same size. At the end of this stage, we therefore have a set of examples from each of the classes that we know the label for.

5.5.4.3. *Feature extraction from the examples*

The objective of this stage is to characterize the examples extracted with the help of a small number of values so as to associate them automatically with the classes they belong to. This characterization is done with the help of the image properties, such as, statistics, color, textures or geometry.

Thus, we obtain, for each example from the database, a pair $(y, \vec{x})$, where y is the label of the class and $\vec{x}$ is a vector whose components are the features extracted.

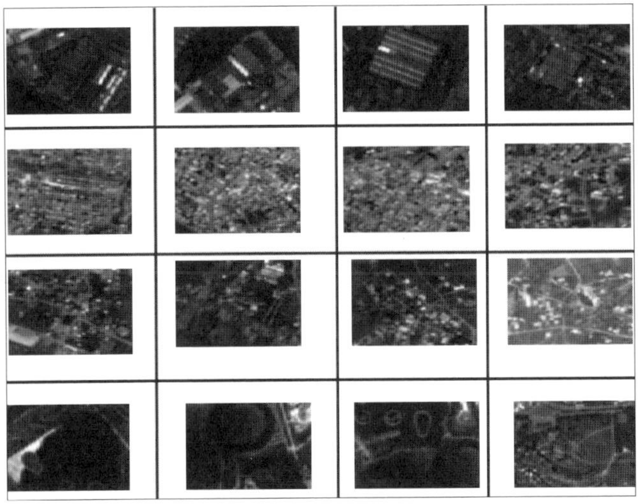

Figure 5.8. *Example of image database extracted from the SPOT5 satellite used for model learning. From top to bottom: examples of industrial areas, dense urban areas, villages and parks*

5.5.4.4. *Feature selection*

During the feature extraction, it is often difficult to know which ones will be those that will allow us to distinguish best between each class with respect to the other. In this case, we will choose to extract a maximum number of features first, to make the selection only afterwards.

The main objective being to best describe a data set with a minimum number of features[2], one of the solutions can be to find intelligent combinations of the existing features in order to synthesize a very small number of these elements: this is called dimension reduction.

The selection of display elements *stricto sensu* consists of keeping a reduced number, but without going through a combination of the extracted display elements. The criteria for deciding if a display element or a combination of display elements must be retained are often founded on measurements of the quantity of information. We approach some of these techniques in section 5.7.

2 The size of the feature vector directly impacts the duration of the search for the class for a given example. We are therefore interested in choosing small sizes.

5.5.4.5. *Model learning*

We wish to find here a function that can connect the characteristics vector to the class label to which it belongs. This function is often called the classification model, or simply, the model. The classification approaches will be characterized in section 5.6.

5.5.5. *Spatial reasoning*

For the composite objects for which the variability (or lack of pertinence) for low level features is high, it is necessary to develop higher level descriptions. Those can rely on spatial relations between features, to deliver a semantic interpretation of objects [MIC 09, VAN 13]. The relations thus modeled take the form of standard language expressions: "near", "along", "between", etc.

5.6. Classification

Classification methods are exploited in image processing for image segmentation and object recognition. We will describe these in this section, while limiting ourselves to *supervised* classification, that is, we have a set of labeled data (class and observation pairs) in order to perform the learning process and establish the models.

5.6.1. *Bayesian approaches and optimization*

The detection of extended objects can be done by classification methods. In their simplest version, they process the pixels independently from one another, but these methods can also be implemented globally as we will see in what follows. These approaches use a probabilistic model of the image, that is, they associate with each pixel s of the image a random variable Y_s. The observed gray level or vector y_s for the pixel, is only one realization of this random variable. The classification then consists of searching for each pixel the realization x_s of a random variable X_s that represents the class. By using a maximum a posteriori criterion, we then search the class x_s that maximizes the probability:

$$P(X_s = x_s | Y_s = y_s) = \frac{P(Y_s = y_s | X_s = x_s) P(X_s = x_s)}{P(Y_s = y_s)}$$

The term in the denominator does not appear in the optimization, so we only need to estimate the likelihood term $P(Y_s = y_s | X_s = x_s)$ and the *a priori* term $P(X_s = x_s)$. For the likelihood, we use a supervised or unsupervised learning process, or *a priori* knowledge on the physical nature of the phenomena observed and of the acquisition sensor. For the *a priori*, it can be known thanks to the probability of the apparition of the classes or ignored assuming that they are equi-probable. This approach, which consists of classifying the pixels independently from one another, is not based on the natural space coherence present in the images. If the distributions representing the probabilities of the gray levels conditional to the classes $P(Y_s = y_s|X_s = x_s)$ overlap, the results will be noisy.

Other approaches have therefore been defined by exploiting the random field in its entirety, that is, the set of the random variables of the image. Considering the global probability $P(X = x)$ with $X = \{X_s\}_{\{\forall s\}}$, we can then take into account models, such as Markovian models [GEM 84], that consider the interactions between neighboring pixels and allow us to obtain a regular solution for the classification. The problem then becomes to minimize a similar energy to the one of equation [5.4], the energy $E_{data}(x, y)$ representing $-\log P(Y = y|X = x)$ and $E_{regul}(x)$ an *a priori* on the regularity of the solution. As mentioned previously, it is difficult to globally minimize this energy that depends on a large number of variables and that is generally non convex. Efficient approaches based on the search for minimal graph cuts have been proposed [BOY 01].

5.6.2. *Support Vector Machines*

Learning methods based on kernels in general and Support Vector Machines (SVM) in particular, were introduced in the field of supervised automatic learning for the classification and regression tasks [VAP 98]. The SVM were successfully applied to text classification [JOA 98] and face recognition [OSU 97]. More recently, they have been used to classify remote sensing imagery [BRU 02, LEH 09].

The SVM approach consists of finding the separating surface between 2 classes in the feature space starting from the selection of the sub-set of learning samples that best describes this surface. These samples are called support vectors and completely define the classifier. In the case where the 2 classes cannot be linearly separated, the method uses a kernel function in order to embed the data in a higher dimensional space, where the separation of the classes becomes linear.

Let us assume that we have N samples available represented by the pairs $(y_i, \mathbf{x}_i), i = 1 \ldots N$ where $y_i \in \{-1, +1\}$ is the class label and $\mathbf{x}_i \in \mathbf{R}^n$ is the feature vector of size n. A classifier is a function of the parameters α so that:

$$(\mathbf{x}, \alpha) : \quad \mathbf{x} \mapsto y$$

The SVM finds the optimal separating hyperplane that verifies the following constraints:

– The samples with the labels $+1$ and -1 are located on different sides of the hyperplane.

– The distance of the vectors that are closest to the hyperplane is maximized. These are the support vectors and the distance is called the margin.

The separating hyperplane is described by the equation $\mathbf{w} \cdot \mathbf{x} + b = 0$; with the normal vector $\mathbf{w}$ and $\mathbf{x}$ is any point in the hyperplane. The orthogonal distance to the origin is given by $\frac{|b|}{\|\mathbf{w}\|}$. The vectors situated outside the hyperplane correspond either to $\mathbf{w} \cdot \mathbf{x} + b > 0$ or $\mathbf{w} \cdot \mathbf{x} + b < 0$.

The decision function of the classifier can therefore be written as:

$$f(\mathbf{x}, \mathbf{w}, b) = sgn(\mathbf{w} \cdot \mathbf{x} + b).$$

The support vectors are situated on 2 hyperplanes parallel to the separating hyperplane. In order to find the optimum separating hyperplane, we set $\mathbf{w}$ and b:

$$\mathbf{w} \cdot \mathbf{x} + b = \pm 1.$$

Given that there must not be a vector in the margin, the following constraint can be used:

$$\mathbf{w} \cdot \mathbf{x}_i + b \geq +1 \text{ if } \mathrm{y_i} = +1;$$

$$\mathbf{w} \cdot \mathbf{x}_i + b \leq -1 \text{ if } \mathrm{y_i} = -1;$$

which can be written as: $y_i(\mathbf{w} \cdot \mathbf{x}_i + b) - 1 \geq 0 \ \forall i.$ The orthogonal distances to the origin of the 2 parallel hyperplanes are $\frac{|1-b|}{\|\mathbf{w}\|}$ and $\frac{|-1-b|}{\|\mathbf{w}\|}$. The modulus of the margin is therefore $\frac{2}{\|\mathbf{w}\|}$ and it has to be maximized. We can therefore write the problem to be solved as:

– find $\mathbf{w}$ and b that maximize $\left\{\frac{1}{2}\|\mathbf{w}\|^2\right\}$;

– under the constraints: $y_i(\mathbf{w} \cdot \mathbf{x}_i + b) \geq 1 \ i = 1 \ldots N$.

The problem can be solved by using Lagrange multipliers, with one multiplier per sample. We can show that only the support vectors will have positive multipliers.

In the case where the 2 classes are not linearly separable, the constraints can be modified using:

$$\mathbf{w} \cdot \mathbf{x}_i + b \geq +1 - \xi_i \text{ if } \mathrm{y_i} = +1;$$

$$\mathbf{w} \cdot \mathbf{x}_i + b \leq -1 + \xi_i \text{ if } \mathrm{y_i} = -1;$$

$$\xi_i \geq 0 \ \forall i.$$

If $\xi_i > 1$, we consider that the sample is poorly classified. The function to minimize is then $\frac{1}{2}\|\mathbf{w}\|^2 + C\left(\sum_i \xi_i\right)$, where C is a tolerance parameter. The optimization problem is the same as in the linear case, but a multiplier has to be added for each constraint $\xi_i \geq 0$.

If the decision surface has to be non-linear, this solution does not apply, and we have to use Kernel functions.

The main inconvenient of the SVM is that, in their classical version, they only apply to 2 class problems. Several adaptations were implemented to be able to process the problems with more than 2 classes [ALL 00, WES 98].

Another approach consists of combining more problems with 2 classes. The 2 main strategies used are:

1) One against all: each class is classified with respect to the set of all the others, which results in N binary classifiers.

2) One against one: we use $N \times (N - 1)$ classifiers to separate all the pairs of possible classes and we make the decision through majority voting.

The SVM are frequently used for the classification of the satellite images using a set of features extracted from the images (Figure 5.9).

5.6.3. *Neural networks*

Neural networks are mathematical functions inspired by the functioning of a biological nervous system. A network is built from a set of inter-connected neurons.

Each neuron is a very simple function that calculates a value (the response of the neuron) on the basis of the values received in the input. These input values are the responses of the other neurons or values at the input of the system.

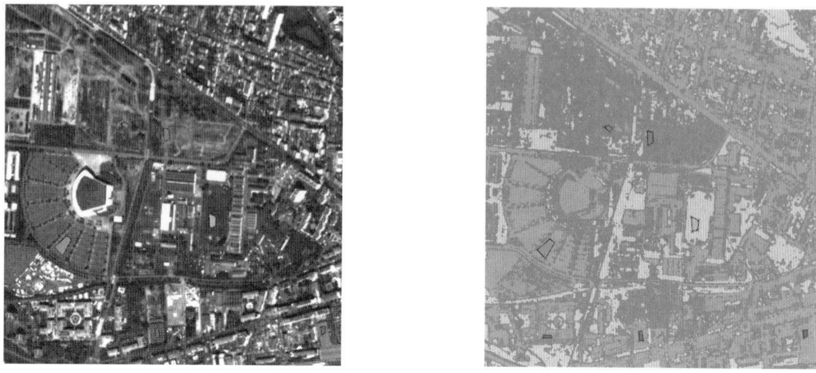

Figure 5.9. *Example of a supervised classification by SVM of a multi-spectral image using manually selected samples. To the left, the original image and the selected samples; to the right, the image of classes (gray: tar, green: vegetation; orange: tile; brown: bare soil)*

The function calculated by a neuron is traditionally a linear combination of the inputs, followed by a non-linear function, f, on which we may implement a thresholding:

$$y = sign\left(f\left(\sum_i a_i \mathbf{x}_i\right)\right)$$

The neuron networks are concurrent approaches to SVM. They remain widely used in remote sensing.

5.7. Dimensionality reduction

5.7.1. *Motivation*

A data set is more difficult to analyze as the number of dimensions that compose it is high. We say, for example, that visualizing an image that has more than 3 bands is difficult, but the problem of large dimensions is above all connected to the particular properties of the data in very large spaces, properties that are very different from the ones that we know for the usual sizes and that are covered by the term curse of dimensionality[3]. Thus, the majority of the classification methods have performances that decrease when the number of dimensions of the space of the characteristics increase (spectral bands, extracted features). If, in order to facilitate the analysis, we choose some of the dimensions from the data set, we risk losing valid information.

It is therefore interesting to have techniques that allow us to reduce the number of dimensions of a data set, all while preserving a maximum of information. These techniques consist of combining the available components in order to generate new ones in the smallest number, while keeping a maximum of information.

5.7.2. *Principal component analysis*

The principal component analysis, PCA, is very commonly used to this end. Representing the initial data structure in an orthogonal reference system, the partial information present in the data can be better distributed following the axes of the new coordinates. It is thus possible to represent the cloud of points in a space that has fewer dimensions than the initial space (Figure 5.10). In practice, we must represent the data in the basis formed by the eigenvectors of the covariance matrix.

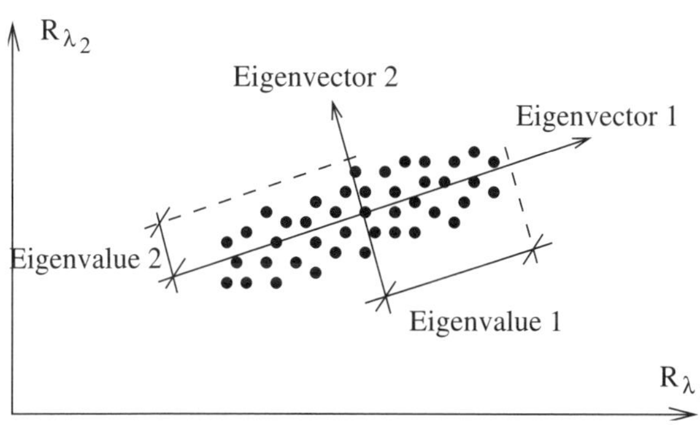

Figure 5.10. *Representation of the rotation of the original mark along the criterion of the main components [LEN 02]*

Similarly, other criteria can be chosen to obtain the new projection axes. We will thus see that criteria of statistical independence or of non-Gaussianity of the data in the target space will give interesting viewpoints.

---

3 Among these properties, let us note for example, that points drawn randomly in a sphere having a unit radius in n dimensions can be found with a probability that tends towards 1 at the distance 1 from the origin when n tends towards infinity.

5.7.3. *Other linear methods*

The change of the orthogonal basis can be assimilated to a rotation in which the new basis will first align itself in the direction of the largest variation of the samples. These directions are predicted starting from the covariance matrix of the data, and its eigenvalue decomposition. With the PCA, the last band then represents the axis of the smallest value and seems to be very noisy. However, this "noise" image always contains a part of the structure of the image (and therefore useful information).

The *noise adjusted PCA (NAPCA)* or *minimum noise fraction* (MNF) allows us to introduce the variability of an observation so as to decompose the multidimensional signal, not only in the sense of decreasing values, but also in terms of decreasing signal-to-noise ratios [LeWoBe-90]. Thus, this transformation can be divided into two stages:

1) Estimation of a multicomponent "noise" image (for example by high-pass filtering) where we apply a PCA. It is not the result of the PCA that interests us, but the transformation matrix estimated by diagonalization of the covariance matrix of the noise image.

2) We use the transformation matrix obtained previously to transform the initial image. We then apply a traditional PCA to the result obtained.

The NAPCA and the MNF are equivalent and they only differ by their implementation.

The independent component analysis (ICA) is based on an independence hypothesis and not on signal decorrelation. This applies to non-Gaussian signals and it involves relaxing the orthogonality constraint shown in Figure 5.10. By inversely applying a corollary of the central limit theorem (which establishes the convergence of the sum of a set of random variables toward the normal law, with a finite variance), ICA aims to inverse a model of a linear mixture of random variables so as to obtain the least Gaussian sources possible [COM 94, CAR 02]. Numerous strategies have been developed in the inversion of this matrix, such as JADE, InfoMax and Fast-ICA, to name just a few.

5.7.4. *Nonlinear methods*

Although in linear models we obtain a new basis that can be represented as a projection matrix whose columns are the basis vectors, in nonlinear methods, we will obtain a functional that allows us to go from one space to the other.

The curvilinear component analysis (CCA), for example, allows us to bypass linearity hypotheses and processes types of structures that are quite varied. The simplicity of the calculations that it engenders and its convergence mode make it easier to use than other nonlinear, more traditional algorithms.

5.7.5. *Component selection*

In some applications, we do not wish to generate new components, because they can be difficult to interpret. In this case, we would rather keep a subset of original components, even if there is a more significant loss of information. These techniques are mainly based on the *forward selection* and *backward elimination* approaches.

Forward selection consists of adding components one by one until a certain criterion of independence between them is observed. We can, for example, start by selecting the maximum variance component (in order to select a maximum of information), then add the most orthogonal component and thus proceed until the number of components desired or until the degree of orthogonality is too low. This solution is similar to the solution of PCA, but it does not create new components. A solution based on entropy and statistical independence instead of the variance and the orthogonality allows us to get closer to an ICA-type solution.

The *backward elimination* is the inverse procedure, which consists of starting with all the components and eliminating them one by one, following a redundancy criterion or an information quantity criterion.

The drawback of these approaches is their sensitivity to the initial choice. One way to make up for this is the *stepwise selection*. This allows us to move in both directions, via suppression or addition of variables at each step.

5.8. Information fusion

Fusion made its appearance in a certain number of applications (that are not exclusively dedicated to satellite imagery) where several sources of information must be combined in order to guarantee the best interpretation and decision possible. Fusion techniques come up as soon as the physical modelings do not allow us to draw an explicit connection between the different measures. We must therefore pass from observations to information without an explicit physical connection, with more general techniques than Bayesian modelings. To see the contribution of fusion techniques with respect to Bayesian models (by using joint probabilities, for example), we must return to certain properties that link the observation and the information.

– Does observation represent the world *exhaustively* ?

In certain theories, this exhaustiveness hypothesis is called the “closed” or “open” world hypothesis. This question can become particularly sensitive in an analysis of a series of images where certain classes cannot be observed all the time, and others appear from a particular date onward.

– Is there an *exclusive* link between observation and information?

In a classification, if the classes are not rigorously separable, it is because this link is not exclusive and two classes can be locally represented by pixels (or attributes) of the same value.

– Is the link between observation and information total?

Depending on the resolution of the sensors, the link between observation and information is often partial. This is why unmixing methods have been introduced (section 6.4). Quite often, the representation of a vegetated surface is accompanied by the signature of the ground, which can be seen from between the vegetation. However, is vegetation a category in itself or is it rather part of a mixture of gluten molecules, collagen molecules and others, according to the specific proportions per type of vegetation or per maturing degree?

Perfect information would be exhaustive, exclusive and having a total, known link. Such information is practically inaccessible and the different theories below will propose certain hypotheses in order to better consider information fusion.

– A piece of information is said to be uncertain if it is exhaustive, exclusive and with a total but unknown link.

– A piece of information is said to be ambiguous if it is no longer exclusive and the link between observation and information may not be total.

5.8.1. *Probabilistic fusion*

Fusion in a probabilistic framework presupposes an uncertain piece of information. The link is estimated, for example, via maximization of *a posteriori* probability, and using the probabilities inherent to different observations. For example, if we consider two sources of information that must be jointly processed Y_1 and Y_2, we will have to maximize the probability $P(X_s = x_s | Y_{1;s} = y_{1;s}, Y_{2;s} = y_{2;s})$ to calculate the probability of having the class label x_s on the site s.

5.8.2. *Fuzzy fusion*

The fuzzy model considers information having a partial link between observation and information. We therefore have fuzzy measures that represent the "chance" so that a piece of information $X_s = x_s$ to $Y_s = y_s$ represents a total link. This measure, represented by $\mu_{X_s=x_s}(y_s)$ in the interval $[0, 1]$, thus appears as a *membership function* that must not be confused with the conditional probability $P(X_s = x_s | Y_s = y_s)$.

These functions do not have axiomatic constraints imposed on the probabilities and they are therefore more flexible for modeling. This flexibility can be considered as a disadvantage since it does not help the user define these functions. In the majority of the applications, this definition is made either from probabilistic learning methods, from the heuristic method or from the neuromimetic methods that allow us to learn the parameters of particular forms of membership functions; finally, the definition can be made by minimizing classification error criteria [BLO 03]. The only drawback of fuzzy sets is that they essentially represent the vague character of information, the uncertainty being represented implicitly and being made accessible only by the deduction using the different membership functions.

As for the probabilistic fusion, the fuzzy fusion first goes through the definition of a joint measure. The fuzzy joint measures must be commutative, associative and monotonous. However, there are two families that come up for the definition of the identity function.

– T-norms are the functions that associate two fuzzy measures:

$$\mu_{X_s=x_s}(Y_{1;s} = y_{1;s}, Y_{2;s} = y_{2;s}) = \mathcal{T}(\mu_{X_s=x_s}(Y_{1;s} = y_{1;s}),$$
$$\mu_{X_s=x_s}(Y_{2;s} = y_{2;s}))$$

where the identity is ensured as follows: $\mathcal{T}(u, 1) = \mathcal{T}(1, u) = u$.

Among these T-norms, we find the Zadeh measure, or min measure:

$$\mathcal{T}(u, v) = \min(u, v)$$

or also the product T-norm : $\mathcal{T}(u, v) = u \times v$.

– The T-co-norms are functions whose identity is ensured by:

$$\mathcal{T}(u, 0) = \mathcal{T}(0, u) = u$$

Among these T-co-norms, we find the max function: $\mathcal{T}(u, v) = \max(u, v)$.

The fusion thus takes place through the application of a T-norm or a T-co-norm between the fuzzy memberships of the different information sources $\{Y_1, Y_2, \ldots\}$. The decision making is then implemented according to the maximum attained by these measurements:

$$\{\mu_{X_s=x_s}(Y_{1;s} = y_{1;s}, Y_{2;s} = y_{2;s}, \ldots), \mu_{X_s=x'_s}(Y_{1;s} = y_{1;s}, Y_{2;s} = y_{2;s}, \ldots), \ldots\}$$

The choice of a T-norm or a T-co-norm induces a behavior that is consensual or not in the decision-making process. Indeed, T-norms will favor the consensus of the different observations Y_i relatively to a value of x whereas the T-co-norms favor one of the observations Y_i that is the more connected to x, meaning that the odds that there is a total link between Y_i and X are the highest.

An application of the fuzzy reasoning to the interpretation of satellite imagery is given in [VAN 13].

5.8.3. *Evidence theory*

The *evidence theory* [SHA 76] proposed by Dempster and Shafer considers, just like in the case of Bayesian fusion, that a piece of information is uncertain with an unknown link between the observation and the information. This link is, however, supposed as known throughout the subsets of the space of observations. This degree of additional freedom allows us to differentiate between equally probable events (e.g. an equivalent decision between the classes x_s and x'_s if $P(X_s = x_s|Y_s = y_s) = P(X_s = x'_s|Y_s = y_s) = \frac{1}{2}$) and non-distinguishable events (from a set point of view $x_s \cup x'_s$ or a logical point of view x_s OR x'_s).

Evidence theory does not define the probabilities but the mass functions $m_X(Y)$ whose normalization allows the composition of simple hypotheses. For example, for a two-class classification $X \sim \{x_1, x_2\}$, we have:

$$m_{X=x_1}(Y = y) + m_{X=x_2}(Y = y) + m_{X=x_1 \cup x_2}(Y = y) = 1$$

whereas the probabilistic model only normalizes on the basis of *simple hypotheses* :

$$P(X = x_1|Y = y) + P(X = x_2|Y = y) = 1$$

The evidence theory fusion is defined in various ways [LEH 97, TUP 99], but we will only present the most widely used one here, the one that involves conjunctive fusion.

This Dempster–Shafer rule introduces a strategic notion in information fusion that is that of the conflict between the sources. It reveals, through a K coefficient taking values in $[0, 1]$, the agreement of the two sources of information ($K = 0$) or, on the contrary, their complete disagreement ($K = 1$). In such a case, *ad hoc* strategies must be adapted so as to account for the confidence that we have in one source over the other, and this is called a *source weakening* [SME 94]. The conjunctive fusion between two observations Y_1 and Y_2 is thus written as:

$$\begin{aligned} m_{X=x}(Y_1, Y_2) &= m_{X=x}(Y_1 = y_1) \oplus m_{X=x}(Y_2 = y_2) \\ &= \frac{1}{1-K} \sum_{x_1 \cap x_2 = x} m_{X=x_1}(Y_1 = y_1) m_{X=x_2}(Y_2 = y_2) \end{aligned} \quad [5.6]$$

the conflict being $K = \sum_{x_1 \cap x_2 = \emptyset} m_{X=x_1}(Y_1 = y_1) m_{X=x_2}(Y_2 = y_2)$.

An evolution of evidence theory was proposed by removing the exclusivity hypothesis [SMA 06]. This paradoxical reasoning allows us to specify the class mix stages, and must not be confused with the imprecision that does not allow us to distinguish one class from the other. Here, the two classes are intricate during the observation (the example of the landcover of an agricultural parcel is particularly useful in this case). If a vegetation indicator such as the normalized difference vegetation index (NDVI) (see section 6.3.1) is at the decision threshold between a bare ground or a covered ground, is it not more rigorous to consider that it is a bit of both?).

5.8.4. *Possibilistic fusion*

Possibility theory is often considered to be similar to the fuzzy model since it equally considers information that has a partial link between the observation and the information. However, it considers an uncertain information (again, like the probabilistic or the evidence theory model). The possibility functions thus defined are very similar to the fuzzy functions, but account for the knowledge that can be extracted from certain intervals in the observation space. These possibility functions can be seen as an envelope that contains the possible probability densities of the observations [DUB 07].

The fusion in the possibilistic frame can be founded on the same operators as those used in fuzzy fusion with T-norms or T-co-norms. However, conjunctive and disjunctive notions can be used for a more nuanced fusion as the fusion rules used in the evidence theory [DUB 92].

5.9. Conclusion

This chapter introduced different families of image processing methods used in remote sensing to improve the received signals, extract relevant information, data fusion, etc. Given the vastness of the field of image processing, this chapter has been an introductory and general one, with a partial choice on the methods presented.

The field of computer vision and image processing constantly evolves and the advances made allow for the definition of more efficient approaches for remote sensing imagery, whether we are talking about parsimonious approaches or non-local approaches, for example. The tools presented in this chapter will be the basis for other chapters (such as the classification methods for change detection in Chapter 8), and some of them will be detailed in the following chapters (i.e. Chapter 6 on the processing of optical data).

Chapter 6

Passive Optical Data Processing

6.1. Introduction

This chapter introduces a series of techniques specific to the processing of passive optical images, e.g. multi- and hyperspectral images. On the basis of the techniques discussed in the previous chapter, we will now study operations such as (1) merging panchromatic images with multispectral images (or *pansharpening*), (2) extracting indices specific to a given application and (3) extracting two types of product – land use maps and subpixel mixture maps.

6.2. Pansharpening

In Chapter 2, the different resolutions of remote sensing images have been presented. In the case of satellite-mounted multispectral sensors at high or very high resolutions such as SPOT5, Formosat-2 (see Figure 3.1) or QuickBird, several spectral bands covering the wavelengths of the visible (400–700 nm) and the near-infrared (700–800 nm) are generally available. Figure 6.1(a) illustrates an example of the image acquired by the multispectral sensor mounted on the QuickBird satellite. Besides the multispectral sensor, these satellites are generally equipped with an additional sensor, called panchromatic, that simultaneously observes the wavelengths of the visible and provides a single grayscale image, the panchromatic band. Because of the quantity of light received, this band has a higher spatial resolution (generally two to four times higher than the corresponding multispectral

---

Chapter written by Devis TUIA.

bands). The central image in Figure 6.1 shows an example of an image acquired by the panchromatic sensor mounted on the QuickBird satellite.

These two types of images complement each other, because the first offers a high spectral resolution (several bands covering different ranges of wavelengths are available), whereas the second offers great spatial detail, but only consists of a single channel in gray tones. The two can be used for different purposes, the multispectral image being useful for discerning objects depending on their spectral signature (see section 3.4), while the panchromatic image is rather useful in tasks of recongnition of texture and fine structures.

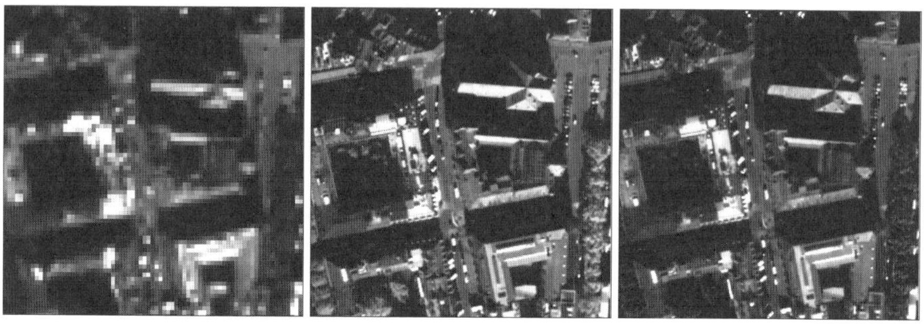

Figure 6.1. *a) Multispectral image acquired by the QuickBird satellite, at 2.4 m resolution, four bands. b) Panchromatic image, at 0.6 m resolution, unique band. c) Result of pansharpening using the PCA method. For a color version of this figure, see www.iste.co.uk/tupin/RSImagery.zip*

This complementarity can be used to synthesize images that combine the advantages of the two types of sensors: through resolution merging, or *pansharpening* [WAN 05, ALP 07], one can obtain the image that would have been acquired by a sensor with the spectral resolution of the multispectral and the spatial resolution of the panchromatic. This has great advantages: beyond the obtention of an image at a better resolution to solve a specific problem, pansharpening brings a general methodology for synthesizing the information of the sensor that does not (yet) exist and whose development is not possible (most often for economic reasons).

There are several methods of pansharpening. In the following sections, we will present two methods:

– the *spectral* methods based on projection-substitution operations;

– the *space-scale* methods.

6.2.1. *Spectral methods: projection-substitution*

Projection-substitution methods are based on the hypothesis that the different channels of a multispectral image contain significant similarities. These methods exploit the projection of the data to a space, where the redundancy of information is minimal: examples of such projections are the principal components analysis (PCA) discussed in section 5.7.2 [CHA 91], the intensity-hue-saturation (IHS) transform [CAR 90] or the simulation of a low-resolution panchromatic image [LAB 00]. In the projected space, one of the extracted components separates the low-frequency spatial information from the specific color information in each channel. In other words, one of the components resembles a low-resolution, smooth panchromatic image.

Once this component has been isolated, it is replaced with the high-resolution panchromatic image and the process is inverted. By doing so, we replace the low-resolution image with another image, a very similar one, but that provides the spatial details (high frequencies). Figure 6.2 illustrates the difference between the panchromatic image and the first principal component extracted from the multispectral image shown in Figure 6.1. The other components, which bring complementary information that is relative to the multispectral bands, remain unchanged. The image that results from the inversion will therefore have the color information and the researched space detail. Figure 6.1(c) shows the result of the pansharpening using the PCA as a projection method [CHA 91].

Summing up the process, the projection-substitution methods consist of the following:

1) Oversampling the multispectral bands at the resolution of the panchromatic image: this operation consists of matching the size of the images, either by replicating the values (by nearest neighbors) or by using an interpolation method. The oversampled bands compose the matrix $\mathbf{x}_{\text{low}}$.

Figure 6.2. *a) Panchromatic image acquired by the QuickBird satellite at 0.6 m resolution, unique band. b) First principal component obtained with the PCA method at 2.4 m resolution*

2) Applying the projection method (e.g. PCA): find the projection vectors $\mathbf{w}$ and project the images $\mathbf{x}^{\text{PCA}} = \mathbf{x}_{\text{low}} * \mathbf{w}$.

3) Substituting the intensity component with the panchromatic image ($\mathbf{x}_1^{\text{PCA}} =$ Pan).

4) Applying the inverse transformation ($\mathbf{x}_{\text{high}} = \mathbf{x}^{\text{PCA}} * \mathbf{w}^{-1}$)

The spectral distortions that result from the fusion deserve special attention. Typically, using an IHS transformation produces significant color distortions, since the angle between the intensity (I) and the bands is different from the angle between the panchromatic image and the bands. To reduce color distortion, one solution is to force this angle to remain the same.

6.2.2. *Space-scale methods: multiresolution pansharpening*

The multiresolution pansharpening methods use a different logic: in these methods, only the high-frequency information of the panchromatic image is injected into the multispectral image bands. In order to do this, a space-scale decomposition (i.e. typically wavelets [MAL 89] (Figure 6.3) or Laplacian pyramids [BUR 83] (Figure 6.4)) is used for extracting the spatial details (high frequencies) of the panchromatic image, which are then injected in the multispectral bands via band replacing or addition to the color bands. To avoid the creation of artifacts, the injection can be weighted depending on the local spatial correlation [AIA 02].

Figure 6.3. *a) Panchromatic image acquired by the QuickBird satellite at 0.6 m resolution, unique band. b) Wavelet decomposition of the image (from upper left, clockwise: low frequencies at the first level of the decomposition, high frequencies in the vertical, horizontal and diagonal directions)*

The multiresolution fusion methods consist of the following:

– Reducing the resolution of the panchromatic image and then oversampling it, in order to obtain a model of the panchromatic image, as it would have been acquired by a sensor with lower resolution. This can be obtained, for example, by applying a Laplacian pyramid up to the resolution of the multispectral bands [AIA 02] (Figure 6.4).

– Obtaining the spatial detail by differentiating the panchromatic image and the model of the panchromatic obtained (Figure 6.5).

– Oversampling the multispectral bands to the resolution of the panchromatic image, as in the projection-substitution methods.

– Adding the spatial detail in the spectral bands:

- through direct addition (in the case of Laplacian pyramids);

- via (weighted) substitution of the components at high frequency and inversion of the transformation (in the case of wavelets);

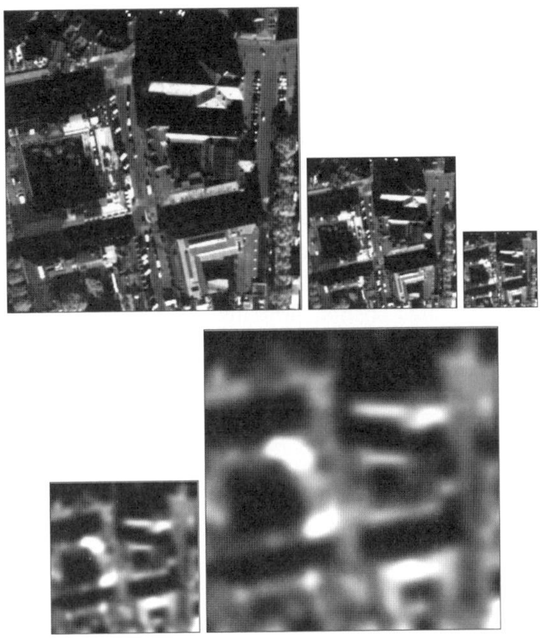

Figure 6.4. *Example of a Laplacian pyramid. From left to right: panchromatic image; first level of decomposition; second level of decomposition; first level of synthesis; second level of synthesis (corresponding to the low-resolution model of the panchromatic image)*

Figure 6.5. *Example of a Laplacian pyramid (follow-up): high-resolution spatial detail obtained by substracting the panchromatic and the result of the synthesis. Dark gray tones correspond to small differences*

6.3. Spectral indices and spatial indices

Through its spectral richness, passive optical imagery allows us to study the characteristics of the materials observed by remote sensing systems. In section 3.4, the optical properties of the different types of surface have been introduced: for example, the vegetation presents a reflectance peak in the near-infrared wavelengths, preceded by a low reflectance in the frequencies of the visible. These kinds of properties have been exploited to define a series of indices, either spectral (therefore sensitive to the spectral properties of the surfaces) or spatial (relative to the arrangement of the objects with respect to their spatial neighborhood). Without aiming to be exhaustive, this section presents a series of indices used in specific application fields: the monitoring of the vegetation, the study of water properties, the detection of clouds and the description of urban structures.

6.3.1. *Vegetation indices*

As was mentioned in section 6.3, the vegetation presents a specific behavior at wavelengths around 700 nm: a major increase in reflectance, called *red edge*, can be seen around these frequencies. The red edge often functions as a marker for the presence of vegetal surfaces and is largely exploited in remote sensing via the normalized difference vegetation index (+ italics, since it is the explanation of NDVI).

$$\text{NDVI} = \frac{(R_{\text{R}} - R_{\text{NIR}})}{(R_{\text{R}} + R_{\text{NIR}})} \qquad [6.1]$$

where R_{band} corresponds to the reflectance in the corresponding band, although the index is often calculated directly starting from digital numbers (DNs) (corresponding to the digital value of a pixel). The normalization by the sum of bands tends to reduce the lighting effects or the slope effects. Its values are comprised between $[-1; 1]$, but are most often between $[-0.5; 0.5]$. Figure 6.6 illustrates the NDVI index, calculated on a QuickBird image at 2.4 m resolution. The NDVI is very informative in low-density covers, but tends to saturate in the presence of a thick vegetal cover. It is also sensitive to atmospheric conditions: a compensation of atmospheric effects is therefore desirable before calculating the NDVI, especially if we wish to make comparative studies between several acquisitions.

Figure 6.6. *a) Multispectral image acquired by the satellite QuickBird, with a resolution of 2.4 m. b) NDVI index. For a color version of this figure, see www.iste.co.uk/tupin/RSImagery.zip*

Several other vegetation indices have been proposed in specialized literature, particularly using the soil line[1] (perpendicular vegetation index (PVI)) or by correcting the atmospheric effects (atmospherically resistant vegetation index (ARVI) and its variations). However, in light of the additional parameters that must be calibrated, the NDVI remains the reference index. For more information on these indices, see [CAL 01].

With the arrival of hyperspectral sensors, research around the vegetation indices has met increased interest: the increased spectral richness of these sensors allowed for the development of indices making use of very specific wavelengths, which are known for characterizing specific phenomena [STA 10]. For example, indices specific to certain pigments were developed, such as the *blue green pigment index* (BGI), the *blue red pigment index* (BRI) or the *carotenoid reflectance index* (CRI), described in [GIT 02, ZAR 05]:

$$\text{BGI} = \frac{R_{450}}{R_{550}}; \qquad \text{BRI} = \frac{R_{450}}{R_{690}}; \qquad \text{CRI} = \frac{1/R_{510}}{1/R_{550}} \qquad [6.2]$$

1 The values of reflectance for soils (from a humid soil to a dry soil) describe a straight line in the space defined by the red and near infrared bands.

where R_λ is the reflectance at a given wavelength (ex: $\lambda = 550$ for 550 nm). Stagakis *et al.* [STA 10] presents a comparative study of the different vegetation indices that are extracted from the hyperspectral images for the estimation of the leaf area index and the concentration of chlorophyll.

6.3.2. *Water-related indices*

With regard to the water, chlorophyll-sensitive indices are often used: for example, in the models of chlorophyll estimation discussed in [HOR 10], the ratios between the radiance ρ at 443, 490 and 550 nm are used:

– in a model specific to the MODIS sensor (*moderate resolution imaging spectroradiometer*, a sensor equipped with 36 spectral bands covering wavelengths between visible and thermal infrared):

$$\begin{aligned} \mathrm{Chl}_{\mathrm{MODIS}} &= 10^{(0.2830-2.753\rho_{SM}+1.457\rho_{SM}^2+0.659\rho_{SM}^3-1.403\rho_{SM}^4)} \\ \rho_{SM} &= \log_{10}\left(\frac{\rho_{443}}{\rho_{550}} > \frac{\rho_{490}}{\rho_{550}}\right) \end{aligned} \qquad [6.3]$$

– in the Clark model

$$\mathrm{Chl}_{\mathrm{Clark}} = -1.4\log\left(\frac{\rho_{443}}{\rho_{550}}\right) + 0.07 \qquad [6.4]$$

The ratio between the radiance at 490 and 550 nm is used, after standardization, in the index K490, which is used to detect turbidity[2] in the water column:

$$K490 = K_w(490) + 0.15645\left(\frac{n\rho_{490}}{n\rho_{550}}\right)^{-1.5401} \qquad [6.5]$$

where $K_w(490)$ is the diffuse attenuation coefficient of the water.

The majority of these indices are based on radiance ratios between 400 and 500 nm. This is mainly due to the generally accepted hypothesis that the reflectance in the near infrared is null at open sea (which is true for pure water). On the contrary, in the presence of suspended particles, this reflectance varies and can be used to measure the concentration of particles. In [DOR 11], an index based on the ratio between the reflectances at 765 and 865 nm is used to differentiate water samples with increasing particle concentrations. In [GIT 11], the bands on the red edge (684, 700 and 720 nm) are used for estimating the chlorophyll in a coastal area characterized by turbid waters.

2 The turbidity is the cloudiness of a fluid due to scattering particles.

6.3.3. *Indices relative to cloud properties*

The bands in the near SWIR can be used to study the properties of clouds and to differentiate them from other bright objects, such as snow. Clouds are bright, white and cold, three characteristics that can be observed via the bands in the near infrared and thermal infrared. On the contrary, snow has characteristics that are very similar in these frequencies: to differentiate between these two surfaces, the bands of the SWIR can be used, because the snow absorbs in these frequencies [PRE 08]. In the following, we will describe two sets of indices: the first based on the near-infrared bands of the sensor medium resolution imaging spectrometer (MERIS) of ENVISAT and the second based on the bands in the mid-infrared and the thermal infrared bands of the sensor advanced along track scanning radiometer (AATSR) of ENVISAT.

– MERIS normalized difference snow index (MNDSI) : this index is used for the detection of snow in [PRE 08]. It uses the two MERIS absorption bands. High values for bright pixels are attributed to snow. The bright pixels are selected by applying a threshold in the blue band at 412 nm (only the clouds and the snow are bright in these wavelengths).

$$\text{MNDSI} = \frac{R_{865} - R_{890}}{R_{865} + R_{890}} \quad [6.6]$$

– In [GOM 07], several indices extracted from the MERIS bands that are useful for cloud detection are described: the more interesting ones are the brightness index (Bri), which is estimated from the reflectance of the bands that are not affected by atmospheric absorption (B_S), and the whiteness index (Whi), obtained by integrating the first derivative of these same bands (which corresponds to analyzing the flatness of the spectrum).

$$\text{Bri} = \frac{1}{\lambda_{max} - \lambda_{min}} \sum_{i \in B_s} \frac{R_{\lambda_{i+1}} + R_{\lambda_i}}{2} (\lambda_{i+1} - \lambda_i) \quad [6.7]$$

$$\text{Whi} = \frac{1}{\lambda_{max} - \lambda_{min}} \sum_{i \in B_s} \frac{e_{\lambda_{i+1}} + e_{\lambda_i}}{2} (\lambda_{i+1} - \lambda_i) \quad [6.8]$$

where $e_\lambda = |R_\lambda - Bri|$. These indices are calculated separately in the visible and near-infrared bands, thus obtaining four indices whose discrimination ability varies depending on both the type of cloud and the context (either over sea or over land).

– Gomez-Chova *et al.* [GOM 07] introduce indices based on absorption bands (760 and 940 nm) of MERIS. Based on the principle that, when reflected by clouds, the photons traverse a shorter section of the atmosphere, the optical path observed on cloud pixels tends to be abnormally short. Consequently, smaller absorption will be observed in the absorption bands. On the basis of this principle, the authors propose an index for the detection of clouds based on the ratio between the reflectance of the oxygen absorption band (R_{760}) and a band without absorption obtained via the interpolation of the neighboring bands (R_{754}, R_{778}).

– In the algorithms used to extract the products of the sensor AATSR [ESA 07], the thermal band at 11 μm is used for estimating the temperature at the top of the clouds detected by the AATSR cloud detector. The difference in temperature between the bands at 11 and 12 μm is used to detect cirrus clouds. The difference between the bands at 11 and 3.7 μm is used for detecting low stratus (solely nocturnal acquisitions are used, to avoid contamination of the band at 3.7 μm).

6.3.4. *Surface texture: occurrence and co-occurrence*

Texture can be considered to be the local visual aspect of images. It informs us on the local arrangement of the pixels, observed in a neighboring area defined by the user. The patches of an image can be considered to be random variables and we are interested in the characteristics of their distribution (average, variance, entropy, etc.). Texture indices are generally used for differentiating homogeneous areas from the heterogeneous areas. Since they are built on the basis of local moving windows, they are generally rotation and translation invariant (unless the index is specifically built for detecting directional structures).

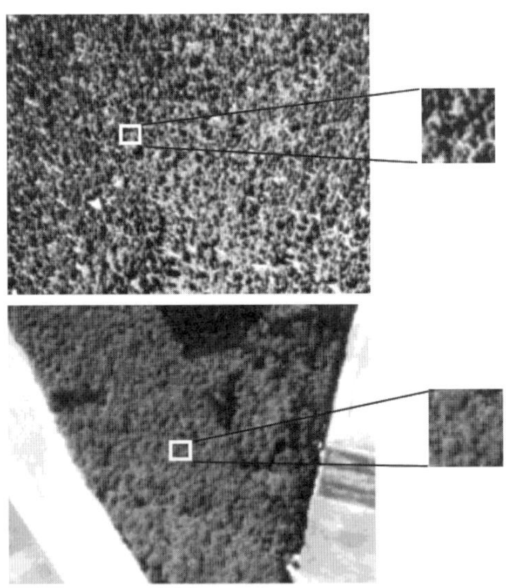

Figure 6.7. *Two tree covers and their differences in texture. Top: an area of coniferous trees; bottom: a deciduous area [CAL 01]*

Figure 6.7 shows an example of texture variation for two tree covers (adapted from [CAL 01]). The first is a cover of coniferous trees, the second is a cover of deciduous trees; it is easy to differentiate them through the spatial layout of the trees; whereas the pixel, taken individually, would make the differentiation more difficult. The texture indices catch this local information and build efficient descriptors for surfaces with similar spectral characteristics, but showing different spatial structures.

In this chapter, we detail two families of texture indices: the occurrence indices and the co-occurrence indices.

– Occurrence: the occurrence indices are calculated directly on the values of the pixels composing the moving window around the pixel considered. If we define the spatial neighborhood of a pixel of coordinates i, j as $V \in \{[i - m/2; i + m/2], [j - m/2; j + m/2]\}$, where m is the size of the moving window, the indices of average and variance are:

$$\mu_{ij} = \frac{1}{m^2} \sum_{m \in V} \sum_{n \in V} \mathbf{x}_{mn} \quad [6.9]$$

$$\text{var}_{ij} = \frac{1}{m^2} \sum_{m \in V} \sum_{n \in V} [\mathbf{x}_{mn} - \mu_{ij}]^2 \quad [6.10]$$

Figure 6.8 shows examples of variance and average occurrence indices for two sizes of moving window (of size 3 and 5 pixels).

– Co-occurrence: co-occurrence considers the spatial information between pairs of pixels in a given window. It studies the autocorrelation between the gray levels in relation to the distance and the angle between the pairs of pixels observed. These relations are summarized in the gray level co-occurrence matrix (GLCM, [HAR 73]). This matrix reports the probability that two pixels show the same gray value for a distance d and a given orientation θ. An example of four GLCM matrices, one by main theta of $45°$, is shown for a toy example in Figure 6.9. The GLCM matrix is often considered in a limited number of quantization levels L (for example we reduce the 255 values at $L = 10$ bins of same width in terms of gray levels).

Once the GLCM has been extracted for all the pixels (there is one GLCM per moving window, distance and orientation), we can calculate indices on the basis of these matrices. The indices can be angular or isotropic, if the matrices per angle are averaged. Two examples of indices are the entropy H and the contrast C:

$$H_{ij} = -\sum_{l=1}^{L}\sum_{c=1}^{L} p_{d,\theta}(l,c)\log\left(p_{d,\theta}(l,c)\right) \quad [6.11]$$

$$C_{ij} = \sum_{l=1}^{L}\sum_{c=1}^{L}(l-c)^2 p_{d,\theta}(l,c) \quad [6.12]$$

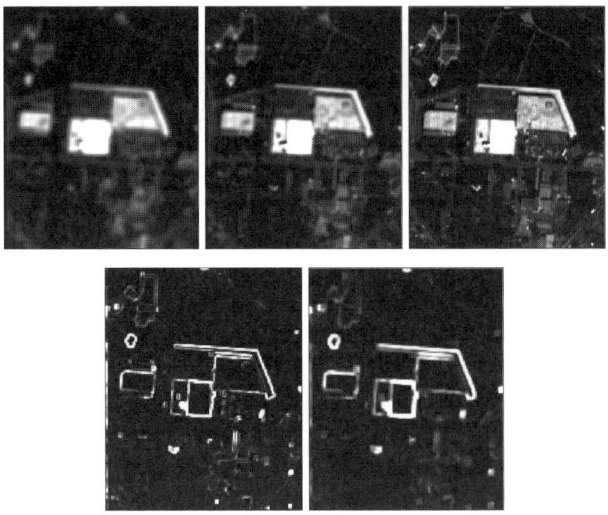

Figure 6.8. *Examples of occurrence indices. In the center, the original image. To the left, average filters obtained by moving windows of size 3 and 5 pixels. To the right, variance filters obtained with the same moving windows*

The entropy is low when few gray values coexist in the moving window and high when there is a large variability of gray values. The contrast is strong when large jumps of gray values coexist in the mobile window. Figure 6.10 illustrates examples of the co-occurrence indices H and C.

6.3.5. *Geometrical indices of surfaces: morphological indices in urban areas*

Extracting indices relative to the geometry of surfaces can be useful when studying urban areas, where the size and shape of the object under study can be discriminative of different types of land use. To obtain this information, recent research has proposed the use of spatial indices computed at several scales. When considering a given index, but extracted at a series of scales, the structures become more easily discernible, because each of the thematic classes reacts in a specific manner depending on the size of the neighboring area considered. In this chapter, we will consider two families of these indices: those resulting from mathematical morphology and those based on attribute profiles. There are other indices, such as the index of urban complexity [YOO 09], which is based on a wavelet decomposition.

34	35	36	35	36
36	36	35	34	35
36	35	35	35	34
34	35	35	35	34
36	34	35	34	36

P(d=1, θ=0) = vertical

	34	35	36
34	2	5	4
35	5	10	4
36	4	4	2

P(d=1, θ=45) = 45° diagonal

	34	35	36
34	2	4	1
35	4	12	2
36	1	2	4

P(d=1, θ=90) = horizontal

	34	35	36
34	0	8	2
35	8	8	5
36	2	5	2

P(d=1, θ=135) = 135° diagonal

	34	35	36
34	4	2	2
35	2	12	4
36	2	4	0

Figure 6.9. *GLCM co-occurrence matrix. To the left, the DNs contained in a subset extracted from an image. To the right, GLCM matrices for the four angular orientations and for adjacent pixels (d = 1). Highlighted, the example for the values 35 and 36 in the vertical direction (four co-occurrences)*

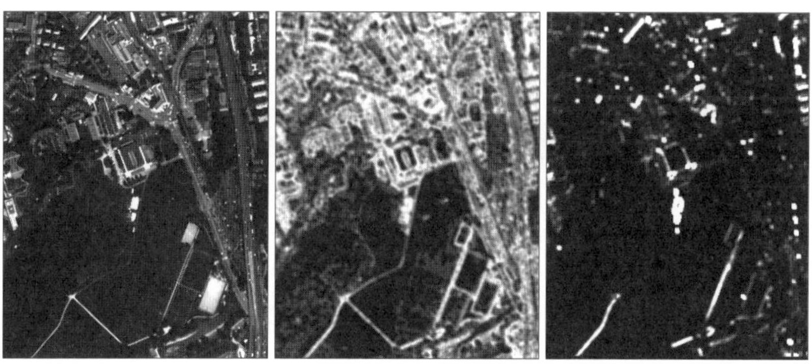

Figure 6.10. *Examples of co-occurrence indices. a) The original image. b) The entropy index for averaged angles and distance d = 3 pixels. c) The contrast index for the same parameters*

– Mathematical morphology: morphological indices are a series of indices assessed in a local neighborhood called structuring element, SE [SOI 04]. The SE is not necessarily a squared window, as in the case of the textural indices considered previously: it can take different shapes, such as a circle, an oriented line or the connected elements that result from a segmentation. Either way, all the pixels contained in the SE are used to evaluate the morphological index. The most used indices in remote sensing are the opening index, γ, and the closing index, φ. For a pixel $\mathbf{x}_i$ and its neighboring area defined by its structuring element SE_i, the opening and closing indices are:

$$\gamma(\mathbf{x}_i) = \max_{j \in SE_i} \{ \min_{j \in SE_i} \mathbf{x}_j \} \quad [6.13]$$

$$\varphi(\mathbf{x}_i) = \min_{j \in SE_i} \{ \max_{j \in SE_i} \mathbf{x}_j \} \quad [6.14]$$

The opening applies two filters, a local minimum filter (called erosion) followed by a local maximum filter (called dilation). By doing so, the opening emphasizes the dark structures, which are smaller than the structuring element. On the contrary, the closing emphasizes light elements, which are smaller than the structuring element. In order to do this, it applies an erosion to the result of a dilation. To obtain a multiscale response, the indices are evaluated with structuring elements of increasing size. Figure 6.11 shows a series of openings and closings for circular structuring elements.

Figure 6.11. *Examples of opening and closing indices. At the center is the original image. Moving to the left, the indices obtained by applying a closing with a circular structuring element ("disk") of size* $[3, 5, 7]$ *pixels. Moving toward the right, indices obtained by applying an opening with the same structuring elements*

– Mathematical morphology by reconstruction: through the application to several scales, the opening and closing filters allow us to acquire information about the size and, when a directional SE is used, the orientation of the objects of interest. Nonetheless, the spatial detail of the original image is lost (see Figure 6.11). To make up for this disadvantage, Pesaresi and Benediktsson [PES 01] proposed the use of morphological indices by reconstruction. The opening and closing by reconstruction filters γ_R and φ_R are iterative filters. Let us take the example of opening by reconstruction, γ_R : a so-called marker image (an eroded image) is dilated, just like in the standard opening. Then, the result of the dilation is compared to the original image: for each pixel, the minimum between the dilated image and the original image is retained, so that no pixel exceeds its value in the original image. The process of dilation/comparison is then iterated until the result of the dilation does not change anything compared to the previous iteration. In the case of closing by reconstruction, φ_R, the marker image is a dilation and we proceed by successive erosions. Figure 6.12 illustrates a series of openings and closings by reconstruction for circular structuring elements. Note how the spatial detail of the original image is now preserved.

Figure 6.12. *Examples of morphological indices by reconstruction. At the center is the original image. Moving to the left, the indices obtained by applying a closing by reconstruction with a circular structuring element ("disk") of size* $[3, 5, 7]$ *pixels. Moving to the right, indices obtained by applying opening by reconstruction with the same structuring elements*

– Differential morphological profiles: another widely used strategy is to extract profiles of the derivatives between indices (*derivative of morphological profiles* (DMP) [BEN 03]): these new indices inform us on what has changed between two consecutive reconstruction indices and show the changes occurring between scales (see Figure 6.13).

– Attribute profiles: first, a thresholding of the image by graylevels is performed [DAL 10]. Then, all the connected objects for which all the pixels have a value that is lower than the threshold are extracted. Finally, several indices are calculated from the objects extracted. For example, if the attribute is the surface of the object, we create a profile where we will review the objects with a surface that is smaller than 10, 20, 50, etc., pixels. Attributes such as the surface of objects, the diagonal of the smallest box surrounding it or the inertia (which informs on the elongation degree of the object) are usually extracted. Figures 6.14–6.16 show these three indices for different attribute thresholding values. Let us take the example of the last image of the three figures: the surface filter shows a maximum value to the light structure in the top part of the image (the surface of that object is larger than 100 pixels), the diagonal filter attributes the maximum value to the forested area (the light element in the top part is smaller than a box of 100 pixels in diagonal). In the inertia filter, the elongated element in the light structure receives the highest value.

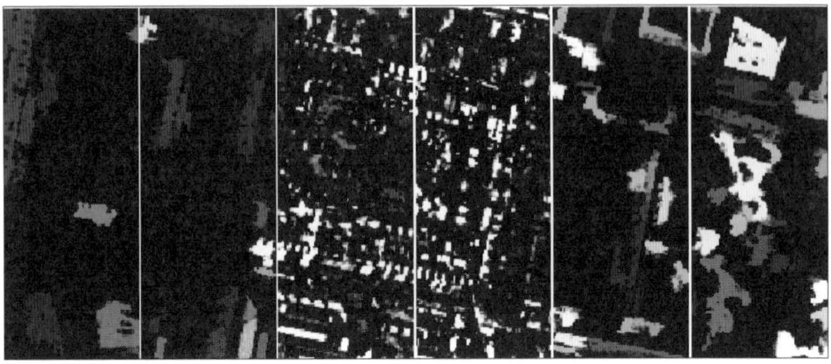

Figure 6.13. *Example of morphological differential profiles. From left to right, indices obtained by differentiation of the indices shown in Figure 6.12. For example, the first block corresponds to* $\varphi_R(disc, 7) - \varphi_R(disc, 5)$ *and the second block corresponds to* $\varphi_R(disc, 5) - \varphi_R(disc\ , 3)$

Figure 6.14. *Examples of surface attribute indices. At the center is the original image. Moving to the left, indices obtained by thresholding the connected elements whose surface exceeds* 50, 100 *and* 200 *pixels for the dark gray levels. Moving to the right, the same indices, evaluated for the light gray tones*

6.4. Products issued from passive optical images

Once the image is brought, the desired resolution and that the valid indices have been extracted, one can finally use it to extract a product. In this section, we will discuss two of the main products, namely the classification maps and the subpixel mixture maps. If the classification has always been the main product (for all the image resolutions), the analysis of spectral mixtures becomes more attractive when we consider hyperspectral imagery.

Figure 6.15. *Examples of diagonal attribute indices. In the center is the original image. Moving to the left, indices obtained by thresholding the connected elements whose diagonal of the box surrounding them exceeds* 20, 50 *and* 100 *pixels for the dark gray levels. Moving to the right, the same indices, evaluated for the light gray tones*

6.4.1. *Classification*

Land use maps are the major product of passive optical satellite images. The simplification of the radiometric information into a limited number of classes is a product of great utility: for example, in urban planning, having a map of the distribution of the buildings in relation to the communication axes and other green spaces can prove to be a crucial piece of information for characterizing the neighborhoods in the light of sanitation works. In forestry, a map of tree types can help forest rangers in planning tree cuts or conservation actions.

In Chapter 5, several classification models were presented. In the following, we will apply them to a very high-resolution data set (Figure 6.17 [TUI 10b]), discuss the validity of a spatial-spectral approach and review post-processing systems.

In the case of spectral classification, only the spectral bands are considered in the process. Each pixel contains a series of d values, one per band: $\mathbf{x}_i \in \mathbb{R}^d$. In an unsupervised context, all the pixels are used for defining spectrally coherent groups via a segmentation algorithm (for example the watershed algorithm or the mean-shift algorithm discussed in section 5.4.4). In a supervised context, a series of pixels, for which the class membership is known (also called training pixels), is available. The training set X_T contains l pixels, each of which is associated with a couple $\{\mathbf{x}_i, y_i\}_{i=1}^{l}$, where y_i is the land cover (or land use) class represented by the pixel, which can be obtained via photointerpretation or through a field campaign.

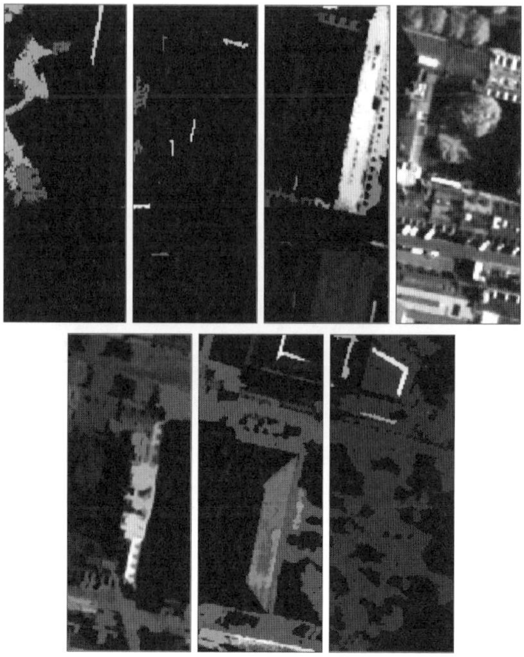

Figure 6.16. *Examples of inertia attribute indices. In the center is the original image. Moving to the left, indices obtained by thresholding the connected elements whose inertia index exceeds* 0.4, 0.5 *and* 0.6 *for the dark gray levels. Moving to the right, the same indices, evaluated for the light gray tones*

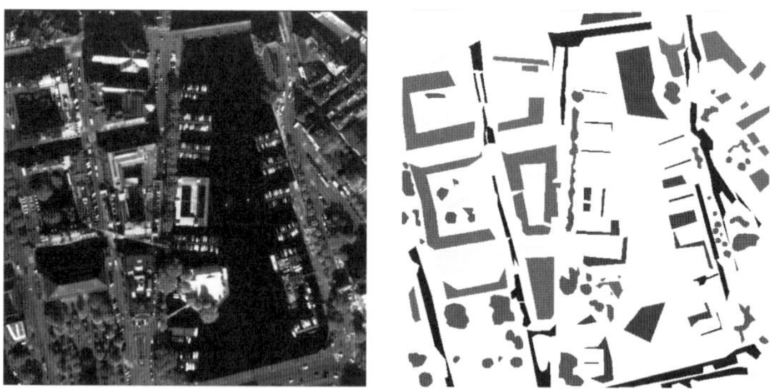

Figure 6.17. *a) Image of the "Bain des Femmes" in Zurich (Switzerland), acquired by the QuickBird satellite and pansharpened with the PCA method [TUI 10b]. b) Ground data for five classes: roads (black), buildings (orange), vegetation (green), water (blue) and shadows (yellow). For a color version of this figure, see www.iste.co.uk/tupin/RSImagery.zip*

The set X_T is used to train a classifier and obtain a decision function, which is used to attribute a new pixel to one of the land cover (or land use) classes. In the remainder of this section, we will consider only supervised classifiers.

Figure 6.18(a) illustrates the result obtained by a spectral classification using the support vector machine (SVM) method (see section 5.6.2). Despite being correct overall, we can still notice some errors, in particular between the classes "Shadow" and "Water" and between the classes "Buildings" and "Roads". The resulting map shows the so-called "salt and pepper" effect, where several pixels are classified in a different class than their spatial neighborhood.

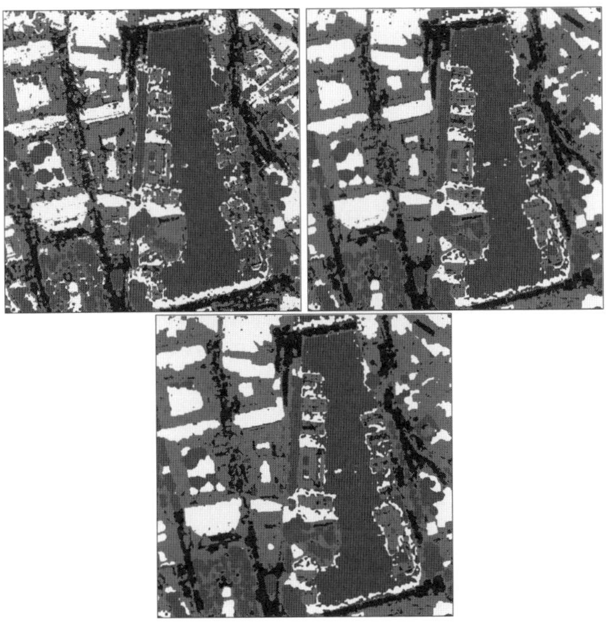

Figure 6.18. *Classification of the "Bain des Femmes" image (shown in Figure 6.17). a) Spectral classification. b) Spatio-spectral classification considering the concatenation of spectral bands, morphological indices (opening and closing) and reconstruction indices (opening and closing by reconstruction). c) Post-processing of the map of the central image by a* 3×3 *majority filter. For a color version of this figure, see www.iste.co.uk/tupin/RSImagery.zip*

To solve these problems, we will consider some of the spatial indices discussed in the previous sections. More precisely, we will add information regarding the shape and the scale of the objects via morphological indices. There are several ways of joining these indices to the spectral information [FAU 13]: the simplest ones concatenate the indices extracted from the spectral bands and then train a classifier on the complete data set [PAC 09, FAU 08, TUI 09]; other more complex approaches consider spatiospectral similarities, where the similarity measure is calculated for each information source separately and is then merged through a dedicated approach [CAM 06, TUI 10b, TUI 10a, LI 13]. Figure 6.18(b) illustrates the result obtained by a spatiospectral classifier: the majority of errors are corrected and the limits of the classified objects become more coherent, with the exception of the class "Roads", for which the spatial discontinuity is still present. This is probably due to the large variability of objects included in this class (cars, road signalization), which prevents a precise delineation of this complex class.

To better understand this qualitative analysis, we consider the numeric performances of two different classifiers. Table 6.1 presents the classification results obtained by the maximum likelihood (ML) and the SVM on the "Bain des Femmes" image:

– the model is trained with $l = 547$ pixels, randomly extracted from the ground data shown in Figure 6.17;

– the parameters of the SVM are optimized with an independent set of validation (an stands for radial basis function (FBF) kernel was used);

– when used, the spatial indices are extracted from the first principal component and then concatenated to the spectral information as additional bands;

– five experiments, obtained by varying the pixels in the set X_T, are reported in order to assess the stability of the different methods (σ in Table 6.1).

Method	Indices	Overall accuracy μ (%)	Overall accuracy σ (%)	Cohen Kappa μ	Cohen Kappa σ
ML	Spectral	62.92	$\pm$ 0.96	0.53	$\pm$ 0.01
	OF+OFR	75.86	$\pm$ 0.71	0.70	$\pm$ 0.01
	AttS + AttD	73.76	$\pm$ 0.67	0.67	$\pm$ 0.01
SVM	Spectral	81.72	$\pm$ 0.89	0.77	$\pm$ 0.01
	OF+OFR	84.44	$\pm$ 1.01	0.80	$\pm$ 0.01
	AttS + AttD	83.66	$\pm$ 0.95	0.79	$\pm$ 0.01

Table 6.1. *ML and SVM classification performances obtained using only the spectral bands (Spectral) or in conjunction with the spatial indices (OC = opening and closing, OCR = opening and closing by reconstruction, AttS = attribute surface profile, AttD = attribute diagonal profile). The parameters used for extracting the indices are those shown in Figures 6.11–6.15*

The results of Table 6.1 show two clear tendencies: (1) the SVM method tends to provide better results and (2) the performance improves when including the spatial indices. The increase is more pronounced for the linear method (ML), which can benefit from a richer source of information and of a higher dimensionality for finding the decision function (the Cover theorem [COV 65] stipulates that the probability of linear separation increases with the number of dimensions). By its nonlinear nature, the SVM uses more flexible functions, which adapt better to the data. Therefore, the performance difference between the spectral methods and the spatiospectral methods is smaller.

Before establishing the final product, it is common to pass through a final *post-processing* stage, for example by applying a local majority filter to the classification map. This filter considers the classes attributed in a local neighboring (typically in a local window 3×3) and attributes to the central pixel the main class attributed in the neighborhood. Figure 6.18(c) illustrates the result of this filtering on the spatiospectral classification map illustrated in Figure 6.18(b): the visual aspect of the result is improved (the salt and pepper effect is further reduced) and the numerical performance increases by an additional 1%.

There are more complex post-classification schemes, for example based on class-specific iterative filtering or applying a fusion of classification maps obtained by different models [FAU 06, LIC 09].

6.4.2. *Subpixel mixture analysis*

Since an image is always issued from the discretization of a source of information, which is continuous and heterogeneous, an open problem is the presence of mixed pixels. Mixed pixels are pixels whose spectral signature is the combination of the reflectance of the different types of land uses it contains. Figure 6.19 illustrates this situation, where the pixel remote considered is composed of three different types of cover. The distinction of the components of the mixture present in each pixel can be very useful, especially in applications such as mining or ecosystem description.

In remote sensing, the field of investigation related to this problem is called *unmixing* [KES 02]. The presence of mixtures is not necessarily solved by an increase in ground resolution, because a mixture at a scale can hide another one at a finer scale. Moreover, mixtures can be inherent to the surface, meaning that the surface observed is intrinsically composed of several interwoven objects, independently from the resolution of the image.

The unmixing problem can be solved by considering linear and nonlinear mixture models. In the first case, we assume that the components of the mixture are found near other components (spatially) and that there is no interaction due, among other things, to atmospheric diffusion. Following these hypotheses, a mixture with p pure elements, or *endmembers*, can be described by:

$$R(\mathbf{x}_i) = \sum_{j=1}^{p} \mathbf{m}_j a_{ij} + \mathbf{n} \qquad [6.15]$$

where $\mathbf{m}_\mathrm{j}$ is the spectral signature of endmember j and a_{ij} is the weight that this pure element has in the mixture observed for the pixel i, thus called *abundance*. The vector $\mathbf{n}$ is an additive noise vector that is specific to each band.

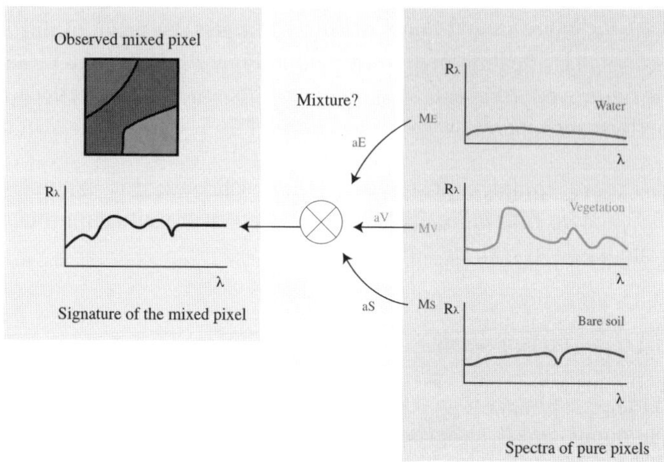

Figure 6.19. *The problem of mixed pixels and the general framework of spectral unmixing*

Nonlinear unmixing assumes that the interaction between light and atmosphere, as well as the interactions between the components of the ground mixture, cannot be ignored. It proposes solutions that are based on the modeling of these interactions, for example by using radiative transfer models [BOR 94]. Since nonlinear unmixing models are very complex and not very intuitive, we will not detail them in this chapter. We will focus on linear models and present algorithmic examples of endmember extraction and is called abundance estimation.

An unmixing workflow consists of the following phases:

– Estimation of the number of endmembers: the first stage in the extraction of endmembers is the definition of the number of pure components that can be differentiated in a mixed data set. This number is generally considered to be smaller than the number of dimensions (bands) of the image. For this reason, the number of endmembers can be estimated with one of the dimensionality reduction techniques presented in section 5.7.5. A classic method is the estimation of the virtual dimensionality (VD) proposed in [CHA 04c]: in this technique, the number of endmembers corresponds to the number of dimensions, for which the correlation matrix shows values that are smaller than those of the covariance matrix. This corresponds to the minimum number of distinct signals that characterize the hyperspectral image.

– Extraction of endmembers: once the number of endmembers has been fixed, a source separation algorithm [KES 02] is used for extracting the spectral signatures of the endmembers. This can be avoided if the endmembers are known in advance, for example if we wish to estimate a mixture of known components, whose signatures are extracted from a database. In the more frequent case where the nature of endmembers is unknown, geometric and statistical methods are generally used [BIO 10, PLA 11].

Geometric methods consider the properties of the convex hull, which is assumed to contain all the pixels in the image. Two strategies are generally followed: the first strategy searches an endmember in each peak of the simplex that contains the data, which corresponds to finding the most extreme pixels present in the data [NAS 05]. The second strategy considers the volume of the actual simplex and tries to minimize it [BIO 09]. The first line in Figure 6.20 shows four endmembers of the image *Cuprite*[3] found with the vertex component analysis method (VCA, [NAS 05]). Geometric methods have the advantages of being computationally efficient and very intuitive, but they can sometimes be inaccurate, especially in the case of strongly mixed spectra (which corresponds to a poorly populated surface of the simplex).

Statistical methods follow an approach based on learning methods. Taking the example of the kernel methods considered in Chapter 5, density estimation methods such as the stands for support vector data description (SVDD) [BRO 09] can be used to define the data support. Then, the support vectors can be considered as the pixels that define this support of the data, i.e. the endmembers, and from which all the others are a mixture. The second line in Figure 6.20 shows four examples of endmembers defined on the basis of the *cuprite* image with this method (SVDD, [BRO 09]). Statistical methods can have the inconvenient of higher computational loads, especially if the images considered are of a large size.

– Abundance estimation: once the endmembers have been defined, we proceed to the estimation of the abundance vector $\mathbf{a}_j$, containing the relative presence of the endmember j in each pixel. The most direct way of obtaining the abundance vector is to reverse equation [6.15] using least squares. In this case

$$\hat{\mathbf{a}} = \mathbf{M}^{\dagger}\mathbf{R}(\mathbf{x}) = (\mathbf{M}^{t}\mathbf{M})^{-1}\mathbf{M}^{t}\mathbf{R}(\mathbf{x}) \quad [6.16]$$

where $\mathbf{M}^{\dagger}$ is the pseudo-inverse of the matrix $\mathbf{M}$ containing the signatures of the endmembers. $\mathbf{R}(\mathbf{x})$ is the matrix that contains the reflectance of the pixels to

3 http://aviris.jpl.nasa.gov/data/free_data.html.

be unmixed. The constraint of a unitary sum can be integrated with the help of Lagrange operators. Figure 6.21 shows the abundance maps estimated for the four endmembers in Figure 6.20 for the VCA method [NAS 05]. Abundance estimation is a very active research field, where inversion methods are being developed (for example, to guarantee the positivity of the coefficients [CHA 06], or by using statistical optimization methods [DEB 06]). Nonlinear [BRO 09] and sparse estimation methods [IOR 11] are among the most recent developments. A review of these methods can be found in [BIO 10].

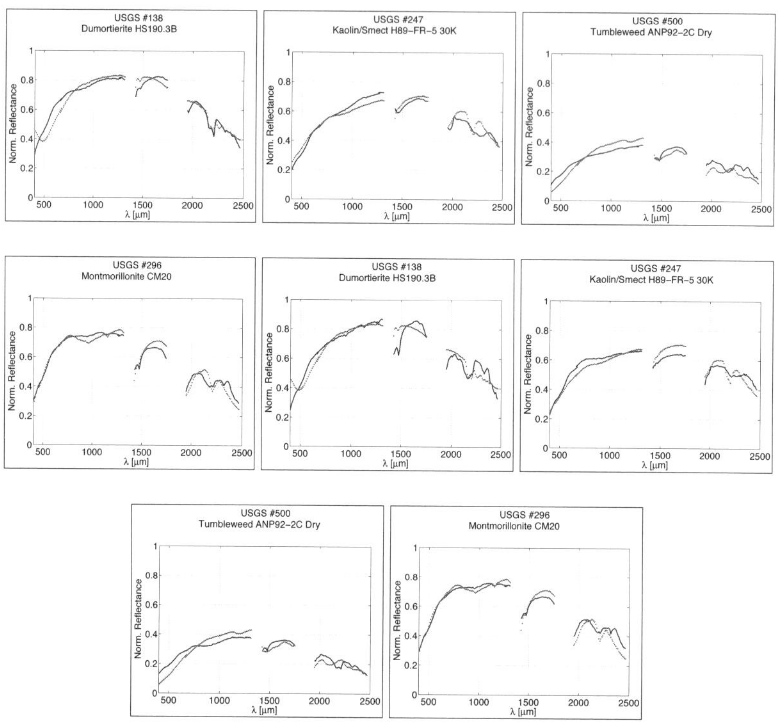

Figure 6.20. *Extraction of endmembers. For each method, a total of 14 endmembers has been extracted (solid line). Then, the most likely spectrum in the USGS library was researched (dotted lines in all the images). The first line shows the endmembers extracted using the VCA method [NAS 05] and the second line shows the endmembers extracted using the SVDD method [BRO 09] for the same minerals*

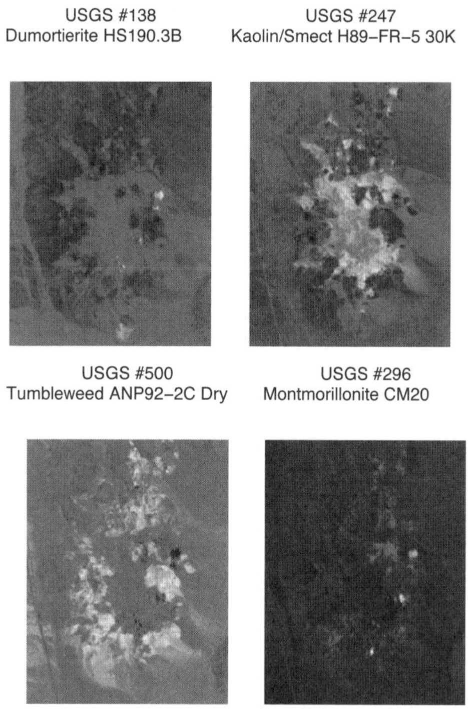

Figure 6.21. *Abundance estimation for the endmembers found by the VCA method (Figure 6.20) and using least squares. The light tones correspond to a larger presence of the endmember in the mixture of the pixel. For a color version of this figure, see www.iste.co.uk/tupin/RSImagery.zip*

6.5. Conclusion

In this chapter, we have reviewed applications of multi- and hyperspectral signal processing. First, we have discussed pansharpening methods that aim at obtaining images with high spectral and spatial resolution from sources that have only one of these two properties. Then, we have presented examples of contextual filters and indices specific to the given applications. In the following section, we have discussed different approaches to the classification of land use at very high resolution, revisiting the methodologies presented in Chapter 5 and focusing on the added value of using contextual information in linear and nonlinear classifiers. Finally, we have introduced the field of unmixing, whose objective is to describe the mixture of materials present in the pixels. Hyperspectral imagery allows us to apply source separation methods and to obtain abundance maps, reflecting the relative presence of each material in the pixels.

The application of machine learning methods to data of high dimensionality, such as the latest-generation optical images, remains a very active research field, where the knowledge in physics, optics, machine learning and vision is interweaving. The fusion of numerous sources of information as well as the development of algorithms capable of processing the huge archives of data acquired (and to be acquired) are challenges for developments in the future and ensure a constant renewal of the field.

Chapter 7

Models and Processing of Radar Signals

7.1. Speckle and statistics of radar imagery

7.1.1. *Physical origin*

The origin of the speckle effect is due to the coherent nature of radar illumination with the emission of polarized waves, examined in Chapter 4. This coherent illumination gives to the SAR image this granular aspect that is so uncomfortable to look at. To explain it, let us consider a physically homogeneous surface, like a field of grass. Each image pixel contains a large number of elementary scatterers. All these elementary scatterers add their contributions to the field emitted by the pixel in a coherent way (see Figure 7.1). Each scatterer presents a geometric phase shift $\Delta\varphi$ arising from its respective range from the sensor. The total pixel response in amplitude and phase is the vectorial addition of these elementary contributions (Figure 7.1).

The resulting electromagnetic field z is thus the addition of a "large number" N_{diff} of complex elementary contributions $\alpha_k e^{j\phi_k}$, with α_k and ϕ_k, the amplitude and phase of the elementary field coming from the kth scatterer, respectively, (ϕ_k is the intrinsic phase signature of the scatterer linked to its electromagnetic properties increased by the geometric term $\Delta\varphi$):

$$z = \sum_{k=1}^{N_{diff}} \alpha_k e^{j\phi_k} \qquad [7.1]$$

Chapter written by Florence TUPIN, Jean-Marie NICOLAS and Jean-Claude SOUYRIS.

Although at a macroscopic scale, adjacent pixels correspond to the same physically homogeneous surface, at the wavelength scale (a few cm), their internal structures are different enough to produce independent phase ϕ_k (a variation of $\lambda/4$ is enough to produce a π deviation). In the case of constructive interferences within the pixel (corresponding to close phase values), then the radiometry will be high. For the opposite case (destructive interferences), the pixel will have low radiometry (Figure 7.1). The speckle effect is due to these uncontrolled radiometric variations from one pixel to another. Due to its unpredictable aspect (although this mechanism is not random), it will be modeled by random variables, and we will see that it can be considered as a multiplicative noise.

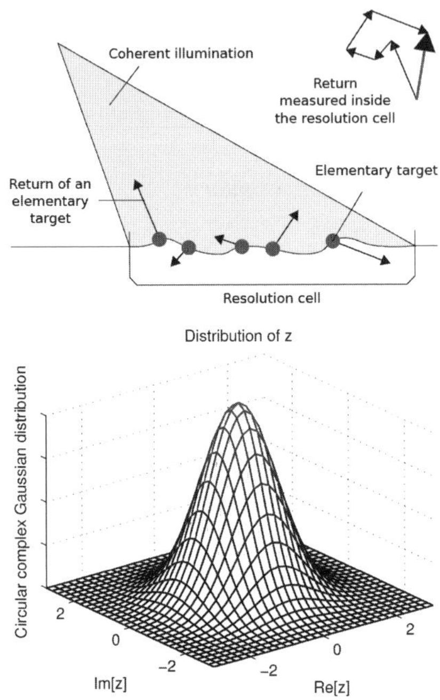

Figure 7.1. *Physical origin of speckle. Top, the coherent addition of elementary contributions of scatterers within a resolution cell. Bottom, the distribution of the complex field z, which is a Gaussian circular distribution for fully developed speckle*

7.1.2. *Statistics of fully developed speckle*

Goodman in [GOO 76] has defined the fully developed speckle with the following hypotheses:

1) the responses of each scatterer are independent of the others;

2) the amplitude α_k and the phase ϕ_k are independent and identically distributed (in other words, all scatterers produce responses of comparable intensity; we are not considering here the specific case of a predominant scatterer within a cell);

3) phases ϕ_k are uniformly distributed between $-\pi$ and π.

Under these hypotheses, the radiometric distribution of a uniform surface can be deduced from equation [7.1]. Based on the central limit theorem (or law of large numbers), we obtain a circular Gaussian distribution with zero mean defined by:

$$p(z|\sigma^2) = p(\Re(z), \Im(z)|\sigma^2) = \frac{1}{\pi\sigma^2} \exp\left(-\frac{|z|^2}{\sigma^2}\right) \qquad [7.2]$$

where $\sigma^2 = E(|z|^2)$ is connected to the backscattering coefficient of the surface, which will be called "reflectivity" from now on and denote by R. Equation [7.2] can be obtained from the Gaussian distribution followed by the real part $\Re(z)$ and the imaginary part $\Im(z)$, with zero mean and identical standard deviation $\frac{R}{2}$.

The equation of the probability density function (pdf) of intensity $I = \Re(z)^2 + \Im(z)^2$ is deduced from this:

$$p(I \mid R) = \frac{1}{R} \exp\left(-\frac{I}{R}\right) \qquad I \geq 0 \qquad [7.3]$$

This is an exponential law. Consequently, whatever the reflectivity of the considered region, null values are predominant. It can also be shown that the mean $E(I)$ (moment of order 1) is equal to R: $E(I) = R$, and that the standard deviation σ_I is also equal to R. This means that the variations of the radiometric values are directly connected to the reflectivity of a region. The more it increases, the more the apparent heterogeneity increases. Therefore, in radar imagery, variations are measured by the "normalized" standard deviation called "variation coefficient" and defined by:

$$C_I = \frac{\sigma_I}{E(I)}$$

The value of the variation coefficient on a homogeneous region in an intensity image is 1.

The pdf $p(A \mid R)$ of the amplitude $A = \sqrt{I}$ is obtained from the relation:

$$p(A \mid R)dA = p(I \mid R)dI = 2p(I \mid R)AdA \qquad [7.4]$$

giving:

$$p\,(A \mid R) = \frac{2A}{R}\,\exp\,(-\frac{A^2}{R}) \qquad A \geq 0 \qquad [7.5]$$

The amplitude of the image is distributed according to a Rayleigh law[1], with a mean $E(A) = \sqrt{\frac{R\pi}{4}}$ and with the standard deviation $\sigma_A = \sqrt{R(1-\frac{\pi}{4})}$. The variation coefficient C_A is approximately 0.523 for a homogeneous area of an amplitude image. Again, when $E(A)$ increases, σ_A increases proportionally. This means that the variations of the gray levels of a region increase with its reflectivity. Thus, the areas with strong radiometries appear "noisier" than the dark areas.

In opposition to the data that we will see in the following section 7.1.3, the images $I = |z|^2$ and $A = |z|$ are called single-look data.

7.1.3. *Speckle noise in multi-look images*

The speckle effect strongly limits the understanding of radar images. Therefore, multi-look images are generated to improve the radiometric resolution of radar signals. The generation of multi-look images is based on incoherent sums of single-look images. Multi-look images are generally built on the basis of intensity images: the L-look intensity image is obtained by averaging L uncorrelated 1-look intensity images. These L images are obtained by dividing the radar synthetic aperture into L sub-apertures. Each sub-aperture creates an uncorrelated image. However, the resolution worsens (multiplication by L). Done in the spatial field, the multi-look operation associates a unique pixel with any batch of n_a azimuth pixel and n_d range pixels:

$$I_L = \frac{1}{n_a n_d} \sum_{k=1}^{n_a n_d} I_k \qquad [7.6]$$

where I_k is the intensity of the kth pixel. The number of looks is given by $L = n_a\, n_d$, if spatial correlation between nearby pixels is ignored. Yet, spatial correlation does

1 According to equation [7.3] the probability density of I reaches its maximum in 0, whereas according to [7.5], the probability density of $A = \sqrt{I}$ is null in 0! This apparent paradox reminds us that we must not mistake the probability density with a probability. The differential form [7.4] ensures the equality of the probabilities connected to the events $A = a$ and $I = a^2$.

occur, as the radar signal is always over-sampled to some extent. The intensity variation coefficient of the multi-look image is given by:

$$C_{I_L} = \frac{\sqrt{var(I_L)}}{E(I_L)} = \frac{\sqrt{(var(I_k)/L)}}{E\ (I_k)} = C_I/\sqrt{L} = 1/\sqrt{L} \qquad [7.7]$$

The decrease in the variation coefficient of a factor $\sqrt{L}$ expresses the amount of speckle reduction.

If we do this multi-look averaging starting from an intensity image that verifies an exponential law (equation [7.3]), then the multi-look image follows a generalized Gamma law (denoting by Γ the Gamma function):

$$p_L(I|R) = \frac{L^L}{R^L\Gamma(L)} I^{(L-1)} e^{\left(-\frac{LI}{R}\right)} \qquad [7.8]$$

To build the multi-look amplitude images, two techniques are available:

– calculating a multi-look intensity image and taking the square root: this is what we call the image $\sqrt{\text{I}}$;

– doing the image average in amplitude: this is a classic operation in image processing, which guarantees certain properties of linearity, among others.

In the first case (image $\sqrt{\text{I}}$), we can analytically calculate the pdf because the pdf is known for the "L-looks" intensity images: we then obtain the Rayleigh-Nakagami law:

$$p_L(A|R) = \frac{2L^L}{R^L\Gamma(L)} A^{(2L-1)} e^{\left(-\frac{LA^2}{R}\right)} \qquad [7.9]$$

In the second case, the probability law does not have a simple analytical equation and needs several approximations.

It is important to note that these types of laws (Rayleigh-Nakagami for the amplitude images, generalized Gamma law for the intensity images) are defined in $\mathbb{R}^+$ and therefore have specific characteristics: in particular, the R parameter, the average (order 1 moment) and the mode (a value corresponding to the law's maximum) are not equal, contrary to the Gaussian case. As we have seen, in the multi-look case with an L factor, the variation coefficients take on the values $\frac{1}{\sqrt{L}}$ in intensity and around $\frac{0.5}{\sqrt{L}}$ in amplitude for a physically homogeneous area.

These speckle laws, both in amplitude and intensity, are illustrated in Figure 7.2, which gives the aspects for different values of the number of looks L. We see that the more the number of looks increases, the more localized the probability law. The case $L = \infty$ corresponds to a homothetic distribution (a simple gain) that we often associate with a degenerate normal law (null variance).

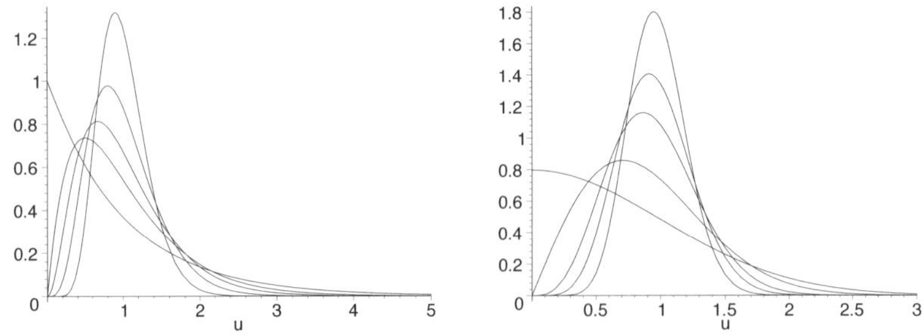

Figure 7.2. *The laws of fully developed speckle. To the left, the intensity laws: Gamma laws with μ=1 and L=1, 2, 3, 5 and 10. To the right, the amplitude laws: Rayleigh-Nakagami laws with μ=1 and L=0,5, 1, 2, 3 and 5. In this example, we must note that the Gamma law (with L=1) and the Rayleigh-Nakagami law (with L=0,5) have their mode in 0.*

7.1.3.1. *Texture modeling*

The distributions seen previously correspond to the case of fully developed speckle, that is, the underlying scene has a local constant reflectivity R. This could be the case of a roof, a field, a lawn area, etc. However, there are other parts of the image that have a reflectivity displaying fluctuations that can be modeled by considering R not as a constant parameter, but as a random variable following a certain distribution.

There are two equivalent approaches that are generally used. The first consists of considering that the previous distributions were, in fact, conditional distributions $p(I|R)$ and $p(A|R)$. In this case, if we wish to introduce a variability on R, it is necessary to model $p(R)$ and then to calculate:

$$p(I) = \int_0^{+\infty} p(I|R)p(R)dR \qquad [7.10]$$

The second approach consists of considering the speckle effect as a random variable S (following the previous laws with a reflectivity parameter equal to 1), and to consider a multiplicative model $I = RS$. The law obtained will finally be the same because:

$$p(I = RS) = \int_0^{+\infty} p(R)\, p(S = \frac{I}{R})\frac{1}{R}dR \qquad [7.11]$$

These reasonings are equally valid in amplitude when using the associated distributions. For the multiplicative model, we have the following relationship between the variation coefficients of the image C_I, the texture of the scene C_R and the normalized speckle C_S:

$$C_R^2(1 + C_S^2) = C_I^2 - C_S^2 \qquad [7.12]$$

The value of C_S is $\frac{1}{\sqrt{L}}$ for a multi-look intensity image. We find $C_R = 0$ (no texture) if $C_I = C_S$, and in this case intensity data have variations strictly due to the speckle effect.

There are several hypotheses that have been studied for the R distribution:

– $R = R_0$, area with constant reflectivity, then I follows a Gamma law with mean R_0 (according to previous results); this hypothesis is well suited for physically homogeneous areas.

– R follows a Gamma law, then I follows a K law; this law corresponds to a process of birth and death of the elementary scatterers within a resolution cell; it is well adapted to vegetation areas;

– R follows an inverse Gamma law, then I follows a Fisher law [TIS 04]; the interest of this distribution lies in the modeling of a wide range of textures, allowing us to model an urban environment that has strong back-scatterers as well as natural areas.

Numerous other models have been proposed, either directly on I, or on the R distribution (log-normal laws, Weibull, Rice, etc.). However more complete, the difficulty with these models comes from the increase of the number of parameters defining the law and their estimation. A manipulation framework of these distributions defined in $\mathbb{R}^+$, called “log-statistics”, was developed in [NIC 02] thanks to the Mellin transform, and to the Mellin convolution, which are adapted tools for multiplicative noise (we could see, in passing, that equation [7.11] expresses a Mellin convolution). It allows us to define quantities specific to the laws on $\mathbb{R}^+$: log-moments and log-cumulants, which have properties similar to the moments and traditional cumulants of the laws defined on $\mathbb{R}$. More specifically, this framework allows us, through the diagram of the log-cumulants of order 2 and 3, to put the laws in relation with one another. It also provides very advanced tools for estimating the parameters that come up in the distributions. The first parameter that must be known is the reflectivity R of the studied area. The estimator in the sense of maximum likelihood is given by the empirical average of intensities. As regards the number of looks, we can either consider that it is given by the agency providing the images in terms of physical parameters, or we can estimate it on the data (see the following section 7.1.4). In this case, the log-statistics allow us to obtain estimators with a low variance. This is also the case for the other parameters that may intervene in the distributions.

Another form of multi-looking is given by the multi-temporal combination of data. This leads to a very significant reduction of the speckle effect.

7.1.4. *Estimating the number of looks in an image*

As we have seen above, the number of looks L is the image quality parameter which characterizes radiometric resolution on a stationary area of the image [MAS 08]). In order to determine it, we select a homogeneous part of the image, with no apparent texture ($R = R_o$ and $C_R = 0$). We then have $C_I = C_S = \frac{1}{\sqrt{L}}$, and the calculation of the variation coefficient leads to estimating the equivalent number of looks. In a second phase, knowing the equivalent number of looks will allow us to determine the texture coefficient for any part of the image using equation [7.12].

In spite of a simple formulation, the number of looks is a difficult parameter to estimate. We recommend renewing the estimation for several homogeneous parts of the image. What is more, the slight oversampling of the image (pixel size < spatial resolution) leads to a spatial correlation between neighboring pixels, which results in a lower number of equivalent looks than expected.

Finally, we should keep in mind that the previous distributions were determined with the fully developed speckle hypothesis. The advent of high-resolution sensors ($\leq 1m$) makes it more likely to have a limited number of scatterers inside a resolution cell (and not the "large number" required for fully developed speckle conditions). The alternative laws mentioned previously (K laws, Weibull laws, Fisher laws, etc.) are then more adequate to describe the statistics of the observed scene [TIS 04].

7.2. Representation of polarimetric data

Let us recall the form of the scattering matrix that was presented in Chapter 4 (equation [4.53]):

$$\left[\bar{\bar{S}}\right] = \begin{bmatrix} S_{xz} & S_{yz} \\ S_{xt} & S_{yt} \end{bmatrix} = \begin{bmatrix} \hat{z}\vec{E}_0^{s1} & \hat{z}\vec{E}_0^{s2} \\ \hat{t}\vec{E}_0^{s1} & \hat{t}\vec{E}_0^{s2} \end{bmatrix}$$

The indexes i and j of the complex coefficients S_{ij} represent the emission mode (x or y) and the reception mode (z or t), respectively.

We have seen that, for a radar in mono-static configuration, ($\hat{z} = \hat{x}$, $\hat{t} = \hat{y}$, $\hat{k}^s = -\hat{k}^{inc}$), we have $S_{yx} = S_{xy}$ where $\left[\bar{\bar{S}}\right]$ is the backscattering matrix. We will adopt this configuration in the remainder of this chapter, with $\hat{h}$ instead of $\hat{x}$ (horizontal direction), and $\hat{v}$ instead of $\hat{y}$ (vertical direction).

7.2.1. *Canonical forms of the backscattering matrix*

The case of a unique interaction between a wave and a dielectric or conducting surface corresponds to single-bounce scattering (or single scattering). Concerning the polarimetric behavior at first order, the surface can be considered as an infinite perfectly conducting plane illuminated under normal incidence. The transmitted electrical fields reflected from this plane undergo a phase offset of π with a total reflection. The resulting backscattering matrix is written (with a multiplying factor including the phase term):

$$\left[\bar{\bar{S}}\right]_{SS} = \begin{bmatrix} 1 & 0 \\ 0 & 1 \end{bmatrix} \quad [7.13]$$

For any kind of odd-bounce scattering, it is the same $\left[\bar{\bar{S}}\right]_{SS}$. For instance, it is the case for third-order scattering (triple bounce) occurring on a trihedral corner reflector (Figure 7.3(a)).

The case of two successive single scattering events corresponds to double-bounce scattering or double scattering (Figure 7.3(b)). If the two surfaces are locally orthogonal, they will create two successive specular reflections, which scatters the incident energy back to the transmitter – this is known as the dihedral or double-bounce effect. Double-bounce effects are extremely common in the urban areas. The resulting backscattering matrix is written:

$$\left[\bar{\bar{S}}\right]_{DS} = \begin{bmatrix} 1 & 0 \\ 0 & -1 \end{bmatrix} \quad [7.14]$$

It is thus possible to distinguish between odd and even diffusions by using polarimetric measurements. Indeed, the comparison between $\left[\bar{\bar{S}}\right]_{SD}$ and $\left[\bar{\bar{S}}\right]_{DR}$ shows that S_{hh} and S_{vv} are in phase for odd scattering, and in phase opposition for even scattering. Many classification algorithms are based on this property.

Diffraction effects occur on sharp edges. In this situation, the polarimetric behavior is similar to infinite straight wires or dipoles. The main feature of a dipole or wire is that it filters out part of the transmitted polarization (to that extent, it is considered to be a polarizer). Admitting that it is only sensitive to the electric field component parallel to it, we obtain:

$$\left[\bar{\bar{S}}\right]_{Dip} = \begin{bmatrix} 1 & 0 \\ 0 & 0 \end{bmatrix} \qquad [7.15]$$

in the particular case of a horizontal dipole (wire). It thus suppresses the vertical component of the emitted fields, and it is a polarization filter. A similar form is obtained for a vertical dipole ($S_{hh} = 0$, $S_{vv} = 1$).

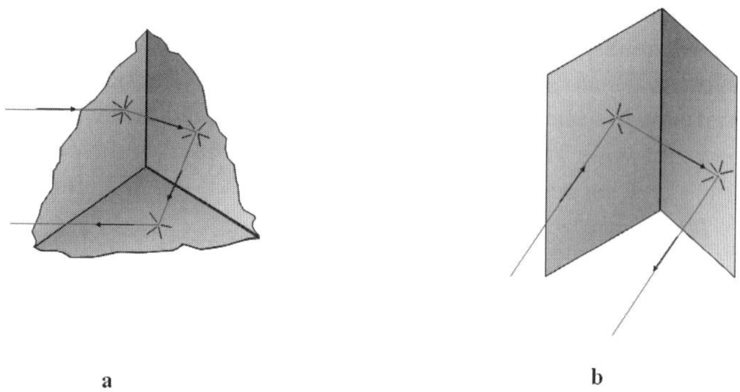

Figure 7.3. *Back-scattering on a trihedral corner reflector a) and on a dihedral corner reflector b)*

7.2.2. *Taking depolarization mechanisms into account*

The interaction processes described by the backscattering matrix $[\bar{\bar{S}}]$ do not allow us to explain the possible creation of a depolarized component during the interaction of the incident wave and the environment (section 4.3). However, fluctuations of some scatterers in space or time imply changes in the polarization state.

In these conditions, the characteristics of the polarization ellipsis ($a_h e^{j\delta_h}$ and $a_v e^{j\delta_v}$), described in section 4.2 and repeated in the equation [4.55], become fluctuating; therefore, it no longer allows us to characterize the state of the polarization. The backscattering wave becomes "partially polarized": a portion of the back-scattered energy is conveyed in a fully polarized form, the remaining part being conveyed in a fully unpolarized form. $[\bar{\bar{S}}]$ is replaced by matrix representations reflecting an "average" behavior of $[\bar{\bar{S}}]$ on the scale of several image pixels or independent radar measurements. The coherence matrix $\left[\bar{\bar{T}}\right]$ is one of them:

$$\frac{1}{2}\begin{bmatrix} \langle (S_{hh}+S_{vv})(S_{hh}+S_{vv})^*\rangle & \langle (S_{hh}+S_{vv})(S_{hh}-S_{vv})^*\rangle & 2\left\langle (S_{hh}+S_{vv})\,S_{hv}^*\right\rangle \\ \langle (S_{hh}-S_{vv})(S_{hh}+S_{vv})^*\rangle & \langle (S_{hh}-S_{vv})(S_{hh}-S_{vv})^*\rangle & 2\left\langle (S_{hh}-S_{vv})\,S_{hv}^*\right\rangle \\ 2\langle S_{hv}\,(S_{hh}+S_{vv})^*\rangle & 2\langle S_{hv}\,(S_{hh}-S_{vv})^*\rangle & 4\left\langle S_{hv}S_{hv}^*\right\rangle \end{bmatrix} \qquad [7.16]$$

where $\langle\rangle$ represents the spatial average operator.

The polarization synthesis technique introduced in section 7.2.2 can be transposed in the $\left[\bar{\bar{T}}\right]$ formalism. This is, for example, implemented in order to reinforce the sensitivity of the radar signal to the bio-geo-physical parameters we are interested in. Sensitivity of the radar signal to the roughness of surfaces (bare grounds, lava fields, etc.) is favored by circular polarizations. Sensitivity to forest biomass rates is reinforced by linear polarizations (horizontal emission, vertical reception) as shown in Figure 7.4. The full perimeter shows bare grounds with different roughness, whereas dotted perimeter shows forest parcels at different growth stages. Bare soil is well distinguished with circular polarization (Figure 7.4(a)), whereas a clear separation is shown with linear polarization for wooded areas (Figure 7.4(b)).

Figure 7.4. *Polarization synthesis effects on wooded zones and bare soil. SIR-C image, Landes forest, L band, 24° . a) Polarimetric data projection onto a circular polarization basis (emission: right circular; reception: right circular). b) Polarimetric data projection onto a linear polarizations basis (emission: horizontal linear; reception: vertical linear)*

The previous considerations lead us to consider the partially polarized wave as the incoherent sum of a fully polarized wave and a fully unpolarized wave. A more in-depth analysis requires us to quantify the respective parts of these two contributions. To do this, we cannot avoid a mathematical approach, by using the Hermitian properties of $\left[\bar{\bar{T}}\right]$ (positive eigenvalues, orthogonal eigenvectors). Thus, we calculate:

– the three eigenvalues: μ_1, μ_2 and μ_3;

– the three eigenvectors: $\vec{e_1}$, $\vec{e_2}$ and $\vec{e_3}$, which we can be written as follows:

$$\vec{e_i} = \begin{pmatrix} \cos\alpha_i \\ \sin\alpha_i \ \cos\beta_i \ \exp^{i\delta} \\ \sin\alpha_i \ \sin\beta_i \ \exp^{i\gamma} \end{pmatrix}.$$

7.2.2.1. *Statistics of polarimetric data*

The vector $\tilde{\mathbf{S}}$ follows a zero-mean complex circular Gaussian distribution [GOO 75]:

$$p(\tilde{\mathbf{S}}|\mathbf{C}) = \frac{1}{\pi^d \det(\mathbf{C})} \exp(-\tilde{\mathbf{S}}^{*t}\mathbf{C}^{-1}\tilde{\mathbf{S}}) \qquad [7.17]$$

where d is the dimension of $\tilde{\mathbf{S}}$ (3 or 4).

When using the empirical covariance matrix Σ locally calculated on L samples:

$$\Sigma = \frac{1}{L}\sum_{i=1}^{L} \tilde{\mathbf{S}}_{\mathbf{i}}\tilde{\mathbf{S}}_{\mathbf{i}}^{\mathbf{t}*}$$

the followed distribution becomes a Wishart distribution:

$$p(\Sigma|\mathbf{C}) = \frac{L^{Ld}|\Sigma|^{L-d}\exp\left(-LTr(\mathbf{C}^{-1}\Sigma)\right)}{\pi^{\frac{d(d-1)}{2}}\Gamma(L)\Gamma(L-d+1)\ \det(\mathbf{C})^L} \qquad [7.18]$$

7.2.3. *Polarimetric analysis based on the coherence matrix*

7.2.3.1. *Entropy*

Respective weights of the polarized and unpolarized components in the backscattered wave can be measured by the entropy H. Indeed, H locally quantifies the "degree of disorder" of the polarimetric response. H is equal to the logarithmic sum of the normalized eigenvalues of $\left[\bar{\bar{T}}\right]$:

$$H = -\sum_{i=1}^{3} P_i \log_3(P_i) \qquad [7.19]$$

where P_i is given by: $P_i = \lambda_i / \sum_{j=1}^{3} \lambda_j$. H tables value between 0 and 1. The extreme case $H = 0$ corresponds to fully polarized backscattering mechanisms, whereas $H = 1$ corresponds to fully unpolarized backscattering mechanisms.

Another indicator is the value $\overline{\alpha}$, which represents quite well an average angle of the eigenvectors:

$$\overline{\alpha} = \sum_{i=1}^{3} \alpha_i P_i. \qquad [7.20]$$

In the $(H, \overline{\alpha})$ diagram, we can then characterize a certain number of targets (forest, urban areas, etc.). In this diagram, the two curves give for each value of the entropy the minimal and maximal values for $\overline{\alpha}$ (Figure 7.5).

– The low-entropy diffusion areas (H lower than 0.5) for which a single backscattering mechanism is predominant:

- **Z9**: $\overline{\alpha}$ is lower than $\frac{\pi}{4}$, which corresponds to a single-bounce backscattering mechanism (with no mechanism introducing a phase rotation of π between hh and vv). In practice, all the mechanisms described by an odd number of backscatterings belong to this class;

- **Z7**: $\overline{\alpha}$ is higher than $\frac{\pi}{4}$ and lower than $\frac{\pi}{2}$, which corresponds to a double-bounce backscattering. In practice, all the mechanisms described by an even number of backscatterings belong to this class;

- **Z8**: $\overline{\alpha}$ is close to $\frac{\pi}{4}$, the proposed mechanism is a dipole backscattering.

– The middle entropy diffusion areas (H between 0.5 and 0.9):

- **Z6**: $\overline{\alpha}$ is low, we have a single-bounce backscattering mechanism with effects that are related to the surface roughness;

- **Z4**: $\overline{\alpha}$ is close to $\frac{\pi}{2}$, we have multiple diffusions (e.g. volume diffusion in the forest canopy);

- **Z5**: $\overline{\alpha}$ is close to $\frac{\pi}{4}$, which can be explained by a certain correlation between the orientation of the scatterers.

– The high-entropy diffusion areas (H higher than 0.9):

- **Z3**: $\overline{\alpha}$ is low, this area, which corresponds to $\overline{\alpha}$ values that are higher than the upper limit, cannot be characterized by this approach;

- **Z1**: this area corresponds to multiple diffusions, as we may see in forest applications;

- **Z2**: this area therefore corresponds to the volume diffusion by a needle-type cloud of particles (with no direction correlation). Let us note that a random noise, with no polarization effect, is represented by the values found to the farthest right of this area.

7.2.3.2. *Dominant/average backscattering mechanism*

The eigenvector associated with the largest eigenvalue of $\left[\overline{\overline{T}}\right]$ gives the dominant backscattering mechanism. This definition does not consider if this dominant backscattering is representative because it does not take into account the weights of the other contributions. [CLO 97] overcame this objection and defined the average backscattering mechanism, which considers all the eigenvectors of $\left[\bar{\bar{T}}\right]$, proportionally to their eigenvalues. The average mechanism is identified by a coefficient $\overline{\alpha}$ varying between 0 and $\pi/2$. $\overline{\alpha} \approx 0$ characterizes single-bounce scattering, $\overline{\alpha} \approx \pi/2$ double-bounce scattering and $\overline{\alpha} = \pi/4$ the dipole-like scattering. The algorithms of polarimetric classification are built on the identification of a backscattering process using $(H, \overline{\alpha})$ values, thus completing the family of canonical mechanisms of section 7.2.1.

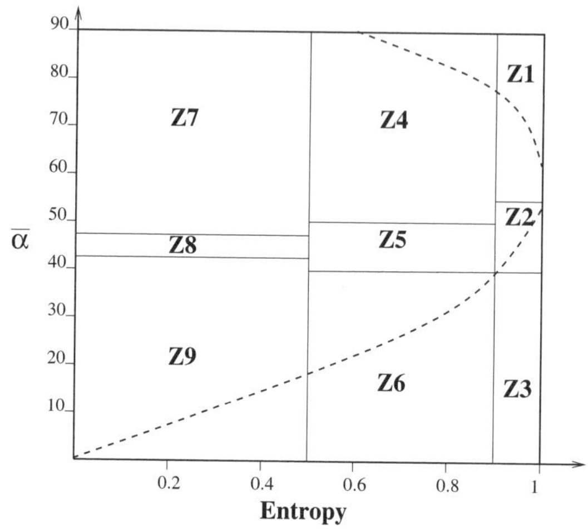

Figure 7.5. *Partition of the "α – H" plane into nine backscattering mechanisms*

7.2.3.3. *Examples of polarimetric images analysis*

Figures 7.6 display *HH*, *HV* and *VV* polarimetric channels of a SIR-C image of the Ulan-Ude region (Russia). Because the mean radiometry of the *HV* channel is approximately 8 dB lower than the co-polarized channels *HH* and *VV*, it has been enhanced for visualization purposes. Although the complementarity of the three channels is not easy to discern at first glance, it becomes obvious in the color combination (Figure 7.6(d)).

In the case of similar radiometric levels, the color is greenish (see for instance, the mountain chain at the bottom of the image). Brighter colors indicate a higher sensitivity to polarization. The blue color dominates in areas with less *HH* signal relative to *VV* polarization. This phenomenon is linked to dominant Bragg scattering and appear on bare grounds. Saturated colors can be observed for slopes facing the sensor or for urban areas near the stream.

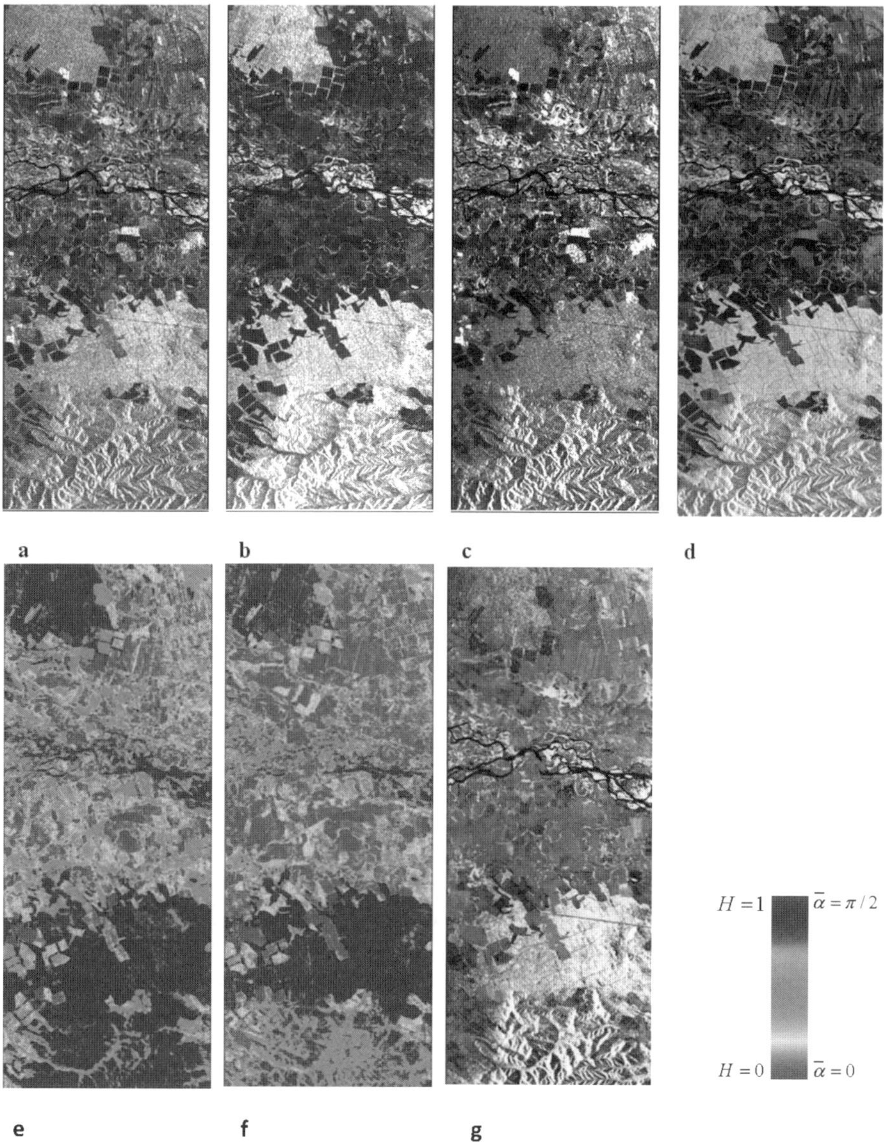

Figure 7.6. *SIR-C polarimetric images (L band, 24° , decameter resolution). The site of Ulan-Ude, Russia, 1994. a) HH channel, b) HV channel and c) VV channel. d) Colored composition HH (R), HV (G), VV (B) in e) H entropy, f) average mechanism $\overline{\alpha}$, g) IHS composition: intensity (span), hue (average mechanism $\overline{\alpha}$), saturation (1-H). For a color version of this figure, see www.iste.co.uk/tupin/RSImagery.zip*

Entropy estimated on a 7×7 pixel analysis window is displayed in Figure 7.6(e) for the SIR-C image. Almost fully unpolarized waves present high entropy (e.g. the mountain in the lower part of the image). These high entropy values generally reflect dominant volume scattering. The high values of the *HV* signal confirm this hypothesis. In addition, the received signal is practically insensitive to transmission and reception polarizations because the *HH*, *VV* and *HV* radiometric values are evenly distributed (Figure 7.6(d)). Similar results can be observed at the C band, although entropy is on average slightly higher than at the L band, making the C band polarimetric analysis more difficult.

The coefficient $\overline{\alpha}$ of the SIR-C image (Figure 7.6(f)) is strongly correlated to entropy. When the back-scattered signal is highly polarized, the three polarization channels are significantly complementary, and it induces low entropy. The concerned areas correspond principally to surfaces producing a simple dominant diffusion, associated with a low $\overline{\alpha}$ value.

Figure 7.7 illustrates airborne imagery of an airport environment. It is made up of buildings with a saturation on the radar signal, and close-cropped vegetation and strips of asphalt with weak radiometry. The merging of channels *HH*, *VV* and *HV* is done in Figure 7.7(a). In this example, an extended polarimetric analysis can be led, using the cartography of the entropy H (Figure 7.7(b)), and average interaction mechanisms $\overline{\alpha}$ (Figure 7.7(c)). The entropy image mainly underlines a section of the track and building areas with low levels, whereas high entropy values generally express a dominant volume diffusion. Going from left to right, i.e. from low incidences toward strong incidences, we note an entropy gradient connected to the incidence variations on the swath. The low incidences preserve the coherence of the wave emitted more than the strong incidences, for which moving away from the specular configuration produces an increasingly diffuse signal, and a decreasingly polarized signal. We will note an entropic break that bevels the section of the diagonal track (it is not visible on the radiometric data in Figure 7.7(a)). This does not seem to be of a physical origin, but it is certainly the result of an incidence effect. At low incidences, the signal sent by the asphalt, although low, remains above the threshold of the image noise. It can thus impose its own polarimetric signature, characterized by a dominant polarized polarimetric component (low H). The intensity of the backscattered signal decreases when the local incidence increases, until reaching the level of the image noise. This noise is fully unpolarized, and is therefore at a high entropy level. The average backscattering mechanism $\overline{\alpha}$ (Figure 7.7(c)) shows strong behavioral similarities with the entropy H. The low H values are generally connected to the surface diffusions, for which $\overline{\alpha}$ is equally low. The presence of a volume component in backscattering triggers an increase not only in H but also in $\overline{\alpha}$. A well-known exception to the correlated behaviors of H and $\overline{\alpha}$ regards strongly polarized mechanisms, other than single-bounce backscattering. The double-bounce scatterings on buildings ($\overline{\alpha} \approx \pi/2$) belong to this category, a category that is however too rare statistically to shake off the impression of similarity between H and $\overline{\alpha}$ observed in numerous images.

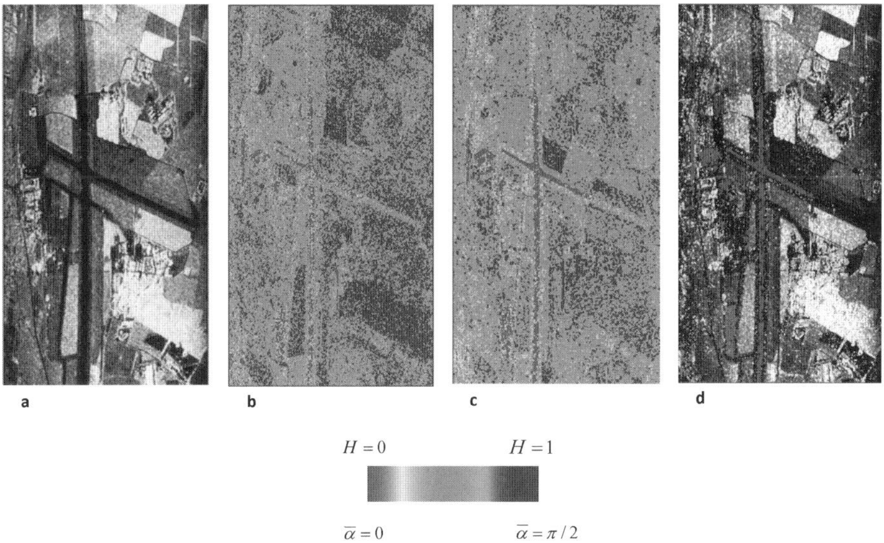

Figure 7.7. *ONERA/RAMSES L band image, 5-m resolution, acquired over an airport area. a) Colored composition HH (R), HV (G), VV (B). b) H entropy, c) average mechanism. d) IHS composition intensity (span), hue (average mechanism $\overline{\alpha}$), saturation (1-H). For a color version of this figure, see www.iste.co.uk/tupin/RSImagery.zip*

7.2.4. *Synoptic representation of polarimetric information*

Due to the vectorial dimension of polarimetric channels with more than three dimensions, it is impossible to display a complete representation for it. In case of high entropy, the phase information is useless, and there is no need to have a complete representation of the information. In [IMB 99], an adaptive visualization system inspired from interferometry is proposed, based on a decomposition into intensity-hue-saturation (IHS)[2]. This system automatically reduces the representation of the polarimetric information to its radiometric part whenever the signal is strongly unpolarized.

2 The intensity associated with every pixel of the image is related to its radiometric content (that is, its "black&white" component), the *saturation* refers to its coloring level and the *hue* to the actual color itself. The *IHS* representation is another way of presenting the tri-chromatic decomposition *red-green-blue* of a color image.

– The intensity channel carries a layer of radiometric information, for example the span image, the incoherent sum of radiometries:

$$SPAN = |HH|^2 + |VV|^2 + 2|HV|^2$$

It thus provides a gray-level background image.

– The saturation channel is controlled by the local polarization state, revealed by the H entropy (section 7.2.3.1) or the polarization degree P (section 4.1.3.3). A high entropy leads to a low saturation (and vice versa), that is to an image that is locally in gray-levels, with an exclusively radiometric content. The law connecting the saturation is usually linear ($S = 1 - H$), although more subtle relations can be involved.

– Finally, because the decreasing entropy colors the image gradually, the Hue channel will associate with each pixel a color related to the local polarimetric behavior (e.g. via the coefficient $\overline{\alpha}$), but only when this is meaningful.

Based on such a representation, the discrimination between polarized and unpolarized areas is easier. In the SIR-C image (Figure 7.6(g)), within the unpolarized maelstrom at the bottom of the image, only the mountain crests retain some wave coherence. On the contrary, the central part of the image presents mostly polarized surface scattering. In the airborne case (Figure 7.7(d)), the IHS representation emphasizes the incidence effect on the polarization of backscattered waves. The bluish points correspond to both low entropy values, and high $\overline{\alpha}$ values, meaning polarized double-bounce backscattering created by built-up areas.

7.3. InSAR interferometry and differential interferometry (D-InSAR)

The geometric aspects of interferometry and D-InSAR will be detailed in Chapters 8 and 9.

Radar interferometry (InSAR) is based on geometric information hidden in the phase of a pixel, accessible with SLC *Single Look Complex* data. Phase information can be divided into two parts. The first part is linked to the electromagnetic properties of the target and is called the intrinsic phase, whereas the second part is linked to the distance between the sensor and the target.

With single-phase information, it is generally not possible to separate these two pieces of information and retrieve the measurement. However, when two measurements of the same pixel ϕ_1 and ϕ_2 are available, it is possible by subtraction to recover the geometric information.

If ϕ_2 is acquired with exactly the same position of the sensor, we will of course have $\phi_2 - \phi_1 = 0$, unless the point observed has slightly moved while remaining in the same radar bin. Generally, the second acquisition is done with a slightly different incidence angle, allowing us to recover not only the potential motion information, but also the topography in the phase difference. Interferometry exploits this property to build digital terrain models (DTM). D-InSAR is concerned with ground movements, by combining two acquisitions with an existing DTM, or by exploiting a set of acquisitions (at least three).

In this section, we will admit that the phase difference contains height or movement information, and we will be interested in the distributions followed by this phase.

7.3.1. *Statistics of interferometric data*

Interferometry is the measurement of two complex data z_1 and z_2, which can be put into a complex vector $\mathbf{Z}$, analogously to what was done in the previous section with the four polarimetric components (equation [7.17]). We may again consider the zero mean circular Gaussian model, and the distribution of $\mathbf{Z}$ is written:

$$p_z\left(\mathbf{Z}|\mathbf{C_z}\right) = \frac{1}{\pi^2 \det\left(\mathbf{C_z}\right)} \exp\left(-\mathbf{Z}^{t*}\, \mathbf{C}_z^{-1}\, \mathbf{Z}\right) \qquad [7.21]$$

$\mathbf{C_z}$ is the covariance matrix of $\mathbf{Z}$, also called the coherence matrix. It is written:

$$\mathbf{C_z} = \begin{bmatrix} \left\langle |z_1|^2 \right\rangle & \left\langle z_1 z_2^* \right\rangle \\ \left\langle z_1^* z_2 \right\rangle & \left\langle |z_2|^2 \right\rangle \end{bmatrix} \qquad [7.22]$$

ρ_{12}, the complex correlation coefficient (or degree of coherence) is written:

$$\rho_{12} = \frac{\left\langle z_1 z_2^* \right\rangle}{\sqrt{\left\langle |z_1|^2 \right\rangle \left\langle |z_2|^2 \right\rangle}} = D\, e^{j\beta}$$

D is simply the coherence, and β the actual offset between the components of $\mathbf{Z}$. Defining the reflectivities $R_1 = \left\langle |z_1|^2 \right\rangle$ and $R_2 = \left\langle |z_2|^2 \right\rangle$, $\mathbf{C_z}$ can then be written:

$$\mathbf{C_z} = \begin{bmatrix} R_1 & \sqrt{R_1\, R_2}\, D\, e^{j\beta} \\ \sqrt{R_1\, R_2}\, D\, e^{-j\beta} & R_2. \end{bmatrix}$$

For $D \neq 1$, $\mathbf{C_z^{-1}}$ is written:

$$\mathbf{C_z^{-1}} = \frac{1}{R_1 R_2\, (1 - D^2)} \begin{bmatrix} R_2 & -\sqrt{R_1\, R_2}\, D\, e^{j\beta} \\ -\sqrt{R_1\, R_2}\, D\, e^{-j\beta} & R_1 \end{bmatrix}.$$

For two complex values z_1 and z_2, equation [7.21] is given by:

$$p(z_1, z_2 \quad |R_1, R_2, D, \beta) = \frac{1}{\pi^2 R_1 R_2 (1-D^2)} \exp\left[-\frac{1}{1-D^2}\left(\frac{z_1 z_1^*}{R_1} + \frac{z_2 z_2^*}{R_2} - \frac{D\left(z_1 z_2^* e^{j\beta} + z_1^* z_2 e^{-j\beta}\right)}{\sqrt{R_1 R_2}}\right)\right]. \quad [7.23]$$

If, instead of considering z_1 and z_2 values, we consider the joint distribution of Σ_z elements, the empirical co-variance matrix:

$$\Sigma_z = \mathbf{Z}\,\mathbf{Z}^{t*} = \begin{pmatrix} I_1 & I_{12}e^{j\varphi} \\ I_{12}e^{-j\varphi} & I_2 \end{pmatrix}.$$

We can then express the joint distribution of the elements of Σ_z depending on the reflectivities R_1 and R_2, and of the complex coherence $De^{j\beta}$ as:

$$p(I_1, I_2, I_{12}, \quad \varphi|R_1, R_2, D, \beta) = \frac{1}{\pi^2 R_1 R_2 (1-D^2)} \exp\left(-\frac{1}{1-D^2}\left(\frac{I_1}{R_1} + \frac{I_2}{R_2} - \frac{2DI_{12}\cos(\varphi-\beta)}{\sqrt{R_1 R_2}}\right)\right). \quad [7.24]$$

We can deduce from this equation the distributions of the anti-diagonal term of the empirical covariance matrix $I_{12}\mathrm{e}^{j\varphi}$, called complex interferogram (Figure 7.8).

In practice, rather than calculating $z_1 z_2^* = I_{12}e^{j\varphi}$, a spatial averaging on L samples is computed (sometimes called somewhat improperly *complex multilooking*, L being the "number of looks") in order to obtain an empirical spatial coherence, denoted by d:

$$de^{j\varphi} = \frac{\sum_{k=1}^{L} z_{1,k} z_{2,k}^*}{\sqrt{\sum_{k=1}^{L} z_{1,k} z_{1,k}^*}\sqrt{\sum_{k=1}^{L} z_{2,k} z_{2,k}^*}} = \frac{I_{12}}{\sqrt{I_1\, I_2}}\,\mathrm{e}^{j\varphi} \qquad L \geq 2. \quad [7.25]$$

This coherence measures the correlation between the complex data used to calculate the interferometric phase. It comprises in $[0, 1]$, the higher values ensuring a good quality of the measured interferometric phase, whereas the lower values indicate the decorrelation between the two images z_1 and z_2. Several factors can cause this decorrelation: the absence of the measured signal (shadow area), the variations of the acquisition geometry, the temporal variations of the ground (vegetation areas, human modifications, etc.), the variation of the atmospheric conditions between the two acquisitions and so on. We will return to these in a future chapter.

Starting from equation [7.24], the law verified by the empirical coherence obtained by the space averaging on L samples is deduced:

$$p(d|D,L) = 2(L-1)(1-D^2)^L d(1-d^2)^{L-2}\,{}_2F_1(L,L;1;d^2D^2)$$

as well as the pdf of the phase φ:

$$p(\varphi|D,\beta,L) = \frac{(1-D^2)^L}{2\pi}\frac{1}{2L+1}\,{}_2F_1\left(2,2L;L+\frac{3}{2};\frac{1+D\cos(\varphi-\beta)}{2}\right) \quad [7.26]$$

where ${}_2F_1$ represents a hypergeometric function. This last equation is written for different values of the coherence D and of the number of samples L in Figure 7.8.

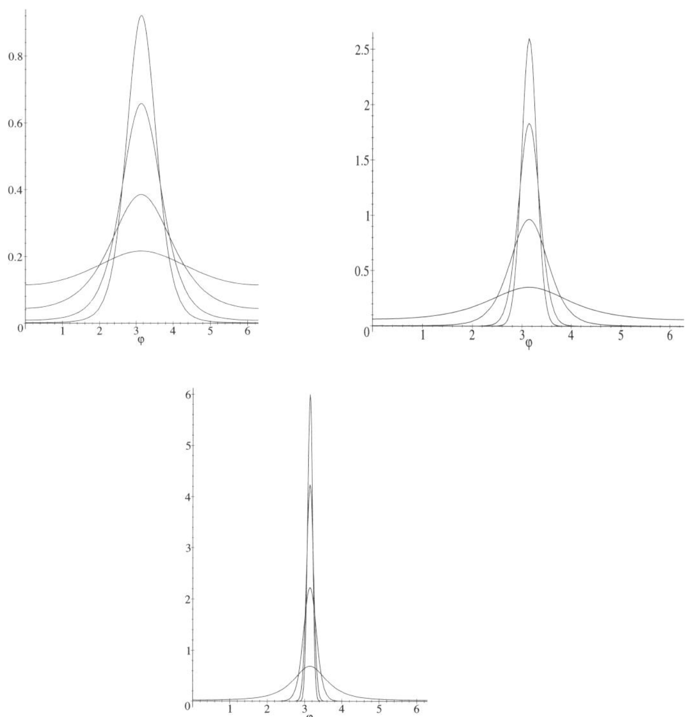

Figure 7.8. *Distributions of the multi-look interferogram (phase difference of the complex interferogram) for D = 0.2 (top left), D = 0.5 (top right) and D = 0.8 (bottom), and for L = 1, 9, 32 and 64 (equation* [7.26]*)* $\beta = \pi$.

The spatial averaging (called *complex multi-looking*) is usually applied to compute the interferogram because of the phase noise (see Figure 7.8). However, this averaging is done at the cost of the spatial resolution of the interferogram. There are several solutions that have been explored for selecting the best samples to be averaged and thus reduce the phase noise without damaging the resolution.

7.4. Processing of SAR data

This chapter aimed to present distributions followed by different radar data. These models are crucially important, as they will be the basis of the processing methods.

As regards phase de-noising, or more precisely, the estimation of relevant parameters (reflectivity, interferometric phase, coherence, etc.), methods will take into account these distributions in order to select the similar samples and combine them. For example, in [DEL 11a] an adaptation of the non-local means is proposed to improve the complex multi-looking.

Classification approaches are generally expressed in a Bayesian framework and rely on the distributions we have studied previously. Criteria such as maximum likelihood or *a posteriori* maximum are used (section 5.6.1). Similarly, detecting relevant objects such as targets will rely on hypotheses tests, defined to use the efficiency of these statistic models (section 5.5.1).

7.5. Conclusion

Radar image processing techniques rely very heavily on the statistics presented in this chapter. Among open issues, it is worth noting that the distributions will never become Gaussian whatever the number of looks. This means that processing approaches relying on additional white Gaussian noise assumptions will never be perfectly adapted to radar imagery, whatever may be the noise reduction technique used.

Another important issue is currently unsolved: geometric aspects are usually not taken into account because they require a DTM or a digital elevation model (DEM). In particular, not taking into account the local slope of the ground or of the imaged objects makes processing less effective. Iterative approaches are necessary to introduce height information. When a DTM or a DEM are available, more and more processing techniques introduce a correction depending on the local slope (e.g. in polarimetry where this correction is essential).

PART 3

Applications: Measures, Extraction, Combination and Information Fusion

Chapter 8

Analysis of Multi-Temporal Series and Change Detection

Change detection refers to the process of highlighting and placing the alterations of the state of a scene's surface, having been observed in several images. These alterations, or changes, can be:

– Slow and tenuous, such as the evolution of the rotations in agriculture or depletion areas. In this case, it is necessary to have a time series of images to detect these *evolutions*.

– Spectacular and fast, such as earthquakes, the damage due to natural or man-made disasters. Two images acquired before and after the event can be enough to detect them. This type of change is generally qualified as *abrupt*, which does not mean it is of a large scale, but rather that it is instantaneously sudden.

With this chapter, we aim therefore to describe and characterize techniques allowing us to detect these changes. We are not concerned with estimating evolution parameters (elevation measures are mainly approached in Chapter 9 and moving measurements are described in Chapter 10).

8.1. Registration, calibration and change detection

The detection of changes assumes a localized comparison of observations. The first step to be consider before the detection of changes is therefore the geo-referencing

---

Chapter written by Grégoire MERCIER and Florence TUPIN.

stage or resampling of images for their co-registration. Chapter 2 has already presented the required methodology to be applied to the geometry of images, so that the pixels of the different images are co-located. When two images have been geo-referenced in the same coordinate system, their pixels become theoretically directly comparable. In practice, even for the geo-referenced data, a residual bias (e.g. coming from the use of different DEM's) must be corrected. Furthermore, we must remember that a geometric correction applied to images needs a re-sampling that could modify the local statistics of the images, see [ING 07b]. Moreover, such resampling induces an increase in the equivalent number of look in radar images.

Moreover, registration between several images is based on distance measurements (such as the standardized correlation or mutual information, see section 2.4). These measurements intend to find the spatial offset that allows us to maximize the similarity (e.g. the correlation) between the images, given that the nature of the field is supposed to have changed throughout the different acquisitions. We must therefore remember that similar measures are used for opposite purposes: maximizing the resemblance for the registration and emphasizing the dissimilarities in order to detect changes.

This is why the scale of analysis is significant for change detection purposes. It is sometimes possible to make this detection at the scale of the pixel (i.e. at the resolution of the sensor) but if the rescaling process requires too significant transformations, it is more reasonable to do the change detection not at the scale of the pixel but at the scale of a neighboring or a larger region. This way, the change detection can be applied directly to the observation data or to the local measurements of these observations or even between the classifications issued from these observations.

Finally, the integrity and the compatibility of the data before their comparison also needs to be considered on a radiometric level. It is related to the calibration stage of the data, which can address atmospheric correction (when using passive data), directional correction (considered in Chapter 3) or inter-calibration [CHA 13]. The data issued from the same sensor are generally inter-calibrated, which allows for a direct comparison. It could be interesting to go back to the physical parameters involved in the measurements before being able to make the comparisons. This is, for example, the case in radar imagery where we need to correct the effects of an azimuth slope before being able to compare the polarimetric data [LEE 00].

8.2. Change detection based on two observations

As a result of 15 years of a political will to maintain a tight control on the environment, several specific requirements have emerged in the field of diachronic remote sensing. International groups (GEOSS[1], CEOS DMGS[2]) or European groups (GMES[3]) have published a set of recommendations seeking to develop the use of services and space technologies, within:

– risk management and prevention (natural or technological);

– natural resource management;

– territorial surveillance;

– sustainable development;

– internal security.

To identify the risk situations, a growing use of massive data coming from remote-sensing is recommended. Particular attention is paid to the immediate availability and accessibility of the data with spatial, spectral and temporal resolutions that are becoming increasingly refined.

The thematic context inherent to 2-date change detection is therefore associated with security, crisis management and rapid cartography. The methodological requirements therefore induce the definition of measurement tools and relatively simple changes from an algorithmic viewpoint; such a viewpoint is immune to some differences between observations, so that the change detections reflect the surface changes without being influenced by the changes induced by the conditions of acquisition and observation.

However, 2-date change detection is also often the basic operator in multi-data change detection, which will be approached in the next section.

8.2.1. *Change measurements between homogeneous data*

A classic scheme of change detection is based on the calculation of a similarity measurement between two images, followed by a thresholding in two classes (change and no change), or by a classification into several classes if different types of change can be emphasized starting from the similarity measurement. In the following sections we will describe the similarity measurements that are currently used in remote-sensing imagery.

---

1 GEOSS: *Global Earth Observation System of Systems*, a network grouping together 55 nations and 30 international organizations.

2 CEOS DMGS: *Committee of Earth Observation Satellites*, whose group *Disaster Management Support Group* was founded in 1997.

3 GMES: *Global Monitoring for Environment and Security*, an initiative launched in 2001 by the European space agency (ESA) and the European Community.

8.2.1.1. *Scalar data in optic imagery*

As we have seen in section 2.4.2.1, the most simple operator that implements a dissimilarity measurement between two images I_1 and I_2 is the difference operator[4]:

$$\mathcal{D}^{\text{diff}}_{(I_1,I_2)}(m,n) = I_1(m,n) - I_2(m,n), \quad [8.1]$$

but as we can easily imagine, this operator is very sensitive to geometric effects and lighting effects that trigger a bias in the values of the pixels and the images. Moreover, this operator is also very sensitive to noise. Therefore, its locally smoothed version is preferred, using a sliding window of an odd size $K \times L$:

$$\tilde{\mathcal{D}}^{\text{diff}}_{(I_1,I_2)}(m,n) = \frac{1}{K\,L} \sum_{k=-K/2}^{K/2} \sum_{l=-L/2}^{L/2} \left(I_1(m-k,n-l) - I_2(m-k,n-l)\right) \quad [8.2]$$

This version limits the noise effect but not the lighting effects.

8.2.1.2. *Spectral data in optic imagery*

The previous difference operator can be very easily extended to the multi-spectral images $\mathbf{I}_i$ (here λ which can be understood as an index in the spectral band):

$$\tilde{\mathcal{D}}^{\text{diff}}_{(\mathbf{I}_1,\mathbf{I}_2)}(m,n,\lambda) = \frac{1}{K\,L} \sum_{k=-K/2}^{K/2} \sum_{l=-L/2}^{L/2} I_1(m-k,n-l,\lambda) - I_2(m-k,n-l,\lambda). \quad [8.3]$$

In order to prevent form lighting effects, we must inter-calibrate the data or define an operator that is not sensitive to affine transformations (of type $y = ax + b$). This is what the MAD operator does (*Multivariate Alteration Detection*) [NIE 98]. This strategy does not sum up the differences in each channel of the spectral signature but performs a linear combination between two observations, by maximizing the variance of this difference. The operation then becomes immune to all the linear transformations and allows us even to compare spectral signatures of different size. MAD is defined by the following equation:

$$\mathcal{D}^{MAD}_{(\mathbf{I}_1,\mathbf{I}_2)}(m,n) = \left(\sum_{\lambda_i} a_{\lambda_i} I_1(m,n,\lambda_i)\right) - \left(\sum_{\lambda_j} b_{\lambda_j} I_2(m,n,\lambda_j)\right) \quad [8.4]$$

4 We preserve here the notations in Chapter 5 and we will denote, in bold, as much as possible, the vectorial data, and with a tilde, the criteria calculated on a local neighboring.

where the vectors $\mathbf{a}$ and $\mathbf{b}$ are assessed so that the variance:

$$\mathbf{a}^t\mathbf{I}_1 - \mathbf{b}^t\mathbf{I}_2$$

is maximized while keeping a unitary variance for each observation:

$$var\left(\mathbf{a}^t\mathbf{I}_1\right) = var\left(\mathbf{b}^t\mathbf{I}_2\right) = 1.$$

The solution is obtained using the notion of canonical correlation analysis [HAI 98]. Figure 8.1 shows the difference between the results issued from the difference between two multi-spectral images to which we have applied an PCA (in this composition using fake colors, the red band follows the axis of the largest variance of the difference image), and the canonical correlation (the red band corresponds to the maximum canonical correlation between $\mathbf{a}^t\mathbf{I}_1$ and $\mathbf{b}^t\mathbf{I}_2$, and the green and blue bands correspond to weaker correlations in a representation that is complementary to the first and then the second axis).

The change measurement defined in equation [8.3] is not fully satisfactory for the multi-spectral data since they do not easily represent the evolution of spectral signature in a synthetic manner (more specifically through a scalar coefficient). The main idea is to compare the observations using a vector difference [MAL 80] :

$$\mathbf{D}^{\mathrm{diff}}_{(\mathbf{I}_1,\mathbf{I}_2)}(m,n) = \mathbf{I}_1(m,n) - \mathbf{I}_2(m,n). \quad [8.5]$$

The difference is thus a change vector (hence the term Change Vector Analysis, CVA) whose dimension is the number of spectral bands in the observations.

The initial idea was to keep just the module of this change vector. A null change vector represents an absence of change. By defining a threshold on the module, we can characterize fine or significant changes. Thresholds can equally be defined band by band, in order to refine the detection of the types of changes – but this way, the analysis is quickly becoming visual and fastidious [LOR 02]. A more formal point of view was developed not by integrating thresholds component by component, but by defining the behavior of angles in spherical representations (or more precisely a hyper-spherical representation in n dimensions) of the change vector [BOV 07]. This mode of representation allows us to build a statistical model of the phase laws and to define the thresholds with a constant false alarm rate (CFAR) It is also possible to distinguish and characterize several types of change.

The change vector analysis can also be defined after the result of a pre-classification in order to implement the change detection within a post-classification analysis [CHE 11].

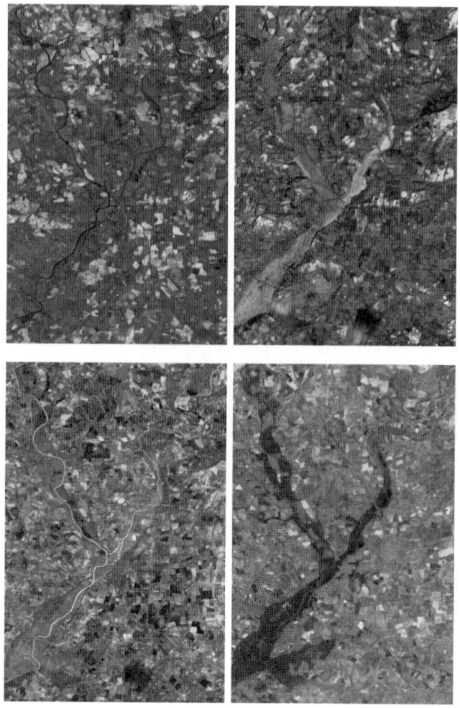

Figure 8.1. *Examples of two SPOT images, acquired during and after a flood that took place in October 2000 in Gloucester, UK. The two images below show the difference between a PCA applied to the image of a spectral difference (equation [8.3]) and the canonical correlation image (equation [8.4]) applied band by band. For a color version of this figure, see www.iste.co.uk/tupin/RSImagery.zip*

We might prefer to resort to more physical information before implementing the change detection. The normalized difference vegetation index (NDVI) defined in section 6.3.1 is often used as a dimension reduction operator allowing us to manipulate scalar data rather than a series of spectral signatures. It is otherwise preferred to the PCA, particularly due to its bio-chemical significance. However, depending on the applications we are interested in (agriculture, forestry along different latitudes, etc), this ranking is sometimes contradictory [COP 04].

8.2.1.3. *Scalar radar data*

When the images do not come from an optical sensor but from a radar sensor, the statistic behavior of observations does not follow a linear model but rather a multiplicative model. This is why the difference operator is to be proscribed. We then have two strategies:

1) We either use an operator log, thus placing ourselves in the framework of log-statistics (which allows us to transform the multiplying model and assimilate the result

to an additional model, despite the fact that the often implicit hypothesis of a Gaussian noise is only an approximation [XIE 02]). The measurement then becomes:

$$\mathcal{D}^{\text{diff log}}_{(I_1,I_2)}(m,n) = \log I_1(m,n) - \log I_2(m,n) \qquad [8.6]$$

which gives the similarity measurement $\tilde{\mathcal{D}}^{\text{diff log}}_{(I_1,I_2)}(m,n)$:

$$\begin{aligned}\tilde{\mathcal{D}}^{\text{diff log}}_{(I_1,I_2)}(m,n) &= \frac{1}{K\,L}\sum_{k=-K/2}^{K/2}\sum_{l=-L/2}^{L/2}(\log I_1(m-k,n-l)\\ &\quad - \log I_2(m-k,n-l)),\end{aligned}$$

which can be associated with a local average as in [8.2].

2) Or we use a ratio, which makes the detector very similar to a contour detection through local averages in a radar image [TOU 88b].

$$\tilde{\mathcal{D}}^{ratio}_{(I_1,I_2)}(m,n) = \tilde{I}_1(m,n)/\tilde{I}_2(m,n) \qquad [8.7]$$

with $\tilde{I}_i$ being the averages calculated locally around (m,n).

8.2.1.4. *Distribution comparison*

The majority of the previously shown measurement measures are only sensitive to the evolution of the average value of the surfaces (first-order moments), even if their local distribution, the distribution of the gray levels, is very different. For example, in radar imagery it is possible to find two regions with close averages but whose roughnesses are significantly different (in particular, if these averages are calculated from small samples). A detector that is sensitive to this change in texture must then integrate statistic measurements of a higher order than the first-order moment, and then compare the local statistic laws using inter-distribution measurements. This is the sense of the following detector:

$$\mathcal{D}^{pdf}_{(I_1,I_2)}(m,n) = d\left(\tilde{f}_{I_1}(m,n), \tilde{f}_{I_2}(m,n)\right) \qquad [8.8]$$

where $\tilde{f}_I(m,n)$ is the density estimation with a local probability around the index point (m,n).

The distance d between two distributions can take various forms but the Kullback-Leibler divergence is used more often. It is expressed as follows for two distributions f_1 and f_2:

$$\mathcal{D}_{KL}(f_1,f_2) = \int \log(\frac{f_1(x)}{f_2(x)})f_1(x)dx + \int \log(\frac{f_2(x)}{f_1(x)})f_2(x)dx$$

Depending on the choice of d, the estimation can be done through a parameter-based law model, a histogram or a development of cumulants [ING 07a]. For different types of distribution, explicit expressions may be obtained depending on the law parameters [DEL 09].

Figure 8.2 compares measurements based on the difference, the ratio and the distance between distributions. Accounting for local distributions seems more satisfying for such change detection. This comes from the fact that the detectors can handle more complex information (basically local histogram in comparison to a simple difference of local averages).

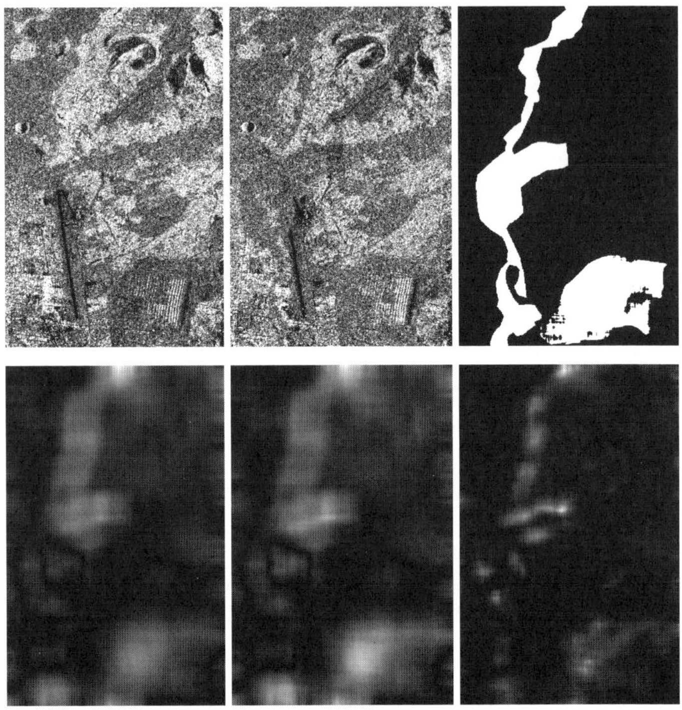

Figure 8.2. *Examples of two RADARSAT images acquired before and after a volcanic eruption in Nyiragongo that took place in January 2002. The two images at the top show the observation before the event, after the event and the shield of the impacted area. The images below represent the local difference, the local ratio and the distance between local distributions. The local estimation is perceived via a sliding window of a size* 35×35

As we have seen in Chapter 2, the exploitation of the joint probability of the two images and particularly the calculation of the mutual entropy can also be a good similarity criterion (equation [2.9]). It is particularly used in the case of heterogeneous data (e.g. optic and radar data).

When we use very high-resolution images (VHR) the consideration of the local neighboring becomes dangerous. The increase in sensors' space resolution comes with a reduction in the truthfulness of the stationarity hypothesis of the image. Thus, taking into account the local neighboring of a pixel induces a mixing of pieces of information that are of specific nature and characteristics. The change measurement yields many false detections. An alternative solution can be found in the use of Hölder distances, which depend on a ℓ parameter:

$$\mathcal{D}^{Hol}_{(I_1,I_2);\ell}(m,n) = \begin{cases} \sqrt[\ell]{\dfrac{I_1(m,n)^\ell + I_2(m,n)^\ell}{2}} & \forall \ell \in \mathcal{R}^* \\ \sqrt{I_1(m,n) \times I_2(m,n)} & \text{si } \ell = 0. \end{cases} \quad [8.9]$$

We thus find the harmonic average for $\ell = -1$, the arithmetic average for $\ell = 1$ and the quadratic average $\ell = 2$. The geometric average is obtained when $\ell = 0$. By considering the geometric and arithmetic averages together, we can make a decision with a CFAR on the hypothesis of a Fisher distribution to model the amplitude of radar images [QUI 12]. The advantage of these Hölder measurements is that they can be easily extended to the analysis of temporal series with a more substantial number of observations.

8.2.1.5. *Hypothesis test for distribution model data*

The procedure via the hypotheses test defines a unified framework that can be applied as soon as a distribution model of the data is available [DEL 12] (whether scalar-optic-, radar-, vectorial, complex, matrix-based, etc.) The main principle is to test the two following hypotheses on the basis of two noisy observations:

$\mathcal{H}_0$: the two observations come from the same scene parameter;

$\mathcal{H}_1$: the two observations come from two different parameters.

The test allowing us to obtain the strongest probability of a good detection with a fixed false alarm rate is the likelihood ratio:

$$\mathcal{L}(I_1, I_2) = \frac{p(I_1, I_2|R_{12}, \mathcal{H}_0)}{p(I_1, I_2|R_1, R_2, \mathcal{H}_1)} \quad [8.10]$$

$$= \frac{p(I_1|R_{12}, \mathcal{H}_0)p(I_2|R_{12}, \mathcal{H}_0)}{p(I_1|R_1, \mathcal{H}_1)p(I_2|R_2, \mathcal{H}_1)} \quad [8.11]$$

by following the notations of Chapter 5, where I_1 and I_2 are the observations and R_1, R_2 and R_{12} are the parameters of the related scene. In practice, these parameters remain unknown, and we calculate the generalized likelihood ratio that replaces these parameters by their estimations in the sense of maximum of likelihood starting from these data:

$$\mathcal{L}_g(I_1, I_2) = \frac{\sup_{R_{12}} p(I_1|R_{12}, \mathcal{H}_0)p(I_2|R_{12}, \mathcal{H}_0)}{\sup_{R_1} p(I_1|R_1, \mathcal{H}_1) \sup_{R_2} p(I_2|R_2, \mathcal{H}_1)} \quad [8.12]$$

$$= \frac{p(I_1|R_{12}^{MV}, \mathcal{H}_0)p(I_2|R_{12}^{MV}, \mathcal{H}_0)}{p(I_1|R_1^{MV}, \mathcal{H}_1)p(I_2|R_2^{MV}, \mathcal{H}_1)} \quad [8.13]$$

The similarity measurement is then defined as $-\log(\mathcal{L}_g(I_1, I_2))$. In the case of additive Gaussian noise, it is given by $(I_1 - I_2)^2$, and in the case of Rayleigh distribution by $\ln(\frac{A_1}{A_2} + \frac{A_2}{A_1})$ where A_i stands for the amplitude. Through this approach, we find a formula that is based on the amplitude ratios of radar images and that is adapted to the nature of the distributions.

The appeal of this hypothesis test approach lies in its large generality. The similarity measurement that it defined can be derived, for example, in the case of polarimetric radar data, with covariance matrices C_1 and C_2 [DEL 09]:

$$\mathcal{L}(C_1, C_2) = \frac{p(C_1, C_2|\Sigma_{12}, \mathcal{H}_0)}{p(C_1, C_2|\Sigma_1, \Sigma_2, \mathcal{H}_1)}$$

with Σ_1, Σ_2 and Σ_{12} as the "true", unknown, covariance matrices of the scene. Hence, the generalized likelihood report:

$$\mathcal{L}_g(C_1, C_2) = \frac{\sup_{\Sigma_{12}} p(C_1, C_2|\mathcal{H}_{\prime}, \Sigma_{12})}{\sup_{\Sigma_1,\Sigma_2} p(C_1, C_2|\mathcal{H}_1, \Sigma_1, \Sigma_2)} = \left(\frac{2^K\sqrt{|C_1||C_2|}}{|C_1 + C_2|}\right)^{2L} \quad [8.14]$$

8.2.1.6. *Comparison by descriptor extraction*

One way of bypassing the calibration of data and registration problems is to go through a keypoint extraction phase and descriptor extraction such as the scale invariant feature transform (SIFT's) (see Chapter 5). Whether these invariant points between different images match or not is a fact that can then be used for building change maps.

8.2.2. *Change measurements between an image and a map*

The change measurements between an image and a map are useful for updating the map. It is not a comparison between two classifications (a part that will be tackled in the following section) but rather a validation of the map (and more generally of a geographic database) through observations.

This validation induces some specific problems that do not exist in the comparison of the two classifications. Here, the data from the map are essentially vectorial, the vectors then defining the marks of buildings, roads, vegetation areas. In the validation phase of the database in relation to the observations, we must therefore define the signatures that characterize attributes, and secondly, define a distance or a membership measurement helping us assess how adequate the database is in comparison with the observations.

There are two examples that can be approached here:

– Updating the map of urban areas starting from a very high resolution optical and/or radar image [POU 11].

The features used are mainly the roads (detected by a radiometric approach, in optical image as well as in radar image), the buildings (extracted through a radiometric classification in optic imagery) and the vegetation (detected by a threshold on the NDVI). The different features are then characterized by their signatures: poly-lines for the routes, polygons for the buildings. Finally, the goodness-of-fit is done by considering geographic considerations and the physical effects induced by the presence of a feature (shadow effects, geometric effects in radar, etc.). These measurements are then injected as mass functions in a Dempster-Shafer formalism (section 5.8.3).

– Updating forest cover maps in a tropical context [RAS 13].

The features used are mainly directed by the thematic context (brightness, greenness, humidity, parameters that are defined under the name of *Tasseled Cap Transform* in [KAU 76] in agriculture). The initial database works here as reference for the classification of the time series. The distance from class centers is then used as a change detector and for non-classified surfaces, and then a CVA is used.

8.2.3. *Change measurement between two classifications*

In this section C_{k,I_1} (respectively C_{ℓ,I_2}) will be denoted as a class k (respectively ℓ) of the image I_1 (respectively I_2). K_{I_1} (respectively K_{I_2}) stands for the number of classes in the classification of the image I_1 (respectively I_2).

When we compare two classified images with the aim of extracting a change detection from them, there are two main approaches that could be taken [COP 04]:

1) Comparison of class characteristics

This approach is in fact based on the measurements presented in section 8.2.1 with the only difference being that the neighboring of a current point at position (m, n) is replaced by a region or the class that the point at (m, n) belongs to.

We must then estimate the average differences, the average ratio or the distances between the distributions of the samples of the image I_1 or I_2 assigned to the classes C_{k,I_1} and C_{ℓ,I_2} with one of the distances $\mathcal{D}$ defined previously.

2) Comparison of class labels

The comparison is done in each point at position (m, n). There are several problems with this type of approach:

i) If the classifications are not supervised, how do we make the link between the different labels that represent similar classes?

$$C_{k,I_1} \quad \Longleftrightarrow \quad C_{\ell,I_2}$$

This relation can be established using hierarchical or incremental classifiers.

ii) If the classifications are not of the same type, how can we make the link of similar classes?

$$C_{k,I_1} \quad \Longleftrightarrow \quad C_{k,I_2}$$

In such a case, we must make an index assignment between the two classifications or include the different classes into meta-classes. When the sensors are of different natures, they do not in fact have the same meaning: a meadow can be characterized by a certain chlorophyll concentration in optics, whereas it is characterized by a certain roughness in the radar.

iii) How can we manage a new class between the observations?

$$C_{K_{I_2},I_2} = ? \qquad \text{when} \qquad K_{I_2} > K_{I_1}$$

When the classifier is not supervised, the encapsulation of new classes can only be done using a hierarchical approach.

iv) How can we quantify or order the changes of a class toward at least two different labels?

$$C_{k,I_1} \quad \begin{cases} \longrightarrow & C_{\ell_1,I_2} \\ \longrightarrow & C_{\ell_2,I_2} \end{cases} \qquad [8.15]$$

Here as well, only a hierarchy approach allows us to order the intensity of the different changes in a non-supervised context.

The difficulty in the change detection based on a classification comparison lies in the fact that the classification makes a tough decision regarding where a class belongs, whereas a change between two observations can be finer. It is therefore often more appropriate to integrate the notions of uncertainty or variability of classes and to refer to the viewpoints offered by the information fusion theories (Chapter 5).

8.2.4. *Changes measurements between two heterogeneous images*

When the physics of the observations do not allow us to make a direct connection between the observations coming from two different sensors, we need classification tools or information fusion tools. We then find the case presented above of classifications that must be processed together with classes that have meanings specific to each type of observation. The fusion techniques potentially allow us to make a connection between the observations. However, whereas they allow us to signify how, more specifically, the observations are in agreement, they do not always allow us to emphasize the changes, should the observations be contradictory [LEH 97, LIU 12]. In this framework, we must not mistake the classification-based change detection and the joint classification of diachronic data.

Wherever possible, it is often much simpler to produce a measurement of the changes from parameters (geometric, physical or bio-chemical changes, such as the NVDI seen later in the text), which can be extracted from the different observations.

8.3. Time series analysis

When we address the subject of time series analysis in remote sensing, we are speaking of processing tens of images at the same time (but only rarely over 100). This represents a very significant volume of data, but the techniques available in the processing of 1D signals (e.g. audio signal) do not apply. The number of samples available per time profile remains very limited, and the problem of the reliability of the estimations in the time axis remains a very current one. Furthermore, given the diurnal rotation, the conditions of observation and the availability of the sensors, the temporal information that must be characterized is not, in the majority of the cases, sampled according to the Shannon theorem. Besides the orthogonal empirical functions (OEFs) that will be approached in the following section, there are few methods that have resorted to the spectral analysis of the time series.

Here, we do not consider the notion of the object in migration nor the spatial increase of a phenomenon.

8.3.1. *Temporal series with scalar data*

As seen previously, we often make a dimensionality reduction starting from multi-spectral data, whether it is by calculating the vegetation index (NDVI) or by making a dimension reduction in order to obtain a temporal vector in each pixel.

Principal components analysis (introduced in section 5.7.2) is used by considering the series of scalar images as an image of temporal signatures. PCA then provides, along the direction of the largest eigenvalue, a band representing the largest excursion of time signatures. It provides a general viewpoint on time series, like PCA applied to spectral data. We must seek the change into a sub-space that is orthogonal to this first eigenvector. Quite often, the image provided by the second component of the PCA is related the change. However, depending on the statistic importance and amplitude of the changes, we must seek these last ones in the third band, or further.

To account for a bi-modal distribution along an eigenvector that would reveal a change class, it may be interesting to adopt an entropy point of view. The entropy of a continuous distribution is at its maximum when the distribution is of Gaussian shape (with a fixed variance) and its maximum value is:

$$\mathcal{H}_{\mathcal{G}}(X) = \frac{1}{2} \ln \left(2\pi e \sigma_X^2\right)$$

σ_X being the variance of the distribution of X, X being the (scalar) source represented by one of the components of the PCA (the second one, in the majority of cases). The more the entropy of a source tends towards 0, the more it grows far away from a Gaussian distribution and more it characterizes a significant temporal evolution.

Here the PCA is applied differently. Previously, the covariance matrix was built so that its dimension was the number (i.e. the dates) of observations in the time series. Here, we absorb the time component and the covariance matrix is then of the same size as the number of observations (i.e. the number of pixels) per image. The diagonalization of such a covariance matrix gives eigenvectors that are orthogonal in space (whence the name of OEFs). Their size is the number of pixels per image, and they can therefore be represented as a map (i.e. a map, of the same size as the images, through an orthogonal function). The projection of the complete time series on each of the orthogonal functions represents a vector of a size of the number of dates in the time series. It is therefore revealing of the *main tendencies* of the series [BJO 97]. The decomposition along the OEFs is quite often used in meteorology or oceanography, fields in which the seasonal cycles are very pronounced. We then can make a spectral analysis of these main tendencies. Furthermore, the initial matrix before the diagonalization can be defined by blocks in order to fuse several sensors in the time series [LOR 03].

8.3.2. *Patterns in long time series with scalar or vectorial data*

The exploitation of remote sensing time series is quickly expanding at the moment. We can distinguish between the approaches that are close to signal processing – which explore the time series in each pixel in order to release different contributions (e.g. general tendency, season oscillations, ruptures) as we have seen previously [VER 10] – and the approaches that have emerged from the computer science community and the search for data that will rely on pattern repetition and which we shall detail here.

Working from the ADAM sequence, which is a geometrically re-scaled, high-resolution image time sequence, a method of structural analysis presented by [LEM 09] allows us to make the best out of the very high-resolution imagery properties and to detect abrupt changes such as ploughing, snowfall, as well as slower phenomena such as the evolution of cultures throughout time. The time space between successive images being quite irregular because of the visibility conditions (i.e. varying from 1 day to 1 month), it is quite difficult to qualify these different phenomena within a classical framework of time tracking; furthermore, it is even more difficult given that the images are not radiometrically inter-calibrated because of the different sensors that are used to acquire them.

We then turn toward a structural approach, or an "object" type of approach, which is induced by the decomposition of the images into regions by building a graph whose nodes represent the regions of the images (after an image by image segmentation) and whose arcs link regions throughout time. This very complex graph that connects several thousands of regions in a time flow of a hundred images is optimized by a representation in accordance with (*Minimum Description Length* (see Chapter 5). By placing the right attributes on the nodes, as well as on the arcs of this graph, all the images are coded in a structure that highlight structural changes. The analysis of the sequence flows from the analysis of this graph. For example, having added a quantified radiometric attribute, we can research the evolution classes of the regions as frequently encountered time branches of the graph. The discretization of the space of the values allows us to improve the problem of time sub-sampling. The phenomenological exploration of the dynamic of the scene flows quite naturally from the detection of the frequent arcs, and the study of the value of the nodes allows us to identify the geometric changes in the scenery.

Furthermore, in [GUE 08] is presented a methodology inspired from the coding theory (information theory and flow-distortion theory, the *Information Bottleneck* principle) for optimally extracting the significant information from a sequence of images and giving it a compression with controlled losses of the space-time events that compose it. These methods are based on the random Gibbs-Markov fields, and they lead to indexing the space-time structures, which is a compact and sparse representation of the information content, and whose transported information quantity is controlled. These works are completed by establishing a Bottleneck multi-information criterion, which gives a more universal aspect to information extraction from the time series, and which is built within the Kolmogorov information theory.

There are non-supervised approaches [JUL 11] that allow us to detect the frequent sequential time motifs. There are measures on the surface covered by the pixels that present these patterns, and their spatial connectivity (calculated, for example in 8-connectivity) then allow us to select the more plausible patterns. Crop supervision apps or coastal deformation analysis apps have been developed (see Chapter 10).

[PET 12] proposes the *dynamic time warping* (DTW) similarity for manipulating irregularly sampled time series, and calculates an optimal alignment between radiometric profiles. Figure 8.3 illustrates the DTW principle: rather than assessing the quadratic distance $\sum_i |s_i - t_i|^2$, we first evaluate $|s_1 - t_1|^2$ while ensuring that $|s_1 - t_2|^2$ or $|s_2 - t_1|^2$ are not smaller. More generally, before using the elementary distance $|s_i - t_j|^2$, we make sure that $|s_i - t_{j+1}|^2$ or $|s_{i+1} - t_j|^2$ are not smaller in that case, it is the smallest one that is used. Adaptations of this similarity to the image time series are introduced, for processing multi-dimensional profiles, imposing association constraints (e.g. seasonal constraints) and for processing the missing data. The main difficulty when we process offset or expanded time series is being able to define an average time series, so that we use different time series in the same classification operation, for K-means in particular. This has been particularly defined in [PET 12].

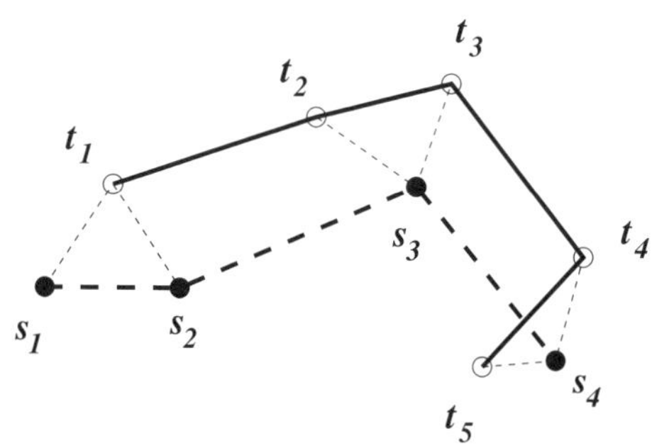

Figure 8.3. *Illustration of the distance calculation by dynamic time warping. The points in the first series are indexed by t_i and connected by continuous arcs whereas the second series (to be compared with the first one) is indexed by s_i and reconnected by dashed lines. The dotted lines show the possible associations*

8.4. Conclusion

In this chapter, we have approached the problem of change detection based on remote sensing data. We have seen that this problem can be more or less complex depending on the sources of information used: images issued by the same sensor and the same acquisition conditions, images of the same sensor in different conditions (different acquisition angles, different resolutions, etc.), images from different sensors but of identical modality, images acquired by different modalities, heterogeneous data, etc. With the multiplication of the sensors, the increase of acquisition rates, and the availability of the archives, change detection approaches should be able to simultaneously account for different sources and to integrate the new information in a dynamic process.

The exploitation of long time series is often connected to a specific field of application such as detection and motion sequence (Chapter 10) or the assimilation of meteorological or oceanographic data (Chapter 11). The democratization of the access to data archives and the constellation of satellites (such as Cosmo-Skymed, Pleiades) will trigger new requirements for efficiently processing these masses of data. Data search and assimilation will most definitely bring some answers.

Chapter 9

Elevation Measurements

9.1. Optic stereovision

9.1.1. *The principle of stereoscopy*

Stereoscopic vision is based on the difference perceived between two images taken from two different points of view. It has existed since the emergence of photography. In 1856, the first photographic camera emerged, equipped with two objectives and allowing us to take several, significantly shifted pictures of the same subject. Stereoscopic observation allows us to obtain a three-dimensional (3D) vision of the object being photographed. At first, we had terrestrial stereoscopic observation, then airborne stereoscopy and then space stereoscopy. The flight plans predict overlapping areas between the exposures taken along the strips (longitudinal overlap) and between the strips (lateral overlap). Regarding space stereoscopy, high-resolution optic systems are rare (HRS on SPOT-5, Alos, Cartosat) and rather non-existent on current satellites. However, all satellites with very high resolution (lower than 1 m) can use their agility (see section 2.2) to acquire couples along the track, because of a stereo or multi-stereoscopic operational mode.

Since the first Earth observation satellites such as SPOT-1 in 1986, several stereoscopic acquisition techniques have been implemented. Since the sensors are not capable of doing simultaneous front and back sights, because stereoscopy is diachronic, we must wait for the satellite to pass next on an adjacent route in order to get the second image of the couple (see section 1.3). The main inconvenience of this

---

Chapter written by Michel ROUX, Olivier DE JOINVILLE, Florence TUPIN and Jean-Marie NICOLAS.

couple, whose two images were taken at different dates, is that the lighting conditions may have varied.

The impractical diachronic method was replaced in the late 20th and early 21st Centuries by agile satellites, such as IKONOS in 1999, and the putting into orbit of some stereoscopic sensors (HRS on SPOT-5 in 2002, ALOS, Cartosat P5). We must note that the recent, more agile satellites allow the acquisition of multi-stereoscopic data: more than 12 images of the same area in a single journey with Pléiades, for example. The HRS instrument on board the SPOT-5 does the simultaneous acquisition of stereoscopic couples on large surfaces (120 km $\times$ 600 km). The main applications are in telecommunications, institutional cartography, airport and defense applications. It has 2 telescopes, one pointed forward and the other toward the back of the satellite's vertical, with a 20° sight angle. The exposure interval along the track is of 90 s.

Reconstructing an object or the shape of terrain means searching the position of the intersection points of pairs of homologous rays. In Figure 9.1, the position of point M will be estimated in 3D, knowing the exact position of the intersection of the two homologous rays leaving from the image to the left of the couple (point m_1) and from the right image (point m_2), respectively, through the two optic centers of the two images (C_1 and C_2). If the two rays cross at the surface of the ground, the altitude of M will be right.

The quality of digital 3D models obtained by stereoscopy will therefore depend on:

– the quality of the couples of homologous points that we can establish between the two images, and

– the precision of the exposure parameters that allow us to trace the optic rays coming from the points in the image.

9.1.2. *Epipolar geometry*

9.1.2.1. *Matrix sensor*

The epipolar geometry is a property verified by the image couples acquired with a matrix sensor, which ensures that the exposure parameters, whether they are intrinsic or extrinsic, are the same for all the points of an image. This property does not exist for satellite images acquired with a *pushbroom* sensor, for which the optic center and the sight will evolve during the acquisition of the entire image.

Epipolar geometry is entirely based on the co-planarity of a set of points: if we consider a point M in the scene, then the point M, the two optical centers C_1 and C_2 and the projections m_1 and m_2 of the point M in the images 1 and 2 are co-planar.

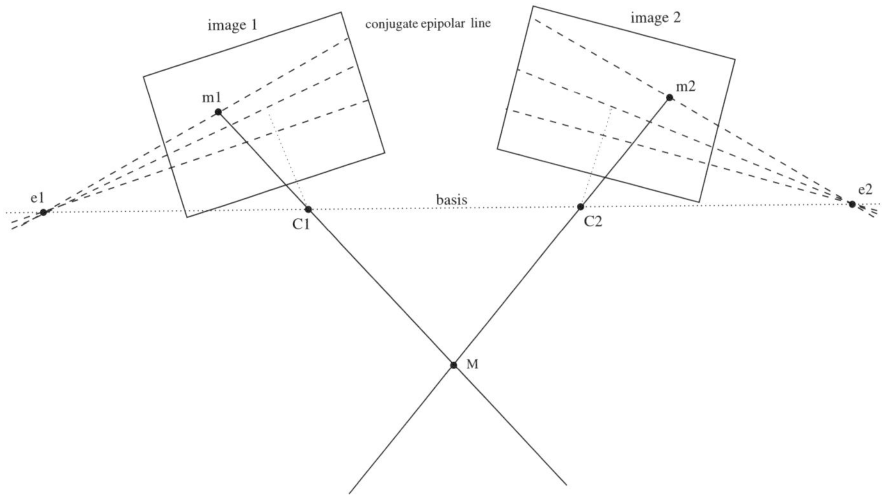

Figure 9.1. *Epipolar geometry: the co-planarity of point M, optical centers C_1 and C_2 and images m_1 and m_2 allows us to define straight lines, through the intersection with the two image planes; these lines are called "epipolar" and m_1 and m_2 are situated on them. These straight lines cut the straight line of the optic centers C_1 and C_2 at the epipoles e_1 and e_2, which are key points in the tri-dimensional reconstruction of the image*

In the search for homologous points, one widely used strategy is to consider a point m_1 of image 1, and to search for its correspondent in image 2. The epipolar geometry allows us to transform this problem from a 2D search (in the complete image 2) to a 1D search because the correspondent of point m_1 is necessarily situated on the straight line, intersection of the focal plane of image 2 and of the plane formed by points m_1, C_1 and C_2. This straight line is called the epipolar line, and is conjugated to the point m_1. Symmetrically, we can search the homologous point of an m_2 point of image 2 solely along the conjugated epipolar straight line, intersection of the focal plane of image 1 and of the plane formed by the points m_2, C_1 and C_2.

The set of epipolar straight lines of an image forms a beam that converges in a point e, the epipole, which is the intersection between the focal plane of the image and the baseline, the straight line going through the two optic centers. In the more particular case where the focal plane is parallel to the baseline, the epipolar straight lines are parallel, and the epipole is rejected *ad infinitum*.

Starting from an acquisition geometry (focal planes that are not parallel to the baseline), we can resample the couple of stereoscopic images in a focal plane that is parallel to the baseline, so that the lines of the images correspond to epipolar straight lines.

In this particular geometry of the images, the search for the homologous points will consist of searching, for a given point m_1 of the first image, for its correspondent m_2 on the same line of the second image. This geometry allows us to improve the calculation time because it then becomes possible to use gliding window algorithms for calculating the correlation [DEV 97].

9.1.2.2. *Push-broom sensor*

Numerous satellite sensors are of the *push-broom* type. This means that the sensor is in motion during image acquisition. The notion of epipolar geometry, in the strict sense, therefore does not apply for these images. However, it is possible to resample the images in a *quasi-epipolar* geometry while using polynomial rational functions [ZHA 05, WAN 10, ZHA 08]. The experiments show that errors of the order of a quarter of a pixel may be obtained on the vertical parallax of the images in epipolar geometry.

Other approaches do not need resampling of images in epipolar geometry. The search for the correspondent of a point m_1 of the first image is implemented by vertically sampling the ray of light issued from this point, and by reprojecting these 3D points in the second image. This method supposes that the two images have the same orientation in relation to North. This is generally the case for satellite images.

9.1.3. *Searching homologous points*

9.1.3.1. *Similarity measurements*

We have already seen similarity measurements in Chapter 2 on registration and in Chapter 8, where it is used for change detection (section 8.2.1). Here, we will reiterate the ones that are frequently used in stereovision. Let us note that in this chapter we must maximize the similarity (or minimize the "distance") in order to find the homologous points.

As we have seen, we can distinguish between two families of similarity measurements when comparing two images (here the neighborhoods centered on the two pixels considered). The first one consists of comparing the pixel values, two-by-two when these two images are superimposed, the second one tries to estimate the probabilistic dependence of the two images, and is based on the gray level distribution within the neighborhoods.

In the first family, the normalized centered cross-correlation is most often used. (see equation [2.7]):

$$\tilde{\mathcal{D}}^{corr}(I_1, I_2)(m,n) = \frac{1}{KL}\sum_k\sum_l \frac{(I_1(m-k,n-l)-\mu_1)(I_2(n-k,m-l)-\mu_2)}{\sigma_1\sigma_2}$$

We have seen that it is equivalent to minimizing the local average quadratic error after the centering and the normalization of the thumbnails.

With regard to the approaches that search for a statistic dependence between the two signals, one frequently used criterion is the maximization of the mutual information (see equation [2.9]).

9.1.3.2. *Search strategies*

Here, we will look at the dense pairing search between the two images. For each point in image 1, the initial strategy consists of seeking its correspondent in image 2. The pairings found can be stored as a disparity map. Let $m_1 = (i, j)$ be a point in the image 1 and $m_2 = (i', j)$ its homologue in image 2 (we suppose here that the images that were resampled into epipolar geometry, the two points therefore have the same ordinate $j' = j$), the disparity of the point m_1 then values $d(i, j) = i - i'$.

The strategies for searching for the correspondence between the two images may vary depending on the constraints used to carry out this search and depending on the parameters of the method.

– Epipolar constraint: the homologous points must be situated on two conjugated epipolar straight lines. For the images resampled in epipolar geometry, the correspondent of a point of one image is necessarily found on the same line of the other image. The strategy then consists of comparing the neighborhood of a point in image 1 with the same neighborhood that we glide along the corresponding line in image 2. In each position, we calculate the similarity measurement between the two neighborhoods, which we preserve as a similarity curve (correlation curve). The homologous point corresponds to the maximum of this similarity curve.

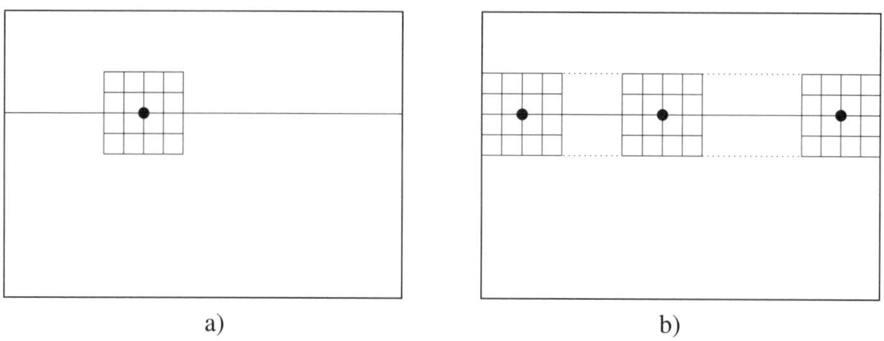

Figure 9.2. *Pairing search in epipolar geometry: for each pixel of a reference image a), which has a small window describing its context, we search its homologue in b), by gliding identical windows along the respective epipolar line*

– Correlation threshold: the pairings for which the similarity measure is lower than a given threshold are eliminated from the disparity map. This constraint is particularly efficient for eliminating the pairings found in parts of image 1 not seen in image 2.

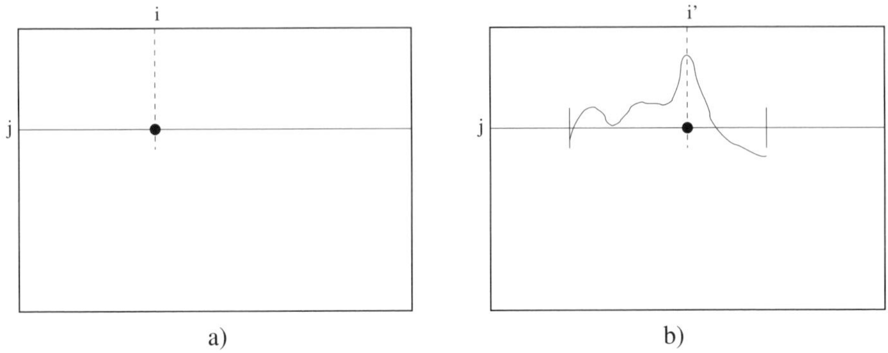

Figure 9.3. *Disparity interval and correlation curve: to the intersection point of the coordinates (i, j) in a), corresponds a correlation function obtained by moving the window. This function presents a maximum at the point i', j. The disparity $d(i, j)$ is measured by $i - i'$*

– Disparity limits: we can limit the search for the homologous point to a given disparity interval. Instead of being done along the entire conjugated epipolar line, searching for the homologous point is limited to an interval that will glide along this line following the horizontal motion of point m_1. This constraint is the same as limiting the distance to the *baseline* of the rebuilt 3D points (see Figure 9.3).

– Order constraint: for opaque surfaces, if we consider two points positioned on the same line of image 1, then their correspondents are placed following the same order on the conjugated line of image 2. This constraint is difficult to apply locally and can only be considered by globally optimizing the correspondence between epipolar lines. Dynamic programming is a good example of a method that specifically considers this constraint of the order [OTH 85].

– Correlation window: the size of the neighborhood used for assessing the similarity of two points strongly influences the quality of the results. A small correlation window (e.g. 5×5 or 7×7 pixels) provides precise – although noisy – disparity maps, and conversely, for larger correlation windows (11×11 or 13×13 pixels). The use of the normalized centered cross-correlation tends to over-estimate the borders of the objects present in the scene of a half-width of the correlation window. There are several approaches that have been proposed in order to limit this *adherence* phenomenon: a Gaussian weighting of the correlation window, a geodesic weighting depending on the contours present in the correlation window, image segmentation and limitation of the correlation to the pixels belonging to the same region.

– Search symmetrization: the previously described procedure (which, for every point m_1 of the first image, seeks its homologue m_2 in the second image) can be made symmetrical by searching for the correspondent m_1' of m_2 in the first image. The correspondence (m_1, m_2) then becomes invalid if the distance between m_1 and m_1' is too large (usually larger than 2 or 3 pixels).

– Hierarchical approach: starting from a decomposition of the couple of images in Gaussian pyramids, the pairing search is firstly carried out at the smallest resolution. The search at the immediately higher resolution is guided by the disparity map at a low resolution: the search of a homologous point is limited to the immediate neighborhood of the point found at the previous stage. The method is iterated until full resolution is reached. Besides gaining in speed, this approach reduces the number of false pairings. However, the fine details of the scene risk becoming lost in the pyramid decomposition.

9.1.3.3. *Global optimization*

We have previously seen that the different constraints show up in the search for homologous points, and that it is difficult to consider them pixel by pixel; a global approach is therefore needed for ensuring a global coherence of the pairings and optimizing the rebuilt surface. We present here two strategies that ensure a global optimization.

– Dynamic programming: epipolar geometry favors the processing of the stereoscopic images line by line. The search for the pairings can then be modeled as a problem of searching the lowest cost path in the graph formed by all the potential couples of pixels between the two conjugated epipolar lines (the connectivity of the graph is directly issued from the connectivity of the pixels in the images). The order constraint seen previously imposes that these paths be monotonous. Dynamic programming provides a linear complexity solution for obtaining the optimum path. The coherence between the lines of the images can be ensured by restricting the disparity along the contours crossing to the epipolar lines [OTH 85, BAI 97].

– Relaxation and energy minimization: the study of the local interactions of the pairings started with simple relaxation techniques, where the safest pairings (the cross-correlation is generally used directly as a safety measure) reinforce, in the iterative approaches, the weak, but coherent, neighborhood pairings [FRA 99, LEL 98].

Using the Bayes relation and Markovian approaches, probabilistic relaxation models the interactions between the neighborhood pairings more rigorously, and it also models the confidence in these same pairings as an energy that needs to be minimized. This energy is traditionally made of a term attached to the data, related to the dissimilarity between the paired points, and a regularization term, which must integrate the surface properties that we seek to reconstruct [NAR 10].

In general, there is no analytical solution to this energy minimization problem, and up until recently, the solutions proposed (ICM, simulated annealing in finite time, etc.) have been suboptimal. Cox and Roy have shown that it is possible to model this optimization problem by searching a minimal cut in a graph [ROY 98, ROY 99, PON 07, PIE 06, PIE 07].

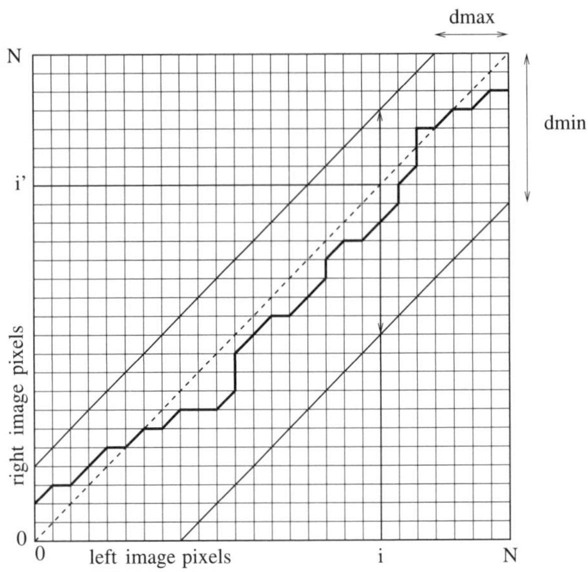

Figure 9.4. *Dynamic programming: the authorized motions correspond to pixel pairings (diagonal motions) and to hidden parts (horizontal and vertical motions), the correspondent i' of the point i is comprised between the values $i - d_{min}$ and $i + d_{max}$, the dynamic programming allowing us to find the optimal route with a linear complexity*

9.1.4. *Reconstruction of the digital terrain and elevation models*

The previously described tools allow us to obtain a disparity map, which preserves the correspondences found in the two images, and a cloud of 3D points, which are the intersections of light rays coming from homologous pixels.

On the basis of these data, we can calculate a digital surface model (DSM), by resampling the points. The DSM gives the elevation of each point as it is seen from the sky, on a regular, geo-referenced grid. The DSM goes, therefore, above the building and the vegetation. These digital models can equally be obtained as a triangulated irregular network that rests on the 3D points, or as a set of level curves.

Obtaining a digital terrain model (DTM), which gives, on the same geo-referenced grid, the elevation of the ground needs to eliminate the above-ground data (mainly the buildings and the vegetation) of the DSM. One approach that is often proposed in the literature is to implement a morphological opening on the DSM with a structuring element of a size that is large enough to eliminate the buildings and the scattered vegetation. The limits of this approach regard the dense forest areas for which the elevation of the terrain remains difficult to access in aerial or satellite optic imagery and the details of the terrain that will be eliminated if they are also smaller than the structuring element [WEI 95]. Classification techniques between ground and above-ground data have also been studied, techniques which use the discontinuities in the DSM in order to separate the two classes, the objects of the classification being the points of the DSM or the regions issued from segmentation [BAI 97].

9.1.5. *Multi-view stereoscopy*

When more than two images of the same scene are available, the generation of a 3D digital model is an information fusion problem that will have to be tackled by considering the complementarity, the redundancy and the contradiction of the sources of information.

Within stereovision, the fusion of information coming from several images may come up at different levels of the processing chain: at the end of the chain, we can merge the digital models obtained from several image pairs, in the middle of the chain, by merging the correlation curves and at the beginning of the chain, via a volumic approach and a reprojection of voxels in the images [ROU 01].

A strong constraint connected to the fusion of data is the requirement to have these data in the same referential. In what follows, we will see what consequences this constraint has on the methods that can be implemented.

9.1.5.1. *Fusion of digital surface models*

A merged DSM can be obtained by combining the set of DSMs that emerged from each couple of images considered, provided that they had been estimated on the same georeferenced grid. The fusion is done at the level of the elevations: for each point (x, y) of the horizontal grid, the median value of all the available elevation values for this point protects against the aberrant values in the initial DSMs.This method is very efficient for filling in the hidden parts that can appear when only a pair of images is used.

9.1.5.2. *Fusion of correlation curves*

The fusion stage can be imagined much earlier, combining the correlation curves that correspond to a point of an image seen in several other images. For this type of method, it is often necessary to define a master image and secondary images.

As we have described previously, the correlation curve is a function of the disparity, which is intrinsic to the couple of images. It is therefore not possible to directly merge the initial curves. One first possibility is to resample these curves depending on the elevation and then to merge them again [ROU 01]. A second possibility is to sample the light ray in elevation emerging from the point considered in the master image (see Figure 9.5) and to calculate the correlation curves directly by reprojecting these 3D points in the secondary images [HER 04].

The fusion operator used is usually the *average* of the correlation curves, which has, however, the disadvantage of supposing that the point considered in the master image is visible in all secondary images. By contrast, a *maximum* type of correlation curves operator would not have this inconvenience, but it will not address the redundancy between the different images and will give a noisier result.

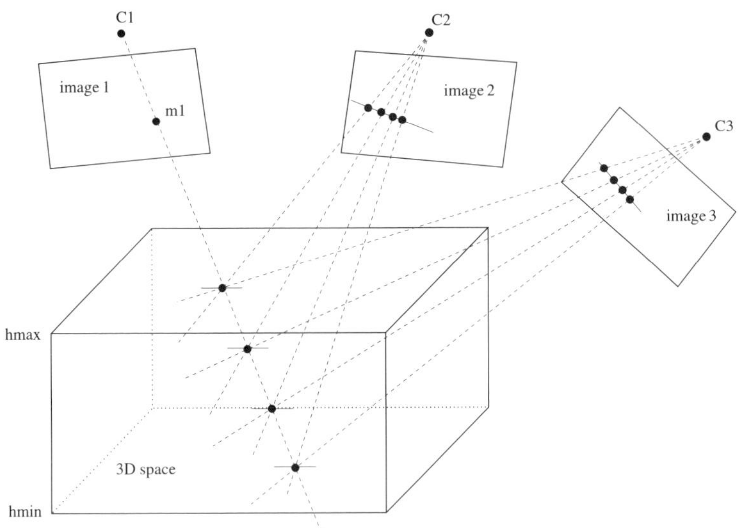

Figure 9.5. *Correlation curve fusion: the ray coming from point m_1 is sampled vertically between the values h_{min} and h_{max}, the 3D points are reprojected in the other images and a correlation curve is estimated for each of these images, the curves are then merged and the optimum altitude is estimated on the resulting curve*

9.1.5.3. *Volumic approach*

This approach consists of explicitly going through a volume modeled by a 3D grid that is sampled regularly. Each voxel of coordinates (x, y, z) is projected in the images available in a position $(u_i, v_i)_{i=1..N}$, where N is the number of images (see Figure 9.6).

A multi-image similarity criterion that considers the radiometry of these pixels or of their neighborhoods gives a measure of the quality of the voxel considered. The implementing of this approach is facilitated by going through the volume plane by plane, orthogonally to the vertical direction, and by resampling the images along a horizontal grid for each elevation.

The multi-image similarity criteria proposed in relevant literature vary depending on how valid they stay when confronted with hidden parts in some images [TSA 83, PAP 00, ROU 01].

One of the advantages of this approach is to avoid having to designate a master image, all the available images being processed in a similar fashion.

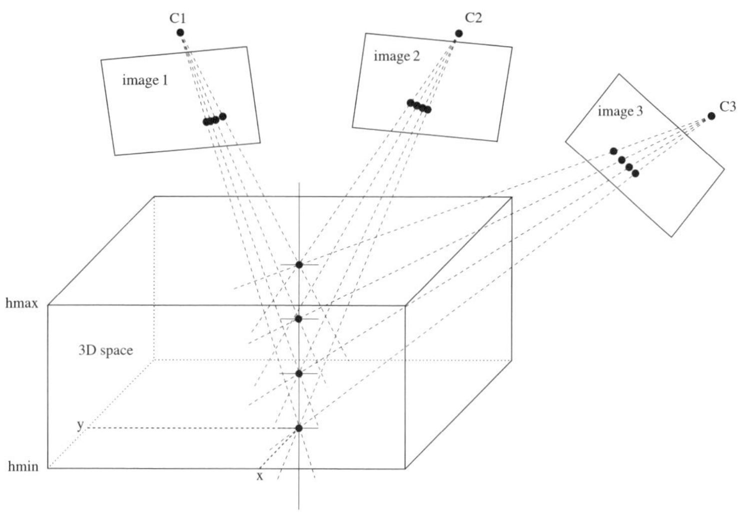

Figure 9.6. *Volume approach: the 3D space is regularly and explicitly sampled, each 3D point being reprojected in the images and a similarity multi-view measurement being estimated; the optimum altitude is then determined for each planimetric position* (x, y)

9.2. Radargrammetry

Radargrammetry refers to the set of stereovision techniques such as photogrammetry, but that use radar images. As in the previous section, we can use the correspondence between homologous points between two radar data to go back to a planimetry and altimetry piece of information, using the acquisition equations in each of the images.

The radargrammetric potential was studied very early [ROS 68, LAP 72], and numerous applications in satellite imagery have been developed since [RAM 86], especially with the radar of the Canadian RADARSAT satellite [TOU 00]. Nevertheless, this relief replication technique is still used much less than interferometry in radar imagery.

9.2.1. *Geometric aspects*

First, we will illustrate radargrammetry in the simple case of parallel trajectories, before moving on to the more general case. If we consider two acquisitions with perfectly parallel trajectories and with a null "*Doppler centroid*" (the boresight of the two antennas is then perfectly perpendicular to the route of the satellite), a point M can be found on the circle of radius S_1M for the sensor S_1, and on the circle of radius S_2M for the sensor S_2 (see Figure 9.7). These circles give images of the point M located in the positions M_1 and M_2 in the two acquired images.

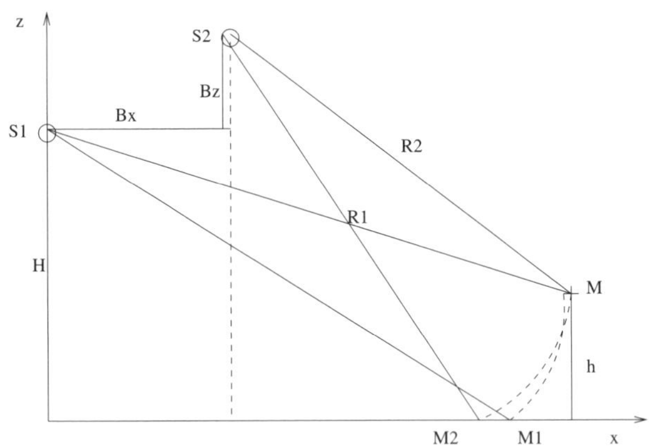

Figure 9.7. *Radargrammetry principle in the case of parallel trajectories with a null Doppler; the disparity M_1M_2 is directly connected to the height h of point M (it is null for $h = 0$, which corresponds to $M_1 = M_2$ and increases with h)*

The relation disparity/height is then very simple:

$$d = h\left(\frac{1}{\tan(\theta_2)} - \frac{1}{\tan(\theta_1)}\right) \qquad [9.1]$$

with d being the disparity ($d = M_2M_1$), h the height of the point M and θ_i the angle of local incidence to the point M_i. On this simplified example, we verify that a ground point will have a null disparity, and that the disparity increases depending on the elevation of the point.

In the more general case where the trajectories of the satellite can, in theory, be anything, the radar equations give a system of 4 equations for 3 unknowns, which can be solved with a least squares approximation. As seen in Chapter 1, the coordinates (p, q) of a pixel of the image provide us with:

– the emission time t of the impulse for which the target is at the center of the beam (with $k = \frac{1}{PRF}$, PRF being the pulse repetition frequency):

$$t = p\,k + t_0 \qquad [9.2]$$

– the sensor-target range r at the time t due to the equation:

$$r = r_0 + q\,\Delta_r \qquad [9.3]$$

where r_0 is the distance corresponding to the first range cell of the radar image and where Δ_r is the range sampling.

The position of the radar at the instant t is given by the vector $\overrightarrow{OS_r}$. The first radar acquisition equation is obtained by writing:

$$\left|\overrightarrow{S_r M}\right|^2 = r^2 \qquad [9.4]$$

This is the equation of a sphere of radius r and centered in S_r. The second equation, relative to the radar boresight, is written as:

$$\overrightarrow{S_r M}.\overrightarrow{V_r} = \frac{f_D\,\lambda}{2}\,r \qquad [9.5]$$

where V_r is the radar speed at the moment t, and f_D is the Doppler frequency. This is the equation of a cone. With the coordinates of the same point seen in both images, we obtain a system of four equations and three unknowns, which can be solved through the method of least squares approximation, in order to find the one solution that satisfies the commonsensical constraints on the position of the target.

As in the case of photogrammetry, we must compromise when choosing the differences between the acquisition angles. The more significant these differences are, the more they increase the precision of the measurement. In return, the search for correspondences becomes more difficult because the signals in the two images look less similar.

Radargrammetry, which exploits the shifts of the pixels in the images, remains on a different principle from the one in interferometry, which uses phase differences between perfectly registered pixels (see Figures 9.8 and 9.9). The registration of two images is possible as long as the difference in the angle is small, that is as long as the baseline (i.e. distance between the two sensors) is small. Beyond a specific basic value, we therefore move in a radargrammetric configuration because we cannot globally calibrate the images in order to do a point by point phase difference.

In this situation, the variations of the altitude h of the points and the baseline between the two sensors create the images of the two points in different radar cells: it is precisely this radar cell change that allows us to recover the value of h in radargrammetry, whereas in interferometry these two images are in the same cell and the altitude of the point is deduced from the phase difference. Consequently, the radargrammetric baseline must be larger than the critical interferometric baseline, which is something we will come back to in the following section. From this analysis, we also deduce that there is no point in using complex data in the radargrammetric approaches because the proper phases do not correspond.

Obtaining the relief based on a couple of radar images goes through the following stages [SIM 02]:

– putting the two images in epipolar geometry in order to reduce the search field of the homologue of a point; the two images are resampled so that they follow the conjugated epipolar lines;

– searching the homologue of each pixel of image 1 in image 2 and calculating a disparity map;

– calculating the 3D coordinates of the pixels on the basis of acquisition parameters.

The rectification stage in epipolar geometry consists of transforming the geometry of images in such a way that the homologous points are on a line, which allows us to significantly accelerate their search. For this reason, a point of the image 1 is projected in image 2 with different heights, producing a curve. This curve is then reprojected in image 1 with a constant altitude. The two curves then correspond to the lines of the rectified images.

9.2.2. *Correspondence*

As in the case of optic imagery, the problem is the detection of the homologous points in the two radar images. The Euclidean distance or the normalized centered correlation between the patches taken in the two images are often used. Even in the case of the radar data where the noise is multiplicative, these measurements can give satisfactory results (particularly if the data have been de-noised). They also allow us to reuse the processing chains developed for optical photogrammetry and therefore to perform radargrammetry at the lowest possible cost.

However, if we wish to consider the specificities of the noise in the radar data, more adequate measurements might perform better. More particularly, a generalized likelihood test is very efficient: a test trying to see if two pixels have a higher chance of being explained by either a common or a different subjacent parameter (see section 8.2.1.5) [?].

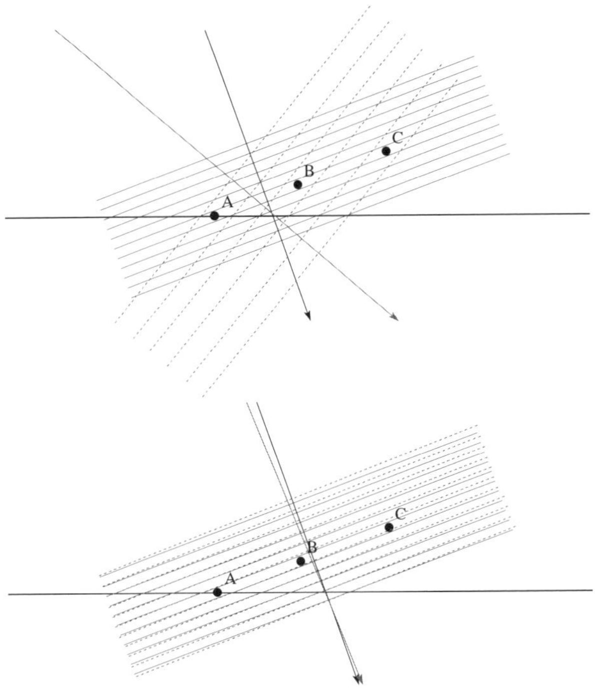

Figure 9.8. *Radargammetry and interferometry principles. The radar cells of the two sensors 1 and 2 are indicated with full and dotted lines, respectively. At the top, the radargrammetry principle: the difference in the incidence angle between the two sensors is significant; the elevation of points B and C then triggers a change in the radar cell for the sensor 2; thus, the shift in the radar cell (disparity) serves to recover the elevation information. At the bottom, the interferometry principle: the difference in the incidence angle is very small, and the points remain in the same radar range cell for sensors 1 and 2. Although for sensor 1, the position of the points A, B and C is at the center of the radar cell, for sensor 2 (dotted lines), the points are shifted within the cell: at the beginning of the cell for point A, at the center for point B, and at the end of the cell for point C; this shifting in the radar cell is directly connected to the elevation of the points and generates a phase variation depending on the elevation*

When acquisition conditions are very different (e.g. an ascending pass and a descending pass), pairing these points can become very problematic. This is related to the sensitivity of the radar signal to acquisition geometry. We then generally restrict ourselves to quasi-parallel trajectories with same side sight.

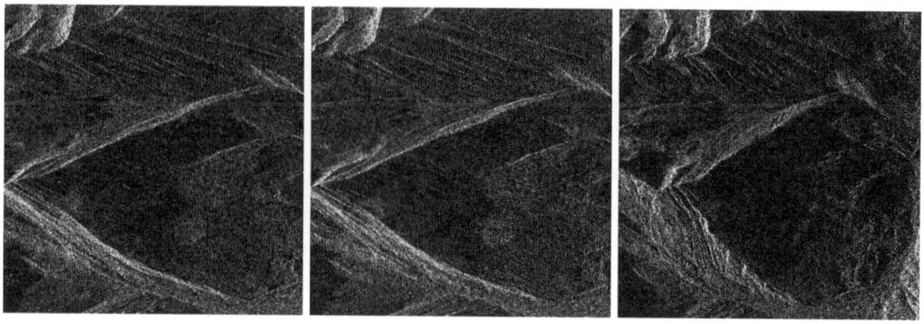

Figure 9.9. *Illustration of radargrammetry and interferometry on the COSMO-SkyMed (CSK) ascending radar images of the Saleina glacier in Switzerland©ASI. The slope is opposed to the sight of the radar (sensor to the left of the image). The two left images have an incidence angle of 26° and are in interferometric configuration (there are no significant amplitude differences : only the phases differ, which allows to build the interferogram of Figure 9.12). The image on the right has an incidence of 49° and is in radargrammetric configuration with the two previous ones: the higher the altitude of the points, the more different are their positions in the horizontal direction on the images*

One of the most significant problems in radargrammetry is the low density of the paired points between two radar images. More particularly, in an urban area and at high resolutions, the radar images are particularly sensitive to the wall/ground corner of the buildings, which backscatter very strongly. However, the really high structures are sometimes not very visible on these data. For example, a flat roof does not respond nearly at all toward the radar and therefore it is difficult to calculate its height in radargrammetry.

Different approaches have been proposed for making the disparity maps more dense. For low-resolution images (more then 10 m), multi-scale approaches allow us to have an interpolation of measures. In the case of HR or very high resolution (VHR) data, the pairings of points are coupled with object detection methods. The use of optic data can also help to make the disparity maps more dense. The approaches used will depend on the desired application and resolution [TUP 05].

9.3. Interferometry

One of the great successes of radar imagery is interferometry. It allows us to build large-scale DTMs, based on the phase difference between two images. The objective of the satellite TanDEM-X, a copy of TerraSAR-X on a close orbit launched in 2010, is to produce DTMs with a relative vertical precision that is lower than 2 m with a spatial sampling of 12 m on the entire surface of the globe, using interferometry. The general principle of interferometry is to exploit the image phase to retrieve the distance between the sensor and the imaged points (Figure 9.8). If, because of the intrinsic phase of a target (called *proper phase*), we cannot do the calculation directly, as soon as we have two acquisitions, the *phase difference* will allow us to retrieve the distance information and therefore to points elevation.

9.3.1. *Geometric aspects*

The principle of interferometry, studying the interferences between two coherent sources, is well known in numerous fields. Because we have two coherent sources, we can analyze the combinations between the two waves thus generated: because the sources are coherent, the waves are combined by considering phase and amplitude at the same time, whereas in the case of incoherent imagery we only add the amplitudes (i.e. the energies). The first experiences in interferometry go back to the beginnings of coherent optics, that is, the beginning of the 19th Century: the wavelengths are submicronic, the most notable experiences were those of the Young fringe, where all we need to do is to pierce two holes in a screen within a distance of millimeters, to be able to see interference fringes on a screen that is positioned within a meter.

The shift toward radiofrequency waves does not pose any problems; on the contrary, because they offer large coherence lengths, they help to build interferograms. However, we must consider the scale change since the wavelengths are a few centimeters long (see the case of current imaging radar systems)[1]. The problem is how to have two coherent sources available in the orbit for building the interferograms. If TerraSAR-X, within the TanDEM mission, has a copy of itself, the scientific community did not wait for two identical satellites to become available in order to make interferograms. In the first shuttle imaging radar B (SIR-B) acquisitions in the 1980s, it exploited one of the characteristics of orbit flights: their stability due to the quasi-spherical shape of the Earth, and the Kepler laws (see section 1.2) in order to make a satellite orbit after a cycle very near from the initial orbit. Using this approach, it becomes possible to acquire coherent datasets and build a synthetic interferogram that best imitates what could have been obtained using two simultaneous sources.

Thus presented, it seems that the problem is oversimple! This is not quite true. On the one hand, the passage of a satellite on an orbit near the previous one is never easy and requires excellent control of the satellite. On the other hand, this revisiting of the site (under identical acquisition conditions: in particular, with the same local incidence) can generally only be done in a relatively high time delay: 11 days for TerraSAR-X and a month for the other SAR satellites (25 days for RADARSAT,

1 Let us note that we can place ourselves in other propagating contexts in order to do interferograms, as in underwater acoustics, for example.

35 days for ERS, 44 days for JERS, 46 days for ALOS)[2]. In the meantime, there can be deep modifications of the studied area completely changing the backscattering coefficient and casting doubt over the possibilities of building interferograms.

Let us note that we speak of multi-pass interferometry when we use two different sensor orbits, the time delay between the two orbits being more or less large. When we can use two antennas simultaneously in reception, for example fixed on an airplane, or on an articulated arm as on the SRTM shuttle of NASA, we speak of mono-pass interferometry. The configuration TanDEM-X of the DLR is nearer to the so-called "mono-pass" conditions, thus avoiding the decorrelation due to the surface changes throughout time and to the variations in the meteorological conditions, as we will see later.

9.3.2. *Topographic fringes*

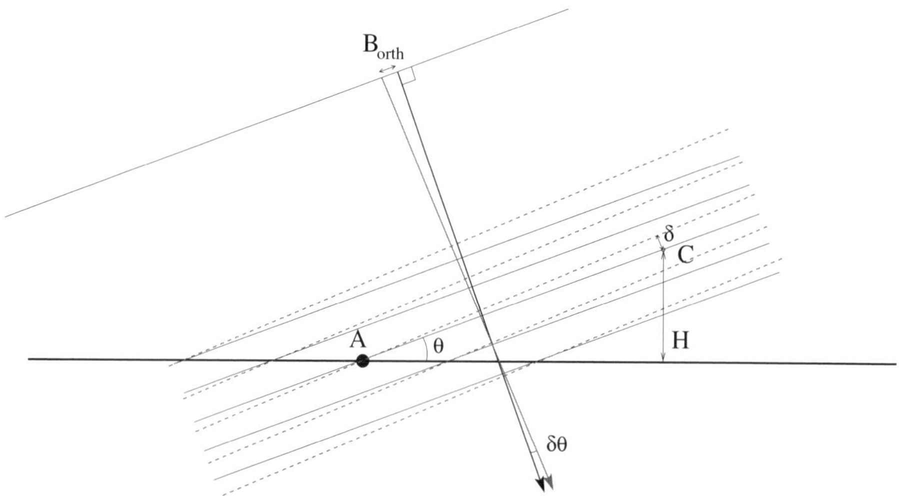

Figure 9.10. *Calculation of the time difference in the interferometric configuration*

Using a simple geometric calculation we will show how a slight difference in the incidence angles of the two sensors generates a difference between the distances and therefore the measured phases (Figure 9.10). Under the hypothesis of a flat ground, the two pieces of data were resampled so that the ground cells have the same size.

2 The complete constellation of the 4 Cosmo-Skymed enables systematic visits every 8 days. Let us also note that, in archive data, we have a revisit of 1 day during the tandem acquisitions ERS1-ERS2.

Point A on the ground is on the range cell edge of the two sensors, whereas point C is on the range cell edge for the "full line" sensor and at the center of the range cell for the "dotted line" sensor.

Let us consider an area imaged along two very near incidence angles: θ for a first sensor ("master" image) and $\theta' = \theta + \delta\theta$ for a second sensor (we can correctly consider that $\delta\theta$ is more or less constant along the swath and the distances between the observation point and the satellites are equal to R). In a preliminary stage, the "slave data" were resampled so that its ground cell is identical to the ground cell of the master image: this stage is done assuming the ground as flat. For every point on the ground (therefore of a null altitude), the positions on the master image and the slave image are identical. We define the orthogonal basis between the two sensors as:

$$B_{orth} = R\,\delta\theta \tag{9.6}$$

A simple geometric reasoning allows us to write (with $\sin(\delta\theta) \simeq \delta\theta$):

$$\frac{B_{orth}}{R} = \frac{\delta}{AC} = \frac{\delta \sin\theta}{H}$$

We therefore have:

$$\delta = \frac{B_{orth}\,H}{R\sin\theta} \tag{9.7}$$

We then see that the time difference δ varies proportionally to the altitude H of the point.

The sensors do not measure the distances directly, but rather they measure the phase of the points. The phase difference between the "master" sensor and the "slave" sensor is linked to the time difference by: $\Delta\phi = 4\pi\frac{\delta}{\lambda}$ in the multi-pass case (and a factor 2 instead of the factor 4 in the mono-pass case). Hence, the relation connecting the phase difference to the height:

$$\Delta\phi = 4\pi\frac{B_{orth}\,H}{\lambda R\sin\theta} \tag{9.8}$$

This quantity is known as modulo 2π, triggering the so-called "fringes" (Figure 9.12). To find the relief, we must "unwrap" the fringes, that is the 2π multiples in each of the points.

Let us note that the interferometry allows us to measure variations in the distance at the scale of the length of λ, that is several centimeters, and this given that the spatial resolution of the sensor can have several meters.

9.3.3. *Orbital fringes*

In the previous reasoning we have assumed that the two images have been resampled so that the ground cells are perfectly equal. We will assume in this section that they have not been resampled: the two images then only have the same range cell (the two images have been acquired with the same acquisition parameters); however, because of the slight variation in the incidence angle, the ground cells will not coincide exactly (see Chapter 2). If we call the master sensor "1" and the slave sensor "2", for a range cell Δr (slant range), the ground cells Δx_1 and Δx_2 are:

$$\Delta x_1 = \frac{\Delta r}{\sin\theta} \qquad \Delta x_2 = \frac{\Delta r}{\sin(\theta + \delta\theta)}$$

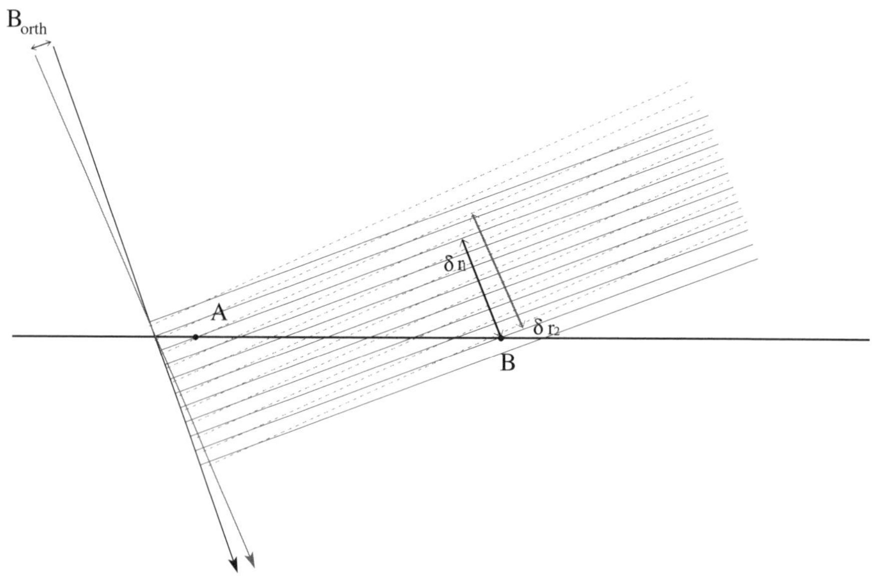

Figure 9.11. *Explanation for the appearance of orbital fringes when the slave image has not been registered on the master image. The range cells (dotted lines and solid lines) are identical, but the ground cells that they define are not*

Let us suppose that the two images, master and slave (with the initial radar range cells), are perfectly registered for point A, and let us see how the difference in the size of the ground cells modifies a point B that is farther away in the swath (Figure 9.11). In relation to point A, point B is shifted with a distance δr_1 for sensor 1 and a distance δr_2 for sensor 2. If we note with x the distance from B to A, we obtain:

$$\delta r_1 = \frac{x}{\sin\theta} \qquad \delta r_2 = \frac{x}{\sin(\theta + \delta\theta)} \qquad [9.9]$$

If the two angles are very near, we can write $\sin(\theta + \delta\theta) \simeq \sin\theta + \delta\theta\cos\theta$, and the difference in the distance between the two sensors measured on the distance axis is written:

$$\delta r_1 - \delta r_2 = \frac{x}{\sin\theta} - \frac{x}{\sin(\theta + \delta\theta)} \simeq \frac{x \cos\theta\, \delta\theta}{\sin\theta} \qquad [9.10]$$

With $B_{orth} = R\,\delta\theta$, we obtain:

$$\delta r_1 - \delta r_2 = \frac{B_{orth}\, x}{R \tan\theta}. \qquad [9.11]$$

This small difference, which can be observed for a flat ground image, will be translated through a dephasing between a master pixel and a slave pixel, which we can observe in a regime of interferometric fringes called *flat earth fringes* or *orbital fringes* (Figure 9.11). Whereas in the beginning, for low distances x, this shift has a weak influence because its only manifestation is the appearance of fringes, the more the distance increases, it will reach the order of the pixel: in this case, the pixels of the two images, master and slave, do not correspond to the same ground area, and pixel comparison no longer makes sense.

This is why it is absolutely necessary that the slave image is correctly registered on the master image in order to be able to state that these are the same objects that are seen in the two images. This registration is done via a resampling of slave data that must necessarily consider the central frequency of the radar (let us remember that for technological reasons, the signal received via the antenna is submitted to an heterodyne processing that allows us to sample it at a frequency closer to the bandwidth, therefore of the order of 100 MHz and not to a frequency close to 10 GHz, which would be unrealistic).

9.3.3.1. *Ambiguity altitude*

We have previously seen that an altitude point H would have a phase difference (equation [9.8]):

$$\Delta\phi = 4\pi \frac{B_{orth} H}{\lambda R \sin\theta}$$

Since the phase is known modulo 2π, if for the height $H + h_{amb}$, we find $\Delta\phi + 2\pi$, it will not be possible to determine if we are speaking of H or $H + h_{amb}$. This altitude that shifts the phase of 2π is called the ambiguity altitude. It corresponds to a time difference of $\frac{\lambda}{2}$. It is given by:

$$h_{amb} = \frac{\lambda R \sin\theta}{2 B_{orth}} \qquad [9.12]$$

As long as the difference in height h between two neighboring points is smaller than h_{amb}, we remain inside a fringe. As soon as we reach this value, it is no longer possible to know if we have increased or decreased the altitude of this value h_{amb}: this is a problem that is perfectly identical to spectrum aliasing that we then find.

The ambiguity altitude plays a key role in the performance of an interferometric system. Let us suppose that we have Q values for coding the phase between 0 and 2π. We will then have Q values for coding an elevation corresponding to h_{amb}. The smaller h_{amb} is, the higher the precision of the height measurement will be. By contrast, as soon as two neighboring points have an altitude difference that is larger than h_{amb}, it will not be possible to find the right multiple of 2π without *a priori*. We see that we must make a compromise between the precision of the interferometric measurement and the difficulties of unwrapping the fringes.

Since the ambiguity altitude h_{amb} is inversely proportional to B_{orth}, we can then formulate the following practical elementary rules:

– When dealing with a strong relief, we will prefer a very low orthogonal baseline, therefore a large value of h_{amb} for limiting the difficulties of fringe unrolling.

– When dealing with a low relief, we will be able to use a higher orthogonal baseline, which will improve the precision of the altitude obtained.

In terms of the precision on the actual phase, it depends on the thermal noise of the sensor but is generally lower than a tenth of a wavelength, so therefore of the order of 1 mm.

9.3.3.2. *Critical baseline*

To be able to attribute a phase difference to a single trajectory difference, the proper phase of the ground cell is exactly the same for the two incidences used. This is the result of the sum of the contribution of the elementary targets present in the resolution footprint. In order for them to be identical, the speckle observed on the master sensor has to be the same as the one noticed on the slave sensor.

The contribution of elementary targets to the speckle considers the local incidence that is slightly different for the two sensors. Considering the footprint as an antenna, we may think that the random process leading to speckle gives two independent observations as soon as the local incidences have a difference that is higher than the value of the antenna lobe of the ground antenna. The angular deviation in the direction θ and for a resolution of the footprint δx is given by $\delta\theta = \frac{\lambda}{2\delta x \cos(\theta)}$. We obtain the value of the critical baseline at a distance ground-satellite R by $B_c = \frac{\lambda R}{2\delta x \cos(\theta)}$. We thus find the relation given in [GAT 94].

9.3.3.3. *The limits of interferometry*

We have seen that interferometry was based on the exploitation of the phase of the radar images. Numerous perturbations can limit the use of this measure.

First, the exploitation of the phase is based on the fact that the two acquisitions "see" the same information and that they are correlated. This correlation is measured by the *complex coherence* between the interferometric signals (see equation [7.25] and Figure 9.12). Several factors can trigger a reduction of the coherence.

First, the noise of the receptor, which is connected to the signal-to-noise ratio (SNR) associated with the receptor. The decorrelation introduced is given by $\gamma_{SNR} = \frac{1}{1+SNR^{-1}}$. There are other factors that can decorrelate the signals. In the case of multi-pass interferometry, there are temporal changes that can intervene between the two acquisitions, which is when we speak of "temporal decorrelation". What is more, as previously explained with the critical baseline, when the acquisition angle difference between the two data increases, the two sensors "see" cells with different ground resolution. We then speak of decorrelation "through the baseline". This loss of correlation increases with the baseline between the two sensors. When the baseline reaches the critical baseline, the two signals are decorrelated and the interferometry is no longer possible. This limit can be lifted for some particular backscatterers, which present an exceptional stability of their proper phase. Called *permanent scatterers* (PS), they are used in differential interferometry in order to exploit very long-time series of images and calculated motion profiles in time (see Chapter 10).

On the other hand, there are some areas that are not seen by the sensor that only measures the thermal noise: this is the case of shadow areas or flat surfaces that do not send the signal toward the radar. We say of these regions that they are incoherent (see section 7.3).

Then, when we try to measure the relief, we suppose that the phase variation is only connected to the altitude. However, the motion of a point between two acquisitions can be translated into a variation in the distance to the sensor, and therefore a phase difference. The measurement of this motion can be done by differential interferometry (Chapter 10). The phase can equally be modified by atmospheric perturbations when the two acquisitions are not simultaneous. We must then correct them to move on.

Finally, as mentioned previously, we must be capable of unwrapping the phase, that is of finding the multiples of 2π allowing us to find a continuous surface. Because of ambiguity altitude, and strong inclinations in the ground, this reconstruction may not be possible. The unrolling difficulty can then equally limit the interferometric measurement.

9.3.4. *Interferogram processing*

The calculation of an interferogram and then of an elevation measurement based on acquired complex data requires a number of operations:

– The first stage is a very precise registration stage between the two data; contrary to the radargrammetric context, the pixels must be perfectly superimposed between the two images, the height information being related to the phases of the two images for the same pixel; this registration stage is done by exploiting the acquisition parameters and by locally maximizing coherence measurements. We can refine this stage by estimating residual phase ramps. These operations allow us to suppress the orbital fringes (the flat earth fringes).

– The second stage is the spectral calibration of images, which consists of only keeping the common parts of the spectrum; the more significant this common part (i.e. the more the baseline between the two images is small), the more coherent the images will be; on the contrary, the more disjointed the spectrums are, the more geometric deformations there are between the two images, and the more responses of the same pixel in the two images risk being different.

– The following stage, which is required in the case of non-simultaneous acquisition, consists of removing atmospheric perturbations that are to be found on the data; they come up when the meteorological conditions between the two images are different, which introduces delays in the wave's journey; there are several methods to remove them, such as the fusion of several interferograms and phase/altitude variation models [BEA 99].

– The calculation of the actual interferogram is done by complex multi-looking (see Chapter 7); there are filtering stages that are generally done to de-noise the phase and facilitate the phase unwrapping.

– The phase being known modulo 2π it is necessary to find multiples of 2π to obtain a continuous surface representing the area imaged. Among the unwrapping approaches, we distinguish the local approaches, which unwrap locally step by step, and the global approaches that perform an optimization on all the data. In this unwrapping stage, the information brought by the coherence on the reliability of the interferometric measure is generally considered. Indeed, as we have seen previously, there are several decorrelation sources that trigger a phase that is useless: the time decorrelation (the scene has changed between the two acquisitions), the geometric decorrelation (the acquisition conditions that are very different trigger a loss in coherence), the shadow areas where only the noise is being measured.

If we wish to sum up the different information contained in the interferometric phase, here are the contributions to be considered:

$$\phi = \phi_{orb} + \phi_{topo} + \phi_{def} + \phi_{atm} + \phi_{res} + \phi_n \qquad [9.13]$$

where ϕ_{orb} represents the contribution of the orbital fringes if the data have not been resampled, ϕ_{topo} the contribution of the elevation, ϕ_{def} that of a potential land slide, ϕ_{atm} that of the atmosphere, ϕ_{res} are the residual contributions connected to a bad knowledge of sensor parameters, and ϕ_n is the noise. In differential interferometry, the objective is not to reconstruct the relief, but the potential motion that might have taken place between two acquisitions. In this case, we delete the component ϕ_{topo}, for example by exploiting another interferogram or an available DTM, and we asses ϕ_{def}. In the case of motion measuring, the interferometric phase is very precise. Let us indeed imagine that the sensor goes exactly through the same spot for the second acquisition. A motion in the boresight of the satellite will lead to a variation δR that will translate into a measured phase difference $\delta\phi = \frac{4\pi\delta R}{\lambda}$. We can therefore measure the motions under a tenth of λ, then lower than a centimeter. The precision is not of the same order when we speak of topography because the relationship between δR and H is $\delta R = \frac{H B_{orth}}{R \sin\theta}$. We will return to differential interferometry in Chapter 10.

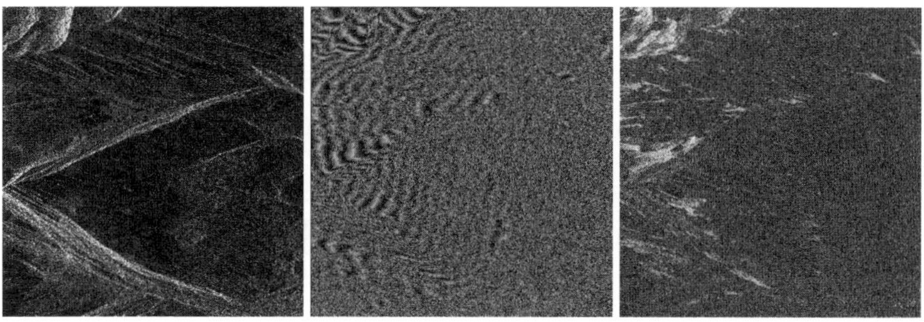

Figure 9.12. *Illustration of the interferometry on the ascending COSMO-SkyMed (CSK) images of the Saleina glacier in Switzerland ©ASI. To the left, one of the original radar images, at the center the topographic fringes obtained after the registration and to the right the corresponding coherence, which presents low values on icy areas and on vegetation*

9.4. Radar tomography

New radar imaging techniques have recently emerged with the help of tomography [HOM 96, REI 00]. We may speak of an augmented interferometry, where several acquisitions with different baselines are simultaneously combined to find not the height of a single point, but the height of all the potential targets present in the scanned volume (Figure 9.13). Everything happens as if a third dimension were added to the trajectory axis and the flight duration axis. We must then make a SAR synthesis in this new dimension to discern the different targets and their heights. The techniques used are signal processing techniques. Tomography has seen a recent development in satellite imaging using the Cosmo-SkyMed acquisitions or TerraSAR-X / TanDEM-X [FOR 12].

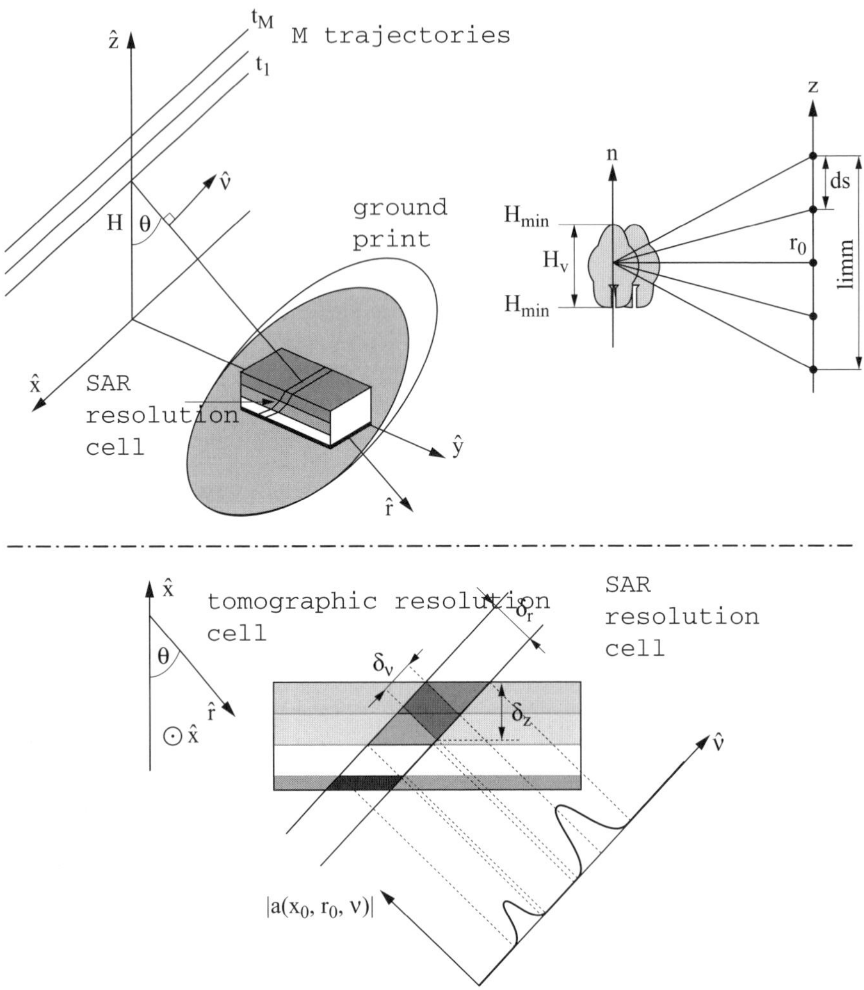

Figure 9.13. *Tomography principle. A geometric configuration of a SAR tomographic measurement (top left), synthetic opening in elevation (top right) and 2D and 3D resolution cells at the bottom (figure by L. Ferro-Famil)*

9.5. Conclusion

This chapter has presented the different elevation reconstruction approaches using satellite images. We have seen that the optic stereovision techniques exploited the shift of the points between two images taken under different acquisition conditions in order to go back to the relief. However, radar interferometry rests on the electromagnetic wave phase received for calculating a path difference and returning to the elevation using images taken with a very small difference between the incidence angles.

Although these techniques are currently completely operational to obtain DTMs, the obstacles we will face are those of digital elevation models in a dense urban environment. Whatever the means, the data obtained are often incomplete in this context with shadow areas and occlusion or covered areas. The combination between several images in stereovision and in radar interferometry or tomography should bring solutions at high and very high resolution.

Chapter 10

Displacement Measurements

10.1. Introduction

The Earth is an active planet: its surface is constantly being reshaped as a result of mass transfers of either internal or external, natural or man-made origin. The displacements of the Earth's surface, whether we are speaking of the ground surface, the glaciers or the water, vary a lot in terms of spatial extension, amplitude and temporal evolution. Their study constitutes an essential part of geodesy, and their quantification is a major topic of geoscience. The objective is not only to understand the current shape of our planet and its evolution, but also to efficiently manage natural hazards. Until the 20th Century, measurements of the Earth's surface displacements were punctual, *in situ* measurements, for example the classic leveling techniques, distance or tilt measurements. The distribution of these measurements on the Earth's surface was not homogeneous, with a concentration in specific areas such as natural observatories (e.g. the Vesuvius volcanological observatory, created in 1845) and a small temporal sampling. At the end of the 20th Century, two major progresses were achieved. On the one hand, the development of spatial geodetic techniques (GPS, optic imagery, InSAR, etc.) has allowed for a drastic improvement of the spatial coverage of the measurements regarding the radar and optic imagery. On the other hand, the generalization of continuous *in situ* measurements has caused splendid improvement in the accuracy of temporal sampling.

With regard to *in situ* or field measurement, the remote sensing measurement is still strongly limited due to its small temporal sampling, which, even though improved, still remains limited when compared to the possibilities currently offered

---

Chapter written by Yajing YAN, Virginie PINEL, Flavien VERNIER and Emmanuel TROUVÉ.

by ground instrumentation. However, it does have its advantages. It is unbeatable in terms of spatial coverage, because it enables us to obtain continuous maps of surface displacement over large areas. This advantage has allowed the detection and quantification of deformation in non-instrumented, remote areas or areas that do not have the necessary financial means and human resources for ground instrumentation. This advantage has also proven very useful for regional studies. Furthermore, because of the archiving system, we can study, a posteriori, areas where a phenomenon we are interested in has been detected, and thus we have access to initial phases. This kind of *a posteriori* study is never possible with ground instrumentation, where data are only acquired with a decision made, and therefore often after the outbreak of the phenomenon.

Because of these advantages, the remote sensing displacement measurement has gained significant development in the past few years, for the detection and quantification of natural as well as man-made deformations. In tectonics, the understanding of the forces at work and the evolution of the Earth's relief is realized by the quantification of the displacement of large units, blocks or tectonic plates, of their level of rigidity and the velocities at their margins. The precise measurements of the displacement field around faults allow us to understand their loading history, to determine locking areas and depths and thus, to feed the models seeking to quantify seismic risk. In recent years, the understanding of the seismic cycle came from *in situ* measurements due to their good temporal resolution [LIN 96, DRA 01]. However, spatial imagery also brings significant constraints on the temporal evolution of displacement fields around faults [JOL 13]. Furthermore, since the first co-seismic study by Massonnet & Rabaute [MAS 93], radar imagery has brought unique information on the geometry of fault rupture areas. In volcanic areas, magma transport and storage in the upper layers of the crust often induce deformations. Consequently, in the field of volcanology, the measurement of surface displacements allows us to infer the geometry and the behavior of the magma plumbing system, and also to detect the arrival of magma at shallow depth, which could potentially precede an eruption. Since the first study by Massonnet *et al.* [MAS 95], remote sensing has become a major asset in modern volcanology [SPA 12, HOO 12b]. It has allowed us to detect magma storage in non-instrumented areas [PRI 02]. Its good spatial coverage has improved our knowledge of the geometry of magmatic intrusions [SIG 10], and it has enabled the detection of deep magma storage zones [DAL 04, OFE 11]. The vertical motions induced by surface load variations (e.g. variations of the water level in a lake, melting of a glacier, accumulation of eruptive deposits) are qualified as isostatic motions. The use of remote sensing to quantify vertical displacements, induced by a known variation in the surface load, allows us to obtain information on the rheology of the shallow layers of our planet [CAV 07, PAG 07, PIN 07]. This information is essential in modeling ground displacement occurring in response to a given evolution of stress. In glaciology, since the study by Goldstein *et al.* [GOL 93], remote sensing has been used to quantify the flow of the glaciers [BER 06, FAL 11, ERT 13], which, because of its sensitivity to the thickness of the glacier and to hydrological conditions underneath the glacier, is a marker of climate change. In this field, the study of the displacement field is also a tool for preventing the glacial risks associated, for example with the accumulation of subglacial water, or serac falls. Remote sensing has also proved to be useful for characterizing landslides [FRU 96, COL 06]. A good knowledge of the displacement and the temporal evolution allows us to better understand the process governing landslides and to better manage the associated risks. In the case of landslides, as for small-size glaciers, the application of remote sensing techniques is, however, more complicated because of the acquisition geometry (strong slopes) and the nature of the displacement, which can be very rapid and strongly localized in space and time. Not all displacements recorded at the Earth's surface are of natural origin, but some are induced by internal mass transfers resulting from human activities (mine excavation, pumping or fluid injection, etc). The quantification of these man-made displacements is also important for preventing future material damages. In this field as well, remote sensing has proven itself interesting [CAR 96].

10.2. Extraction of displacement information

Displacement measurement by remote sensing is based on the comparison of the images acquired with different dates so as to estimate the ground motion that potentially took place between these dates, on a part or in the entirety of the region imaged. This subject is different from that of the change detection (Chapter 8), for several reasons:

– The nature of the sought-after information: the displacement measurement is scalar or vector information (depending on the number of components in the displacement vector), whereas the change detection is symbolic binary information (or N-ary information if we seek several change classes). We then deal with an estimation problem where the sources of uncertainty must be analyzed and propagated in the processing chain, whereas the change detection is more related to the problem of automatic or supervised classification.

– The source of the sought-after information: when we are interested in temporal changes, the albedo or the radiometry of pixels generally provides the information that can be compared directly, whereas the displacement information cannot be obtained from a comparison between the values of the pixels. It comes either from the difference between their positions in the image, considering the geometry of the sensor, or from the phase difference that is related to the difference in the path traveled by the electromagnetic wave in the case of radar imaging.

The displacement measurement therefore needs a step of information extraction exploiting one or the other of these two sources:

– The measurement of the difference in position (offset tracking) by looking for a maximum of similarity can be applied both in optic and radar imagery. This approach, developed in section 10.2.1, is limited by the resolution of the sensor, and allows us to achieve a precision of the order of a tenth of the pixel's resolution. However, it uses the two axes of the image and allows us to obtain a 2D displacement measurement.

– The measurement of the phase difference using the technique called differential interferometry (D-InSAR) is mainly applied to SAR satellite images acquired on repeated orbits that are sufficiently close (i.e. repeat pass interferometry) and when the time evolutions of the surface are sufficiently reduced to maintain the coherence (see sections 9.3 and 9.3.3.3). This approach, developed in section 10.2.2, can reach a precision of the order of a fraction of the wavelength, that is a precision of the order of centimeters or even millimeters by exploiting series of interferograms (see section 10.3.1). With a given acquisition geometry, it only allows us to measure the displacement in the direction of the radar line of sight or range.

The combination of these two sources of information, or measurements, issued from the data acquired in different geometries then allows us to look for higher-level information. Particularly, this second step, illustrated in section 10.3.2, allows us to rebuild the three components (east, north, up) of the displacement vector (3D) or to inverse the physical model describing the phenomenon that causes the displacement observed at the Earth's surface (faults, volcanoes, etc.).

10.2.1. *Maximum of similarity*

The basic objective when searching for the maximum of similarity is to find the position of a point or a pattern in an image. We have seen in the previous chapter that this is the basis of the optic and radar stereo vision. The difference in this context lies in that the search area is less constrained. In stereo vision, having been in epipolar geometry, we have seen that the homologous points were situated on the same line, thus limiting the search to a single dimension (section 9.1.2). We have also seen in Chapter 2 that searching for a maximum of similarity was also the basis of the registration (section 2.4). In displacement information extraction, the application varies slightly from the registration, but the equations used remain the same. We do not seek to identify a global transformation function of one image into another, but try to measure the displacement that has happened to the points between the two images. For the sake of simplicity and in order to reduce the size of the search areas, we suppose that the two images used are issued from a global registration (see section 2.4), carried out by relying on, for example, the fixed areas or on the knowledge of the acquisition geometry and the topography.

10.2.1.1. *General case*

Once the images are registered, we need a local, more intense and more complete similarity search than the one used for the registration in order to extract displacement information. For this, let us consider a master image I_1, which is generally the one with the earlier date, and a slave image, I_2. We thus calculate the displacement in the slave image with respect to the master image. The main principle is to define, in the master image, a local window, called the master window hereafter, that acts as a pattern, and then to look for this pattern in a search area defined in the slave image. Afterward, the master window and the search area will be shifted by a given step, and the search will begin for that new pattern. This step continues for all the regions of the image where the displacement information needs be extracted. Figure 10.1 illustrates this principle.

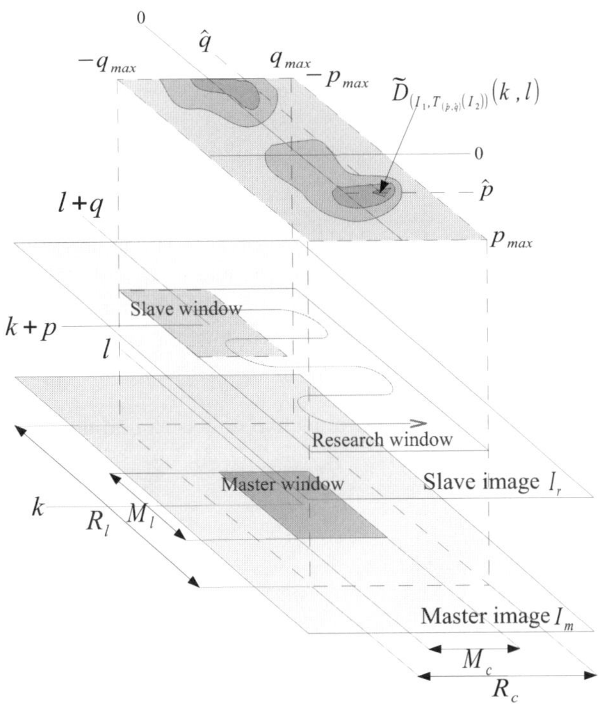

Figure 10.1. *Illustration of the principle of the algorithm of the search for the maximum of similarity; on the first level, the master image and the master window; on the second level, the slave image containing the search area that is larger than the master window and the slave window of the same size as the master window; at the third level, the results of the similarity function for all displacement* (p, q)

To determine the position of the master window in the search area, a similarity criterion must be determined. There are many such criteria in the literature, as we have already seen in previous chapters. Some of them are dedicated to optical images, but they can easily be applied to SAR amplitude images. Others are dedicated to SAR images and their characteristics [DEL 10a].

First of all, let us consider the general case of a similarity function $\tilde{\mathcal{D}}_{(I_1,I_2)}(k,l)$ calculated locally at the point (k,l) between the images I_1 and I_2, that is between the master window I_1 centered at (k,l), and the slave window I_2 also centered at (k,l). Because we search for a displacement between I_1 and I_2, we make several displacements (p,q) on the slave image I_2. Note $T_{(p,q)}(I_2)$ the image I_2 translated of (p,q). We then estimate the similarity $\tilde{\mathcal{D}}_{(I_1,T_{(p,q)}(I_2))}(k,l)$ and look for its maximum value in order to find the right displacement. Note $\Omega_{(k,l)}$ the local neighborhood centered at (k,l) defining the master window:

$$\Omega_{(k,l)} = \Omega_k \times \Omega_l = \left[k - \frac{M_l}{2}; k + \frac{M_l}{2}\right] \times \left[l - \frac{M_c}{2}; l + \frac{M_c}{2}\right].$$

It is quite natural to choose M_l and M_c as odd in order to center the window on (k,l).

With the master window defined, the search area W_r is determined through the knowledge of the maximum displacement in the line $p_{\max}$ and in the column $q_{\max}$ of the point (k,l). Thus, the search window $(R_l \times R_c)$ has the size $R_l = M_l + 2p_{\max}$ and $R_c = M_c + 2q_{\max}$. Similarly to the master window, the search window is generally centered on (k,l) if there has been a global pre-registration (see Figure 10.1).

The problem thus established, the objective now is to find the values $\hat{p}$ and $\hat{q}$ that maximize the similarity function $\tilde{\mathcal{D}}_{(I_1,T_{(p,q)}(I_2))}(k,l)$:

$$(\hat{p},\hat{q}) = \text{argmax}_{(p,q)}\tilde{\mathcal{D}}_{(I_1,T_{(p,q)}(I_2))}(k,l) \qquad [10.1]$$

with the values $\hat{p}$ and $\hat{q}$ being comprised in the intervals $[-p_{\max}, p_{\max}]$ and $[-q_{\max}, q_{\max}]$, respectively. We thus obtain the best correspondence between the master window and the slave window in the search area. The choice of the similarity measurement is therefore crucial.

10.2.1.2. *Similarity functions in optical images*

One of the most often used similarity functions is the cross-correlation, hereafter noted CC (cross-correlation) (see section 2.4.2.1). It can be applied well to both optical images and SAR amplitude images. The CC is given by:

$$\tilde{\mathcal{D}}^{CC}_{(I_1,T_{(p,q)}(I_2))}(k,l) = \frac{1}{M_l M_c} \sum_{(i\in\Omega_k)} \sum_{(j\in\Omega_l)} I_1(i,j).I_2(i+p,j+q). \qquad [10.2]$$

This similarity function[1] has certain advantages: it is directly related to the standard L^2 norm of the difference between the master and slave images. It is maximum in case of equality and then equal to the energy of these images. It can be calculated in the frequency domain using the fast Fourier transform (FFT), which allows us to accelerate the calculation when the size of the windows becomes significant. It is also possible to directly measure the offset between the two images in the frequency domain, on the basis of the FFT phase slope of the CC.

The major disadvantage of this similarity function is that its value depends on the intensity of the images. It favors the high-intensity pixels and does not allow the comparison between the different maximums found. For example, two low-intensity patterns perfectly correlated can have a correlation level lower than two high-intensity patterns that are not as well correlated. Therefore, we generally prefer to use the normalized cross-correlation (NCC):

$$\tilde{\mathcal{D}}^{NCC}_{(I_1,T_{(p,q)}(I_2))}(k,l) = \frac{\sum\limits_{(i,j)\in\Omega_{(k,l)}} I_1(i,j).I_2(i+p,j+q)}{\sqrt{\sum\limits_{(i,j)\in\Omega_{(k,l)}} |I_1(i,j)|^2. \sum\limits_{(i,j)\in\Omega_{(k,l)}} |I_2(i+p,j+q)|^2}}. \quad [10.3]$$

This function helps to solve the CC issue because, through normalization, the similarity result is comprised in the interval $[0,1]$.

Finally, it could be interesting, under certain conditions, to center the values and thus to use a centered normalized CC, usually noted ZNCC for Zero-mean NCC. The authors of [FAU 93] show, for example, that the ZNCC function is more efficient when the Gray distributions (histograms) of the images are visibly different:

$$\tilde{\mathcal{D}}^{ZNCC}_{(I_1,T_{(p,q)}(I_2))}(k,l)$$
$$= \frac{\sum\limits_{(i,j)\in\Omega_{(k,l)}} \left(I_1(i,j)-\overline{I_1}\right).\left(I_2(i+p,j+q)-\overline{I_2}\right)}{\sqrt{\sum\limits_{(i,j)\in\Omega_{(k,l)}} |(I_1(i,j)-\overline{I_1})|^2. \sum\limits_{(i,j)\in\Omega_{(k,l)}} |(I_2(i+p,j+q)-\overline{I_1})|^2}}, \quad [10.4]$$

where $\overline{I_m}$ and $\overline{I_r}$ are, respectively, the means of the master and slave windows:

$$\overline{I_1} = \frac{1}{M_l\,M_c} \sum_{(i,j)\in\Omega_{(k,l)}} I_1(i,j), \qquad \overline{I_2} = \frac{1}{M_l\,M_c} \sum_{(i,j)\in\Omega_{(k,l)}} I_2(i+p,j+q).$$

---

1 This function corresponds to the similarity $\tilde{\mathcal{D}}^{prod}_{(I_1,I_2)}$ of section 2.4.2.1, equation [2.6].

Note that the denominator corresponds to the product of the empirical local standard deviation. The centering and the normalization thus give a similarity measurement [2] comprised in the interval $[-1, 1]$.

There is a large number of similarity functions. Several of them, used in the context of computer vision, are presented in [CHA 03]. We could equally refer to Chapter 8.

10.2.1.3. *Similarity functions in SAR images*

The previous functions can also be applied to SAR amplitude images, but they do not consider certain specificity of these data, particularly the speckle effect, the shape of the impulse response or the possibility to use phase information.

When we only use the images detected (amplitude or intensity), we can try to introduce the characteristics of the speckle effect, as seen in section 4.3 and to use the approach of maximum likelihood. Depending on whether the speckle is correlated between the two images or not (which corresponds to the coherence loss in interferometry), this approach leads to two different similarity measurements. In the case where the speckle is not correlated, by considering the logarithm of the images in order to reinforce the contribution of the low-intensity pixels, we obtain the function noted uncorrelated maximum likelihood (UML) [ERT 09]:

$$\begin{aligned}\tilde{\mathcal{D}}^{UML}_{(I_1,T_{(p,q)}(I_2))}(k,l) \quad = \quad & \sum_{(i,j)\in\Omega_{(k,l)}} \left(\log\left(\frac{I_1(i,j)}{I_2(i+p,j+q)}\right)\right. \\ & \left. -2\log\left(1+\frac{I_1(i,j)}{I_2(i+p,j+q)}\right)\right).\end{aligned}$$

In the case where the speckle is correlated with a correlation level $\rho_{k,l} \in [0,\ 1]$, we obtain the function noted correlated maximum likelihood (CML) [ERT 09]:

$$\begin{aligned}\tilde{\mathcal{D}}^{CML}_{(I_1,T_{(p,q)}(I_2))}(k,l) = & \sum_{(i,j)\in\Omega_{(k,l)}} \left(\log\left(\tfrac{I_1(i,j)}{I_2(i+p,j+q)}\right) - 2\log\left(1+\tfrac{I_1(i,j)}{I_2(i+p,j+q)}\right)\right. \\ & \left. -\left(1+\frac{1}{2N}\right)\log\left(1-\frac{4\rho_{k,l}\frac{I_1(i,j)}{I_2(i+p,j+q)}}{\left(1+\frac{I_1(i,j)}{I_2(i+p,j+q)}\right)^2}\right)\right) \end{aligned} \quad [10.5]$$

We can also refer to the measurements presented in Chapter 8.

---

2 This function corresponds to the similarity $\tilde{\mathcal{D}}^{corr}_{(I_1,I_2)}$ of section 2.4.2.1, equation [2.7].

The use of one or the other of these functions can be guided by the *a priori* knowledge of the acquisition conditions and the imaged surfaces (temporal decorrelation, significant baseline, forest areas etc.) or by an estimation of the correlation level on previously registered areas. In both cases, these functions are not normalized. We must choose a criterion such as the relative height of the similarity peak, in order to have a confidence indicator associated with the measured displacement and to be able to remove aberrant results.

The previous criteria correspond to the displacement measured on the distributed targets whose response is affected by the speckle effect. However, if we are interested in the displacement of the point targets where the response is dominated by a scatterer, we can seek to correlate these points with the impulse response that results from SAR image formation. Once this 2D sinus cardinal response is known, we can thus finely localize the response of these targets in one image or another, and directly deduce their subpixel displacement [SER 06]. This approach noted "SINC", has particularly been tested for measuring the displacement of corner reflectors that follow the motion of a glacier [FAL 11]. TerraSAR-X strip-map images (resolution 2m) have been used for assessing the three approaches: ZNCC, UML and SINC. The conclusions tend to show that the gain in performance of the specific methods, that is UML on distributed targets and SINC on punctual targets, does not discriminate regarding the use of the correlation method that has the advantage of being applied to different types of targets and providing a well-calibrated confidence measurement.

When we have complex data, that is Single Look Complex (SLC) images, we can benefit from all information in order to search for the maximum of similarity by using the module of the complex normalized correlation function:

$$\tilde{\mathcal{D}}^{coh}_{(I_1,T_{(p,q)}(I_2))}(k,l) = \frac{\left| \sum\limits_{(i,j)\in\Omega_{(k,l)}} I_1(i,j) I_2^*(i+p,j+q) \right|}{\sqrt{\sum\limits_{(i,j)\in\Omega_{(k,l)}} |I_1(i,j)|^2} \sqrt{\sum\limits_{(i,j)\in\Omega_{(k,l)}} |I_2(i+p,j+q)|^2}}$$

where I_2^* refers to the conjugate of I_2. We fall on the coherence used in radar interferometry (formula [7.25]) for the estimation of the phase difference with a spatial averaging to reduce the noise and provide an indicator of the phase stability see section 9.3). This approach noted coherence optimization procedure (COP) [GRA 01, STR 02] is much more precise because the coherence drops as soon as the pixels $I_m(i,j)$ and $I_r(i+p,j+q)$ do not cover the same resolution cell on the ground. However, the use of this similarity function needs to take several precautions:

– The interferometric conditions must be fulfilled (small baseline and absence of temporal evolution) so that the "proper phases" related to the speckle effect are the same in the two acquisitions and disappear in the product $I_1.I_2^*$ (see section 9.3.3.3).

– We must perform an over sampling of the complex data (for example using zero-padding in the Fourier domain) so as to search for a subpixel displacement that reduces the decorrelation due to the fact that the resolution cells are not overlapping.

– The choice of the size of the window Ω is based on the compromise between the need to average a sufficient number of samples in order to reduce the variance of the coherence estimation and the phase rotation within the estimation window depending on the size of the fringes.

Finally, when we have polarimetric SAR data, several strategies are possible to benefit from the richness of the polarimetric information in the displacement measurement. The first consists of using a similarity function that considers the different polarimetric channels (HH, HV, VV or after the transformation in the Pauli basis, see Chapter 4) or the coherence matrix. There are several similarity measurements that have been proposed, in the context of the change detection [DIE 02], the filtering [DEL 10b] or the displacement measurement [ERT 13]. Another strategy consists of decomposing the coherence polarimetric matrix using a spherical invariant random vector model in order to obtain a variable of scalar texture that we then try to track between the images. Supposing that this variable follows a Fisher law, a similarity function is then deduced from the maximum likelihood [HAR 11].

10.2.1.4. *Sub pixel displacement*

Searching for the similarity using functions such as correlation, or some of its varieties, gives a displacement of $(\hat{p}, \hat{q})$ pixels in integer values. We can, however, refine this measurement in order to extract subpixel displacement information. Traditionally, an interpolation of the similarity function in the neighborhood of the maximum allows us to obtain this information. The most simple methods that can be used are the interpolation by two parabolas, one in x and the other in y:

$$\mathfrak{D}(x) = ax^2 + bx + c$$

$$\mathfrak{D}(y) = dy^2 + ey + f,$$

or the interpolation via a paraboloid:

$$\mathfrak{D}(x, y) = ax^2 + by^2 + cxy + dx + ey + f.$$

Thus, the subpixel displacement corresponds to the coordinates of the interpolation maximum in the studied neighboring area. In the case of a paraboloid, the function obtained via interpolation may become of the hyperbolic type and therefore may provide an aberrant result. A conformity test of the solution (belonging to the maximum of the interpolation area) allows us to rule out these values.

The implementation of the displacement calculation via a maximum of similarity is illustrated in Figure 10.2. The module and the orientation of the displacement measured corresponds to the motion of the Argentière glacier (Mont-Blanc massif) at the level of the Lognan serac fall. The height of the similarity peak allows us to detect the area where the displacement cannot be measured because of the fall of ice blocks between the two dates. The points where the subpixel interpolation did not work are mainly situated in these areas of poor correlation.

10.2.2. *Differential interferometry*

Differential radar interferometry (D-InSAR) is today widely used to measure the displacements of the Earth's surface, whether of seismic, volcanic or gravitational origin [MAS 98]. Commercial software such as DIAPASON [MAS 97] and Gamma [WEG 05] or free software such as ROI-PAC [ROS 04], NEST from the European Space Agency (ESA) or the EFIDIR_Tools developed by research laboratories allow us to obtain displacement measurements between two dates or to build series of interfergrams in order to monitor smaller and smaller deformations.

The first steps of the D-InSAR processing chain can be regrouped in two parts: on the one hand the SAR synthesis, the registration of the slave image on the master image and the generation of the differential interferogram, and on the other hand the phase filtering and the phase unwrapping. These steps are generally followed by corrections of geometric or atmospheric artifacts presented in section 10.2.3. These corrections may also need to reloop with the previous steps.

10.2.2.1. *Generation of differential interferograms*

The SAR synthesis, the registration of the slave image on the master image and the generation of the interferogram, introduced in section 9.3, are related: using the same "Doppler centroid" or, on the contrary, deleting the disjoint parts of the spectrum [GAT 94]. Once the SLC master and slave images have the same geometry, we may obtain a first estimation of the coherence using spatial averaging (complex multilooking) in order to reduce the noise [GOL 88]. The number of lines and columns of the averaging window can be chosen so as to obtain approximately square pixels on the ground, for example 5×1 or 10×2, for ERS satellite images whose SLC data have a resolution of approximately 4 m in azimuth and 20 m in range on the ground (ground range).

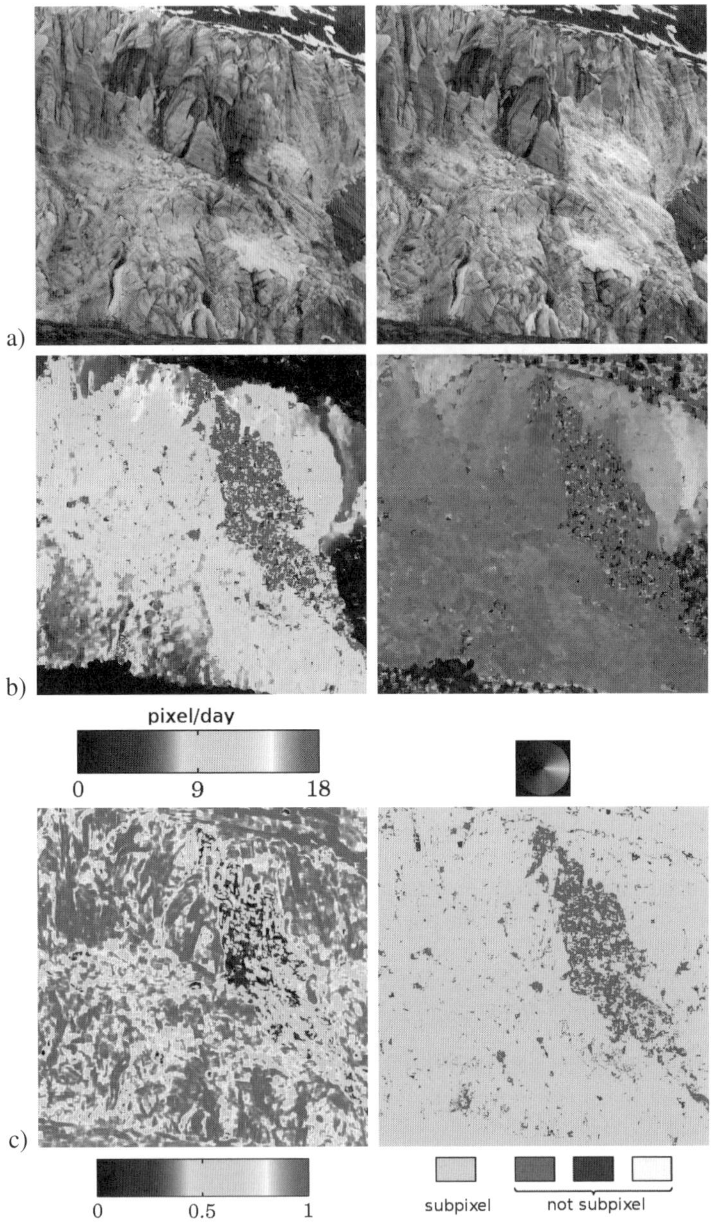

Figure 10.2. *Displacement calculation by maximum of similarity. a) Lognan serac fall (Argentière glacier) on May 29 and 30, 2009 (photograph by Luc Moreau); b) 2D displacement vector transformed to module and orientation; c) height of the cross-correlation peak (NCC) and subpixel interpolation. For a color version of this figure, see www.iste.co.uk/tupin/RSImagery.zip*

The phase difference $\phi_{orb/topo}$ due to the increase in distance in range (so-called "orbital" fringes) and the relief (so-called "topographical" fringes) can be calculated from orbital data and a digital elevation model (DEM). Different from the previous chapter, we do not seek to determine the elevation of the imaged points, but we use auxiliary information in order to calculate the path difference ΔR between the two orbits and the position of the points supposed to be fixed on the ground. This difference is then transformed into a phase via the wavelength λ: $\phi_{orb/topo} = 4\pi\,\Delta R/\lambda$ and can be subtracted during the averaging step, in order to reduce the number of fringes and to avoid the aliasing phenomenon. In a pixel P associated with a window Ω to the initial resolution, the phase $\phi(P)$ and the coherence $Coh(P)$ are then estimated by (see equation [7.25]):

$$Coh(P).e^{i\phi(P)} = \frac{\sum_{(k,l)\in\Omega} z_1(k,l) z_2^*(k,l) e^{-i\phi_{orb/topo}(k,l)}}{\sqrt{\sum_{(k,l)\in\Omega} |z_1(k,l)|^2}\sqrt{\sum_{(k,l)\in\Omega} |z_2(k,l)|^2}} \qquad [10.6]$$

where z_1 and z_2, respectively, designate the master image and the slave image that have been previously registered. After this step, we dispose of an interferogram that should only measure the displacement between the dates. The result of this initial averaging and the subtraction of the orbital and topographic fringes is illustrated in Figure 10.3.

In practice, because of the imprecision of the auxiliary information on the orbits, a residual orbital contribution can remain in the interferometric phase. This residual contribution is usually corrected by adjusting a parametric model such as a plan throughout the interferogram [HAN 01, LOP 09]. Similarly, in certain cases, DEM errors also cause a residual phase term that is proportional to the perpendicular baseline. There are approaches that use a series of interferograms such as the small baseline subset (SBAS approach) and the permanent scatterers (PS) approach, allowing us to correct this residual term (see section 10.3.1).

10.2.2.2. *Phase filtering and phase unwrapping*

The noise, which perturbs the interferometric phase and constitutes an obstacle in phase unwrapping, is essentially induced by decorrelation (temporal evolution, distributed targets seen from slightly different angles, etc.). To reduce this noise, there are three categories of filters introduced in the literature (before, during and after the construction of the interferogram). The filters before the construction of the interferogram consist of separating the signal from the noise in the spectral domain [HAN 01]. The filtering during the construction of the interferogram corresponds to the complex multilooking technique given by equation [10.6]. The increase in the number of looks allows us to highly reduce the variance of the estimation of the phase and the coherence (see section 7.3, Figure 7.8), but at the cost of a decrease in the sampling frequency that can turn out to be problematic in the areas with strong deformation gradient. To further increase the number of averaged samples without scale reduction, it is necessary to apply, after the construction of the interferogram, an "averaging"- type filter [ROD 92, MAS 93]. For this, it is necessary to respect as much as possible the hypotheses of stationarity and ergodicity that allow for the Hermitian product $z_1(P)z_2^*(P)$ by relying on samples from the neighborhood of the point P. To ensure these hypotheses, two ways are preferred:

– using adaptive neighborhood that search for a subwindow or a set of pixels belonging to the same statistical population as the filtered pixel [LEE 94, GOL 98, VAS 04, DEL 11b].

– compensating the local fringe pattern in order to "flatten" the phase in the filtering window. This compensation can be done by estimating a first-order model of fringes given by 2D local frequencies on a rectangular neighborhood [TRO 96] or an adaptive neighborhood [VAS 08].

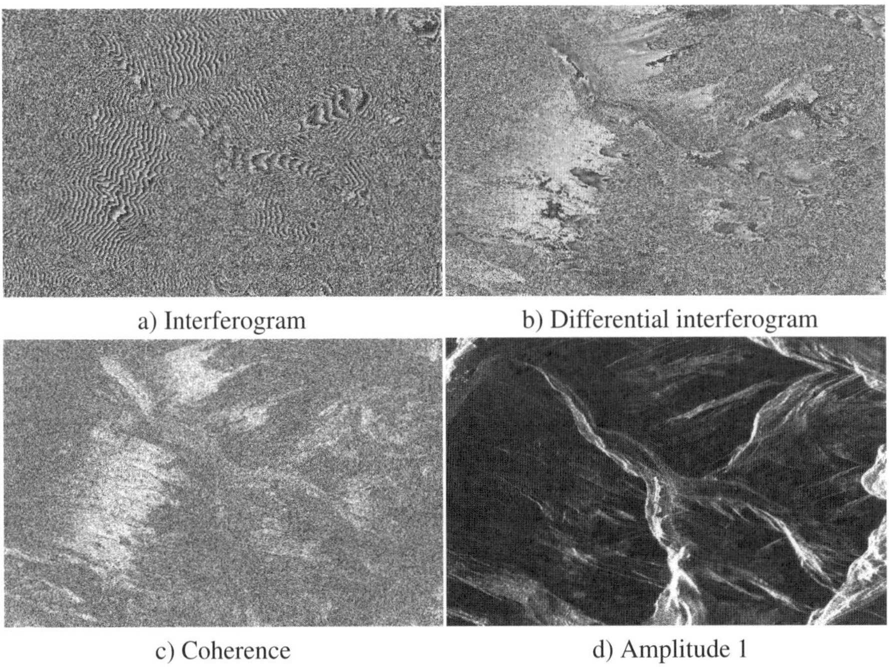

Figure 10.3. *ERS tandem interferometric data, 1995/12/31 – 1996/01/01, Mer-de-Glace glacier, initial averaging* 5×1 *; a)-b): phase before and after subtraction of the topographic fringes; c) initial coherence; d) amplitude of one of the two images*

The filtering then consists of re-estimating the wrapped phase as well as the coherence, by applying the averaging in equation [10.6] on the selected window, and compensating the rotation of the phase using the estimated local frequencies. This step allows us to obtain a "de-noised" interferogram in order to address the phase unwrapping through propagation, and a coherence image that is much more discriminating in order to identify the areas that cannot be unwrapped correctly. The local frequencies can also be used directly for phase unwrapping using the method of least squares [TRO 98]. Figure 10.4 illustrates the filtering and the unwrapping steps on a co-seismic interferogram that measures the deformation of the fault during the Kashmir earthquake in 2005.

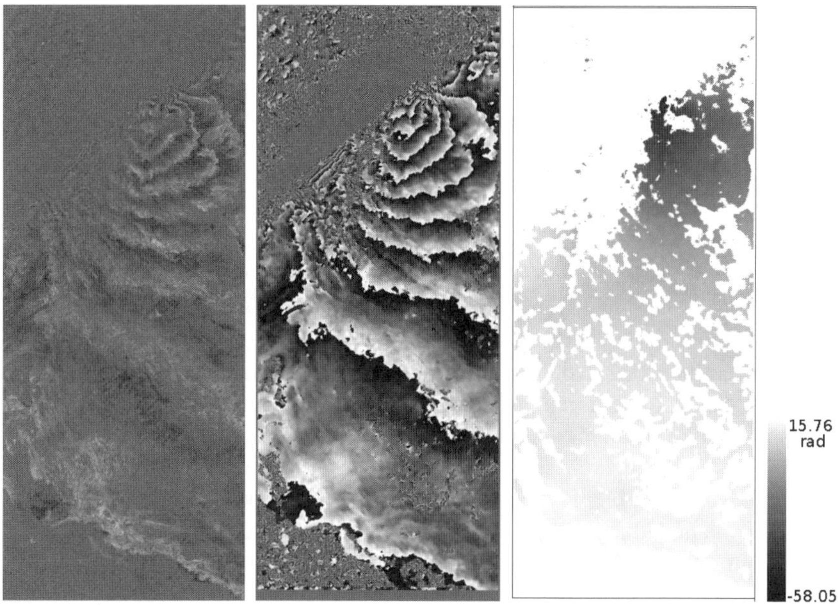

a) b) c)

Figure 10.4. *ENVISAT interferogram (2004/11/06-2005/11/26) of the earthquake in Kashmir (2005). a) Original phase estimated with an averaging (10×2); b) phase filtered through a multi scale approach with an estimation of local frequencies [YAN 13]; c) phase unwrapped using least squares*

The interferometric phase, which measures the difference in the round-trip trajectory of the wave sent by the radar, is only known modulo 2π, so the interferogram only measures the corresponding displacement modulo $\lambda/2$. To remove this ambiguity, phase unwrapping is necessary. In each pixel, we seek to find the right multiple of 2π to be added to the value of the main phase $\phi(P)$ given by the interferogram in order to have the exact value of the phase $\varphi(P) = \phi(P) + 2\pi k(P)$, with $k(P)$ as a relative integer [CHE 00, CHA 04a]. The fundamental hypothesis for phase unwrapping is to consider that the surface to rebuild is relatively regular and the unwrapped phase is continuous; in other words, no noise is present and the Nyquist criterion is respected during the sampling. These conditions imply that the phase varies within π between two adjacent pixels. In the literature, there are two large types of methods for phase unwrapping: the local methods and the global methods. The local methods are based on a propagation of the phase value, pixel by pixel. In these methods, each pixel is assessed individually along paths with necessity of the continuity of the coherent area. The branch-and-cut method [GOL 88] and the minimum cost flow method (MCF) [CHE 00] are in this category. Different from the local methods, global methods seek a solution on the whole image that minimizes the deviation between the phase gradients measured in the interferogram and those of the result. Some methods are based on image processing techniques such as the segmentation, the cellular automata and the Markovian models. Complete work has been proposed for this specific problem [GHI 98]. In the geophysical community, the tool extensively used for phase unwrapping is SNAPHU[3] [CHE 02], which is based on the MCF algorithm.

Phase unwrapping remains a crucial step in differential interferometry that conditions the success of its application. The choice of the method depends on the nature of the interferograms to be processed. There have been several attempts for developing methods that avoid the problem of phase unwrapping [FEI 09], but even until today, this problem remains a delicate subject, as no method seems fully operational. The problems currently encountered in phase wrapping are the discontinuity of the coherent areas and the strong gradient of the displacement that causes the aliasing problem, which corresponds to the appearance of fake fringes in insufficiently sampled areas. For the latter, we sometimes use *a priori* models of the displacement or measurements from other sources of information in order to reduce the number of fringes and make the phase unwrapping easier [SCH 05]. However, this solution cannot always be applied because of errors in the model or the imprecision of the measurements issued from other sources or simply because of a lack of information, especially for an event that has just taken place.

10.2.3. *Corrections*

The measurements from the amplitude correlation provide a "distance" or a "difference in position" between two pixels corresponding to the same area on the ground. As the acquisition conditions of the two images are slightly different, this "distance" is the sum of several contributions:

3 Statistical-Cost, Network Flow Algorithm for Phase Unwrapping.

– a topographically induced stereoscopic effect;

– a distortion induced by the lack of parallelism of orbits and the potential difference of the "Doppler Centroid";

– a displacement of the pixel on the ground between two successive acquisitions.

To deduce the surface displacement, we must carry out some corrections in order to eliminate the first two contributions [MIC 99]. In practice, the measurement in range is corrected using a DEM, which has a resolution equivalent to that of the radar image, and the measurement in azimuth image is corrected by removing a ramp, which is estimated outside of the deformation area.

Regarding the influence of the topography, this is due to the stereoscopic effect (see Chapter 9) and, depending on whether we correct an offset or a phase difference, it can be expressed in the following form (equation [9.7]):

$$\delta R_{topo} = \frac{B_{orth}\,H}{Rsin\theta} \qquad \phi_{topo} = \frac{4\pi B_{orth} H}{\lambda Rsin\theta}, \tag{10.7}$$

with B_{orth} the perpendicular baseline that characterizes the distance between the 2 trajectories (defined by equation [9.6]), H the elevation of the given point, R the target-satellite distance and θ the incident angle. Potential errors of DEM will have less impact in the case of smaller perpendicular baselines.

Regarding the measurement from differential interferometry, besides the topographic and orbital corrections carried out using the DEM and the auxiliary information of the orbits (see section 10.2.2), an atmospheric correction usually turns out to be necessary in order to reduce this source of uncertainty and to be able to quantify displacements of order of centimeters, or even millimeters. The variations of physical properties of the atmosphere between two radar acquisitions cause a variation of the propagation speed of electromagnetic waves. This induces a phase difference of an atmospheric origin that can be wrongly interpreted as ground displacement. This effect can induce several fringes in an interferogram, equivalent to ground displacement of several tens of centimeters. This is therefore the main limitation to the use of InSAR for ground displacement measurement [ZEB 97, HAN 01].

The air refractivity is sensitive to the pressure of dry air, the partial pressure in water vapor, the temperature and the quantity of water in the form of clouds. It can be expressed as follows:

$$N = k_1\frac{P_d}{T} + k_2\frac{e}{T} + k_3\frac{e}{T^2} + k_4W_c + k_5\frac{n_e}{f^2} \tag{10.8}$$

where P_d is the dry air partial pressure (Pa), e is the water vapor pressure (Pa), T is the temperature (K), W_c is the cloud water content (kg.m^{-3}), n_e is the electron density in the ionosphere, f is the electromagnetic wave frequency, $k_1 = 0,776$ K.Pa^{-1}, $k_2 = 0,716$ K.Pa^{-1}, $k_3 = 3,75.10^3$ K^{2}Pa^{-1}, $k_4 = 1,45.10^3$ m^{3}kg^{-1} and $k_3 = -4,03.10^7$ m^{3}s^{-2} [SMI 53, PUY 07, DOI 09].

Any change in the temperature, humidity or pressure along the trajectory of the radar wave can, therefore, cause a phase variation. The so-called "atmospheric" phase that results from this can be decomposed into a turbulent component and a stratified component. The turbulent component comes from the dynamic of the atmosphere, which is too difficult to model to perform accurate corrections. In most cases, it can be considered as random for each acquisition date on the scale of the SAR image scene. It can, however, be strongly reduced, or even eliminated, by stacking the interferograms or by filtering [SCH 03, HOO 07].

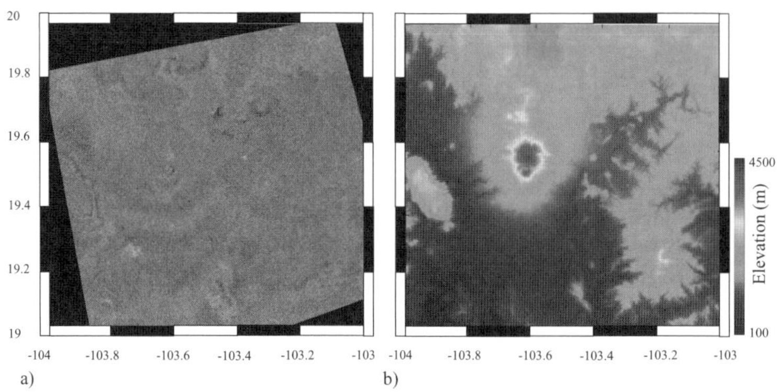

Figure 10.5. *Stratified tropospheric artifact in the volcanic region of Colima, Mexico. a) Interferogram corrected for topographic effects, in ground geometry. Ascending ENVISAT data with a perpendicular baseline of 5 m and a temporal baseline of 385 days. b) DEM (SRTM) for the same geographic area. The interferometric fringes are strongly correlated with the topography. For a color version of this figure, see www.iste.co.uk/tupin/RSImagery.zip*

The stratified part is induced by the variation in the pressure, the temperature and the water vapor content profiles in the troposphere between different radar acquisitions. It corresponds to atmospheric fringes that are, on the first order, correlated to the topography (see Figure 10.5). In the absence of complementary information, this effect can be empirically reduced, by using the correlation between the phase and the elevation observed on the images in non-deforming areas [REM 03, CAV 07]. This method is based on the assumption of the absence of lateral variations of stratification on the scale of the image. It presents a major drawback: a potential partial elimination of the displacement signal, particularly on volcanoes, where the ground displacement induced by a magma reservoir is usually strongly correlated to the topography. Alternatively, when there is a sufficiently dense GPS network on the area studied, it can be used, successfully, to estimate the stratified atmospheric delay and the turbulent delay [WEB 02, JAN 04, LI 06b]. Locally acquired meteorological data can also be taken into account in order to estimate the tropospheric delay [DEL 98]. Some authors also propose to use multispectral data acquired at the same time as the SAR data in order to estimate the water vapor content and to correct the tropospheric effects [LI 06a]. Finally, another interesting approach consists of using temperature profiles, pressure and water vapor content provided by global meteorological models in order to calculate the tropospheric delay [DOI 09, JOL 11]. These models have a temporal resolution of several hours and a spatial resolution of several tens of kilometers. Figure 10.6 shows the variation of the delay versus elevation ratio, a ratio calculated on the basis of global meteorological models in the area of Colima Volcano in Mexico. There is a seasonal variation of amplitude 8.8 rad/km. More precisely, this means that between an ENVISAT image acquired during the dry season and another one acquired during the wet season, on the area of the volcano that extends from 100 to 4,460 m altitude, we can observe 5 fringes that could be misinterpreted as a 15-cm displacement; whereas, on this area, having corrected this artifact, much smaller displacement of approximately 1.5 cm per year, has been identified [PIN 11].

10.3. Combination of displacement measurements

10.3.1. *Analysis of time series*

Today, SAR data are available for more than 20 years. Particularly, for the C band, the most used band, the two ERS satellites are compatible for SAR interferometry, and under extreme configurations, we can obtain an interferogram between an ERS and an ENVISAT image [PEP 05, PER 06][4]. Furthermore, with the two satellites ERS2 and ENVISAT having been operational at the same time, it is possible to put together the displacement time series independently acquired with ERS and ENVISAT when the common time span is sufficient. For given sites, several hundred images are available so that, in the past few years, methods seeking to exploit the time series in order to reduce the errors and increase the accuracy of interferometry have been developed – thus favoring its ability to detect and quantify displacements of small amplitude (of the order of mm/year). Such an accuracy is required for the study of phenomena characterized by a small deformation rate such as the inter-seismic displacement along an active fault or the current isostatic readjustment. The first processing of a series of images consisted of averaging different interferograms covering the same event [SIM 07]. Later, methodological

4 ERS and ENVISAT not having exactly the same wavelength, it is not possible to build a standard interferogram between an ERS image and an ENVISAT image.

developments have allowed going beyond this elementary stage by establishing two different approaches simultaneously: the PS approach and the SBAS approach. Hooper *et al.* [HOO 12a] propose a review of these techniques.

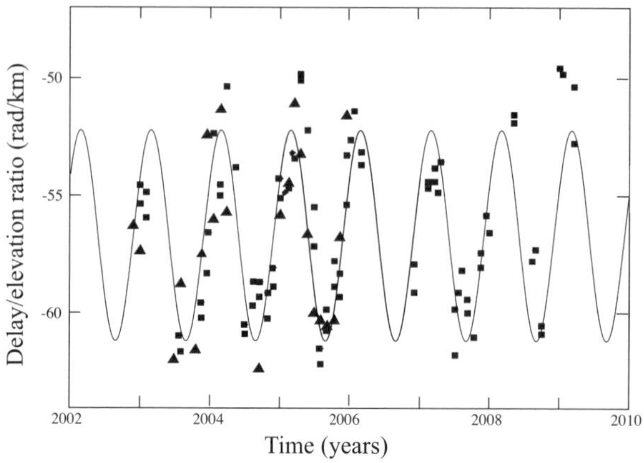

Figure 10.6. *Temporal evolution of the delay/elevation ratio (in rad/km) induced by a stratified troposphere in the vicinity of Colima volcano, Mexico. Ratios represent an average over the 100–4460-m elevation range corresponding to the topography of the volcano. The squares are delay values calculated from the ERA interim global meteorological model at the time of the ALOS acquisitions. The triangles are the values estimated using the NARR meteorological model (according to [PIN 11]) at the time of the ENVISAT acquisitions. The values are calculated at the time of the image acquisition during the descending pass. The sinusoidal curve corresponds to the best adjustment of the seasonal variations obtained by considering all the daily meteorological data (amplitude of 8.8 rad/km)*

10.3.1.1. *Permanent scatterers (PS) approach*

The first method is based on the detection and monitoring of punctual targets: the response of the pixel is dominated by a particular scatterer whose phase presents great temporal stability and remains very little sensitive to the acquisition geometry variations, thus allowing the use of large baselines that cannot usually be exploited on the distributed targets. This is the permanent scatterers or persistent scatterers (PS) method [FER 01]. In urban environments, these scatterers are generally the roofs of the buildings oriented so that they backscatter maximum of energy in the direction of the satellite like a mirror or the result of a reflection on the ground and then on a perpendicular structure. In rural areas, the PS are generally less common, for example rocks with a specific orientation. To make use of these PS, the processing is done at the highest possible resolution, so that neither filtering nor spatial averaging is applied. A common master image is chosen from the available images, and it is combined with each of the other images in the series to produce interferograms. The chosen master image usually corresponds to the one that minimizes the spatial baseline and the temporal baseline of the interferograms produced (see Figure 10.7 (a)). However, the baselines can be very large for certain interferograms, which will limit their coherence. The smartness is then to select and focus on the pixels whose phase remains stable.

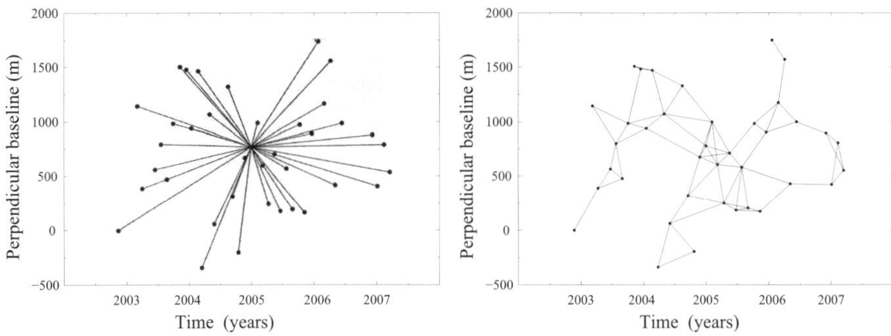

Figure 10.7. *Spatio-temporal distribution of interferograms calculated by two different methods of time series processing. The nodes and the arcs represent, respectively, the 38 images used and the interferograms calculated. a) The case of the PS in which a master image is chosen and only 37 interferograms are produced. b) The case of the small baseline approach (SBAS) for which 71 interferograms are produced, all characterized by small spatial and temporal baselines. The images in this example correspond to the 38 ENVISAT images acquired from November 2002 to March 2007 in the city of Mexico (Mexico) and used for quantifying the urban subsidence (according to [YAN 12a])*

In the classical approach [FER 01], the phase is modeled as a function of the spatial baseline and the temporal baseline, and the displacement is considered to be linear in time. This phase is integrated temporally and then spatially. Different varieties [WEG 05, ADA 09, ZHA 11] of the PS technique have been implemented, modifying or combining one or several previous techniques. *A priori* information on the displacement is often necessary, from which a deformation model is established. In this model, the average displacement rate and the DEM error constitute the two main parameters. The estimation of these two parameters is carried out between neighboring pixels. This approach works in areas where the density of the selected pixels is significant and the displacement is effectively linear in time, as long as a sufficient number of images is used (around 30 images). The PS approach has been successfully applied in urban areas where strong density of stable and bright targets exists [LIU 09, PAR 09, OSM 11, YAN 12a]. Figure 10.8 shows an example of displacement rate obtained using the PS technique, that is the subsidence rate on Mexico City. In natural areas, the application of this technique still remains a great challenge because of the limited PS coverage. However, Hooper *et al.* [HOO 04] have succeeded quantifying the deformation in a volcanic region by using a variety of the PS method, not needing any *a priori* on the temporal function of the displacement [HOO 04].

10.3.1.2. *SBAS approach (small baseline subset)*

The second approach assembles the methods qualified as SBAS, which are based on the minimization of the spatio-temporal decorrelation through the combination of interferograms characterized by a small spatial and temporal baseline [BER 02, SCH 03, USA 03]. This approach is optimal for the pixels whose response is not dominated by a unique scatterer, but by a distribution of scatterers. The small temporal and spatial baseline allows us to maximize the coherence, which facilitates the extraction of reliable phase on a time series. The selected interferograms form a redundant network that connects the images in space and time simultaneously (see Figure 10.7 (b)). The decorrelation noise is reduced by spatial filtering, which causes a loss in spatial resolution. As a result, the high-frequency displacement is usually eliminated during the filtering. After the spatial phase unwrapping, an inversion allows us to obtain the displacement by dates. The SBAS technique has been successfully applied in numerous contexts [SCH 03, CAV 07, LOP 09]. Since its first implementation, this technique has been modified and adapted to the specificity of the phenomena studied. Lopez-Quiroz *et al.* [LOP 09] has applied specific processing techniques in order to adapt to the subsidence measurement of Mexico characterized by large area of deformation, significant rate and gradient of deformation. The version developed within the ANR EFIDIR project by M.-P. Doin called NSBAS, allows us, moreover, to incorporate corrections for the stratified part of the tropospheric artifacts [DOI 11] (the turbulent part is classically eliminated by filtering in the time series). The spatial coverage is generally larger than in the case of the PS method, which makes the correction of orbital effects much easier. However, this method does not provide punctual displacement. Recently, the selection of the interferograms according to a small baseline criterion has been applied on a series of amplitude correlation measurements. Casu *et al.* [CAS 11] have shown that it was possible to benefit from the information redundancy and the high quality of the offset of the amplitude images for image pairs with small spatial baseline, in order to obtain information on the displacement in range and azimuth, on the basis of the offsets of a series of amplitude images with a precision of the order of a 30th of a pixel (compared to a 10th of a pixel with only two images). This is a remarkable variety of the SBAS technique, and it allows us to monitor the events with large displacement.

The PS and SBAS techniques allow us to achieve an accuracy of millimeters/year in displacement rate measurement [HOO 12a]. The SBAS technique provides a continuous description of the deformation structure and its temporal behavior without any *a priori* hypothesis on the displacement, whereas the PS technique enables a quantitative discussion of the disparity of small-scale deformation. These two methods thus provide information on the displacement at different but complementary scales. The accuracy of these two techniques strongly depends on the number of images available, the resolution of the images, as well as that of the DEM. The constraint on the quantity of the images is currently becoming less and less important due to the successive launching of the satellites equipped with high-resolution sensors. Given the advantages and disadvantages of the SBAS and PS techniques, the combination of these two methods opens up new prospects. Certain attempts have already been implemented. Hooper [HOO 08] has combined the SBAS techniques and the PS techniques in order to monitor the time evolution of the displacement associated with volcanic intrusions. The spatial coverage of PS points was improved due to this combination, which facilitates the application of PS techniques in natural areas. In Yan *et al.* [YAN 12a], the subsidence of Mexico City was analyzed by using simultaneously the SBAS and PS techniques in order to measure at the same time the global subsidence and the punctual subsidence associated with isolated objects that have a different behavior from that of their neighbors. In Liu *et al.* [LIU 13], the PS technique was applied on a series of SAR images whose perpendicular baseline is small in order to measure the subsidence of the city of Tianjin (China). Due to the use of small baseline images, the analysis between neighboring points with the PS technique can be done without the DEM, because the DEM error is very small between two neighbouring points. Thus, the impact of the DEM of resolution lower than that of SAR images could be avoided. Given the increasing availability of the high-resolution time series, the combination of these two techniques seems to be able to provide very precise displacement measurements over large areas.

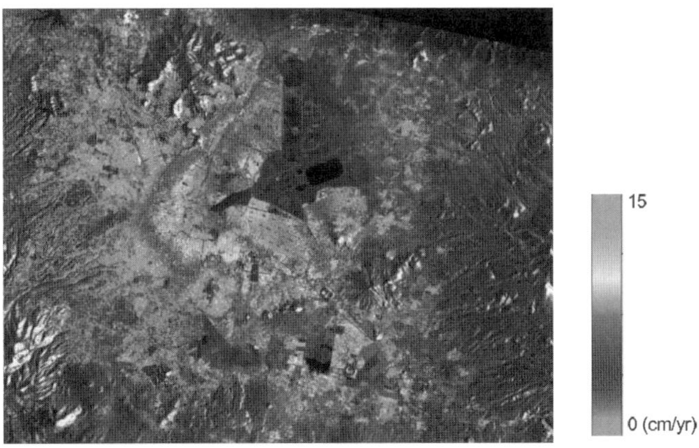

Figure 10.8. *Subsidence rate of Mexico City estimated using the PS technique. The rate is superimposed on the amplitude image. A cycle represents 15 cm per year [YAN 12a]. For a color version of this figure, see www.iste.co.uk/tupin/RSImagery.zip*

Data mining techniques are also being developed with the objective, via extraction of spatio-temporal features, to summarize the time series of satellite images and to encourage the knowledge exploration. The Grouped Frequent Sequential patterns, as proposed and defined in Julea *et al.* [JUL 11], allow us, for example, searching groups of pixels that are sufficiently connected between them and covering a minimum surface. These connectivity and surface constraints allow the extraction of these patterns under standard conditions in terms of execution duration and memory requirement. From the point of view of the duration of the execution and the available memory. These operational aspects are detailed in Julea *et al.* [JUL 12]. Besides the technical interest that these constraints have, the latter also allow the user to extract evolutions that are meaningful with regard to the application. For example, in Meger *et al.* [MEG 11], experiments carried out on interferograms formed using the SBAS approach have shown that it is possible to characterize a seismic fault by highlighting the creep phenomenon along the fault.

In the case of differential interferometry, even after all the corrections detailed in section 10.2.3 and the noise reduction by the processing in the time series, only a relative displacement value can be obtained from the phase, that is to say, a constant shift from the absolute value exists. To obtain the absolute displacement, a reference such as a point on an area whose displacement is known is required. Usually, in practice, the displacement in far-field is considered to be null, which raises the problem for the displacement over large area (larger than 100 km). Yan *et al.* [YAN 13] propose another solution using the displacement measurement issued from the amplitude correlation in order to "register" the displacement obtained by differential interferometry. Similarly, Manconi *et al.* [MAN 12] has combined time series from amplitude correlation and differential interferometry by using the SBAS approach in order to benefit from the advantages of each type of measurement. Other types of geodesic measurement, such as the GPS data, can also be integrated in order to correct the orbital and atmospheric effects and to provide punctual complementary information in 3D, with better sampling frequency [WEB 02, GOU 10]. The integration of these different sources of displacement measurement is often performed during the reconstruction of 3D displacement field.

10.3.2. *Reconstruction of 3D displacement field*

The 3D displacement at the Earth's surface, classically projected on the terrestrial coordinate system axes east, north and vertical (E, N, Up), is commonly used for characterizing the surface displacement induced by earthquakes [WRI 04, PAT 06, YAN 13 or by volcanic activities [WRI 06, GRA 09]. The knowledge of this vector allows us, in particular, to facilitate the interpretation of the displacement field in terms of sources, to calculate maps of deformation rates, as well as subsidence or lifting volumes by integrating the vertical component (Up) of the displacement.

In SAR imagery, the displacements are measured in the direction of the line of sight (LOS or range) and/or in the direction of the satellite movement (azimuth) for each acquisition. These measurements correspond to the projections of the 3D displacement in the directions of each acquisition (Figure 10.9).

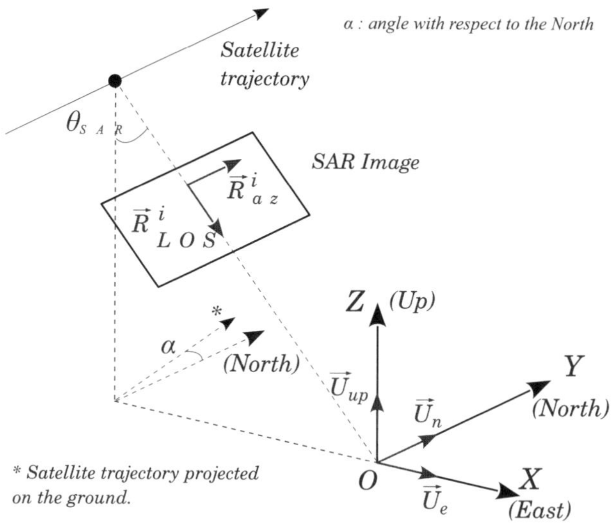

Figure 10.9. *Geometric illustration of 3D displacement and of the displacement in the LOS and in the direction of the radar movement (azimuth) during SAR acquisition (according to [FAL 12])*

We may therefore write:

$$R = PU,$$

where R corresponds to the vector of displacement measured by amplitude correlation or differential interferometry. P corresponds to the projection vector matrix, U denotes the 3D displacement vector to estimate with three components E, N, Up.

To reconstruct the 3D displacement using the displacement measurements obtained from SAR images, at least three different projections are necessary. Diverse acquisition geometries, including variable incident angles (typically for ENVISAT, the incident angles vary from 19° to 44°), different orbital directions (descending and ascending) and different displacement directions (range and azimuth) make this reconstruction possible. Outside of the high latitude areas, the azimuth directions of the ascending and descending passes are nearly collinear. In practice, they are often considered as the same projection. To benefit from the complementary information brought by each of the projections, the common procedure involves combining all the available projections simultaneously and calculating the optimal solution by means of the least squares approach [YAN 12c]. Other combination strategies that consist of either selecting the highest quality projections before the inversion or fusing the inversion results obtained with independent subgroups of projections are proposed in Yan [YAN 11].

In most of the studies, the errors presented in the measurements are considered, in the first approximation, as random and characterized by a Gaussian distribution, which justifies the common use of the least squares approach for the estimation of the 3D displacement. Noting with Σ_R, the error covariance matrix of the measurements R, the 3D displacement estimated by the least squares approach is then given by:

$$U = \left(P^t \Sigma_R^{-1} P\right)^{-1} P^t \Sigma_R^{-1} R. \qquad [10.9]$$

In this approach, the uncertainty associated with the 3D displacement is given by the covariance matrix Σ_U:

$$\Sigma_U = \left(P^t \Sigma_R^{-1} P\right)^{-1}. \qquad [10.10]$$

In many cases, the errors are considered to be independent from one measurement to another, which reduces the covariance matrix Σ_R to a diagonal matrix.

However, the sources of error in SAR imagery are very complex: they come from different perturbations that took place along the radar wave propagation and at the back scattering surface, as well as the noise generated in the electronic processing. Moreover, imperfect corrections (geometric and atmospheric, see section 10.2.3) carried out in the processing chain also induce epistemic errors that remain constant or vary predictably in repeated measurements [YAN 11, YAN 12b]. These diverse sources result in errors with very different characteristics and distributions, making the hypothesis of random independent Gaussian errors questionable. To take random errors and epistemic errors into account at the same time, a fuzzy approach based on the possibility theory [ZAD 78] is preferable. This approach has been applied to displacement measurements by satellite imagery for the first time by Yan and Yan *et al.* [YAN 11, YAN 12b]. The 3D displacement is then estimated by the fuzzy arithmetic based on the principle of the extension of [ZAD 78] according to the equation:

$$U = (P^t \Sigma_R^{-1} P)^{-1} P^t \Sigma_R^{-1} \otimes R \qquad [10.11]$$

where the components U and R are no longer scalar values but fuzzy sets and $\otimes$ refers to the matrix operator of fuzzy multiplication where the sum and the conventional scalar product are replaced by the corresponding fuzzy operations.

This fuzzy approach differs from the probabilistic approach in the manner of representing and propagating the measurement errors in the estimation of the 3D displacement. In this approach, the errors are modeled by possibility distributions, more precisely, here, symmetrical triangular distributions that include a family of probability distribution [MAU 01, DUB 04]. With possibility distributions, the random and epistemic errors are integrated in a unified modeling. According to Yan *et al.* [YAN 12b], the 3D displacement errors obtained with the probabilistic approach provide a lower envelope, whereas the errors obtained with the possibilistic approach provide the upper envelope (see Figure 10.10). The real errors should be situated between these two extremes. The more justified the hypothesis of independent random errors, the closer the real errors are to probabilistic errors, and conversely, the less justified this hypothesis, the closer real errors are to possibilistic errors.

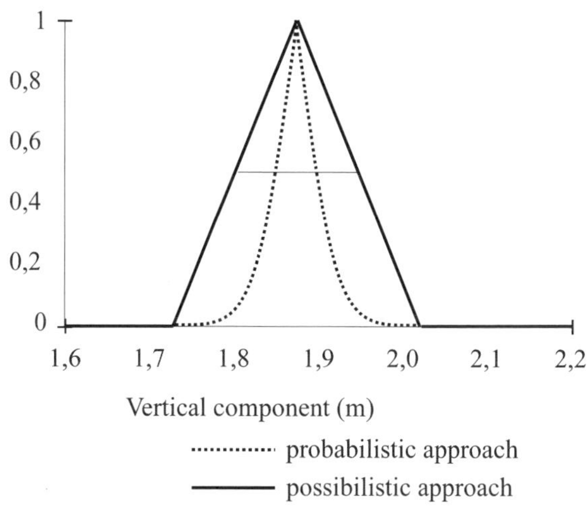

Figure 10.10. *Comparison of possibility distributions resulting from the propagation of uncertainties using the probabilistic approach and the possibilistic approach. For the probabilistic approach, the variance is calculated from equation [10.10] and the possibility distribution is built through equivalence with the Gaussian law. For the possibilistic approach, the input uncertainties were modeled by triangular possibility distributions and propagated according to equation [10.11]. The possibilistic approach gives a more significant uncertainty than the probabilistic approach. The width at half maximum is traditionally used as an uncertainty parameter in the possibilistic approach*

The combination of measurements resulting from amplitude correlation and differential interferometry constitutes an interesting subject in 3D displacement estimation. The measurement in the range direction resulting from amplitude correlation is obtained in the same direction as the measurement resulting from differential interferometry. These two measurements, therefore, correspond exactly to the same displacement. However, in practice, these two types of measurement are considered to be complementary. The measurements resulting from amplitude correlation are often reliable in the near-field where large displacements are observed but they are not very precise in the far-field, where the amplitude of displacement is small. On the contrary, the measurements resulting from a differential interferometry are often available in the far-field where the displacement is small, but they are much rarer in the near-field for large earthquakes, because of the coherence loss and the aliasing problem induced by strong displacement gradients. This is illustrated in Figure 10.11 by taking the Kashmir earthquake (2005) as an example. The spatial distribution of displacement measurements issued from amplitude correlation and differential interferometry for the Kashmir earthquake (2005) is presented in Figure 10.12. These measurements cover at the same time the near-field and the far-field in relation to the fault, approximately 400 km in the North–South direction and 250 km in the East-West direction. In the near-field of the fault, the displacement measurements are obtained from amplitude correlation. In the far-field, the measurements are mainly from differential interferometry (see Figure 10.4), except that in the NW part where there is no available interferometric measurement. The coverage in the far-field is important, which allows us to constrain the modeling of the deformation source in depth, because the surface displacement in the far-field is strongly related to the slip on the fault plane in depth.

The 3D displacement at the Earth's surface, built from these measurements, is illustrated in Figure 10.13. The spatial distribution is smaller than that of the measurements in range and azimuth directions (see Figure 10.12), because of a lack of different projections on certain areas in the far-field. In Figure 10.13, the deformation area, that occupies a strip of approximately 90 km in length and oriented NW-SE, is highlighted. The maximum displacement can be found on the NE compartment that overlaps on the SW compartment. A rupture takes place on the trace of the fault that separates the fault into two main segments. On the south segment, the displacement is relatively small, lower than 2 m in horizontal and 4 m in vertical. On the north segment, the maximum displacement reaches 5 m and 3 m in vertical and horizontal, respectively.

The 3D displacement fields at the Earth's surface are not the information sought after by geophysicists who are interested either in the rheology of the upper layers of the Earth's, or in the depth of this displacement: the geometry and the fault slip distribution, the geometry and the pressure variation of a magma intrusion, etc. In general, all this information is obtained through the inversion of a physical model using the surface displacement measurements resulting from amplitude correlation and differential interferometry. The 3D displacement is not directly used as a source of measurements in the inversion of the model, because of its low spatial coverage and potential errors introduced during its estimation. However, the 3D displacement can also be obtained from the prediction of the physical model, as long as the geometrical and mechanical parameters of the model are available. The integration of the 3D displacement estimated from the surface measurements and the physical model allows for a cross validation between the estimated 3D displacement field and the parameters of the physical model obtained from the surface displacement measurements. In Figure 10.14, the vertical component of the 3D displacement of the Kashmir earthquake (2005) estimated from the surface measurements and predicted by the physical model, as well as the uncertainties associated according to the probabilistic approach and the possibilistic approach, are shown [YAN 12b]. A general agreement between the displacement values obtained by the measurements and by the model is observed, despite the significant processing difference between the two procedures. In the far-field of the fault, there is a very good superposition between the two cases, although the fluctuation of the displacement value estimated by the measurements is significant. In the near-field of the fault, there is a discrepancy due to the defect of the global model used, the maximum displacement is under-estimated by the model near the fault. This comparison allows us to validate the 3D displacement estimation procedure by the least squares approach and the inversion procedure of the physical model. This type of comparison is very useful when another source of measurement, or the ground truth, is not available.

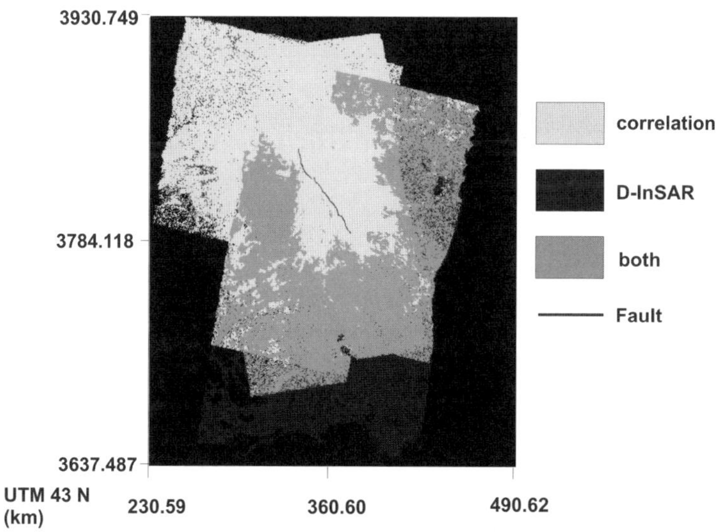

Figure 10.11. *Availability of different types of co-seismic measurement for the Kashmir earthquake (2005) [YAN 11]*

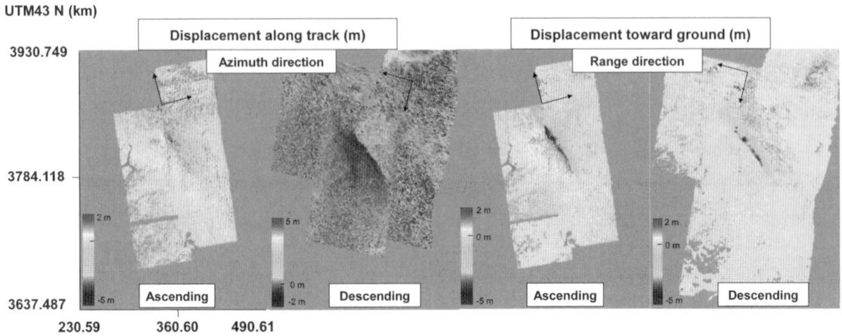

Figure 10.12. *Spatial distribution of the co-seismic displacement induced by the Kashmir earthquake [YAN 13]. For a color version of this figure, see www.iste.co.uk/tupin/RSImagery.zip*

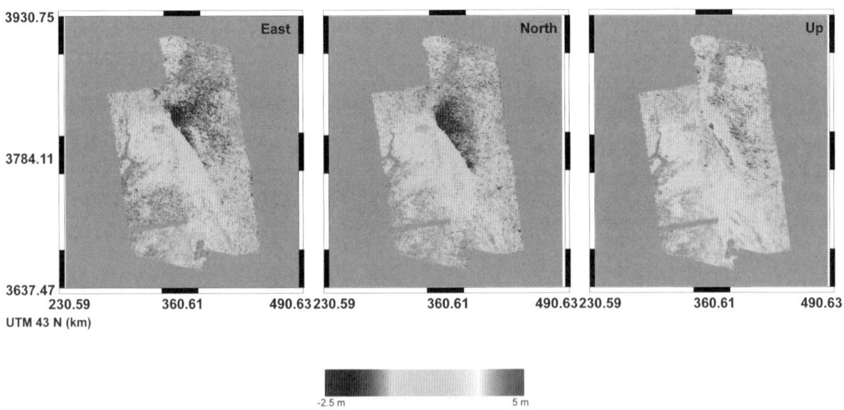

Figure 10.13. *3D displacement at the Earth's surface estimated by least squares approach, using the measurements resulting from amplitude correlation and differential interferometry [YAN 13]. For a color version of this figure, see www.iste.co.uk/tupin/RSImagery.zip*

10.4. Conclusion

In this chapter, we have presented the main techniques for displacement measurement using remote sensing sources, as well as their applications, in particular the use of SAR images that provides double geometric information via sampling, and the phase of the complex data. The arrival of the spatial imagery has caused a true revolution in geodesy by significantly improving our ability to measure the ground displacements and their temporal evolution with great precision over large areas. Spectacular results have been obtained in numerous fields, including the study of subsidence in urban environment, of the co-seismic, inter-seismic and post-seismic motion, of glacier flows, of volcanic deformation, etc. The displacement measurements by remote sensing cover nearly the whole world, with an accuracy within millimeters, which would have been impossible with traditional tools. They provide the predominant sources for the studies of the terrestrial deformation nowadays. Thus, for the latest large earthquakes (since the Wenchuan earthquake in 2008), it is indeed due to SAR and GPS measurements that have been made available to the public, that we were able to model the fault, understand the deformation mechanism and predict the area of the aftershocks so efficiently.

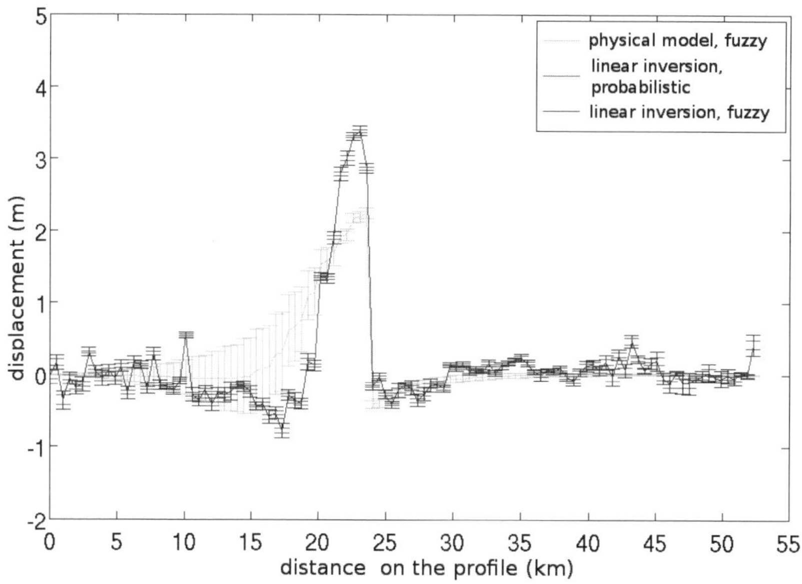

Figure 10.14. *Comparison of 3D displacements at the Earth's surface, obtained from the least squares approach and from the prediction of the physical model in the case of the Kashmir earthquake (2005) [YAN 12b]*

The SAR image processing techniques have been improved rapidly in the past 20 years, reaching processing chains that go from the calculation of the offsets to the reconstruction of the 3D displacement, from the formation of the interferograms to the inversion of a series of images in order to obtain the temporal evolution of the displacement, or even to the PS detection to correct the elevation of targets and to measure their displacements.

Today, these techniques are still being improved in order to integrate as much as possible the benefit of the high spatial resolution and the increasing frequency of the data acquisition for terrestrial displacement measurement. At the same time, efforts to combine different techniques, for example amplitude correlation and differential interferometry, the PS technique and the SBAS technique, seek to make the best use of the information contained in the data, using the complementarity of the different approaches.

Currently, the displacement measurement by remote sensing is still mainly applied to past events. The time series are particularly used for the study of the time evolution of the events that have taken place before data processing. With the next launching by ESA of two Sentinel-1 satellites, SAR data will be acquired nearly everywhere on the Earth at least every 6 days, with a resolution similar to that of ENVISAT. By adding the data issued from other satellites (RADARSAT-2 and the future constellation RADARSAT of three satellites, TerraSAR-X and TanDEM-X, the four satellites COSMOS-SkyMed, ALOS-2, etc.), real-time monitoring by time series will become possible. The availability of a considerable amount of data because of the high spatial resolution and the strong repetitiveness of acquisitions constitutes a true technological challenge. If we do not wish to be quickly limited by the storage capacity and computation time, we will have to adapt our ways of working by relying on data mining and by adapting our tools and means of computation. It will also be important to modify our processing algorithms of time series so that they can integrate the new data gradually, without having to restart from the beginning. Given this challenge, the combination of remote sensing techniques in real-time and physical models (extensively used in meteorology) is possible. It should allow us to predict the evolution of an event such as a magma reload of a reservoir located beneath an active volcano or a rupture of serac. Data assimilation, today extensively exploited in the science of atmosphere and oceanography, will then open up new prospects for the observation and prevention of natural hazards and will lead us toward a new age of remote sensing data usage in displacement measurement.

Chapter 11

Data Assimilation for the Monitoring of Continental Surfaces

The objective of this chapter is to present the basis of data assimilation by limiting the mathematical developments as much as possible. Furthermore, we emphasize here the data assimilation of remote sensing observations in land surface models, although the assimilation techniques presented can be applied to different kinds of dynamical models. We have used the notation described in [IDE 97], which has been traditionally adopted in data assimilation.

Besides the references mentioned in this chapter that the reader can refer to, we consider three documents to be benchmarks in their respective fields for anyone who wishes to study them in greater depth. They are as follows:

– dynamic modeling of natural land surfaces [GUY 99];

– methodological aspects of data assimilation [BOU 99];

– data assimilation applications to hydrology [HOU 10].

11.1. Introduction to data assimilation in land surface models

While global changes and particularly climate changes are about to alter our environment, having a spatialized and dynamic view of terrestrial surfaces is a concern for both the researcher and the manager of water and biotic resources. Remote sensing data, thus, seem an essential add-on to *in situ* measurement that are

Chapter written by Lionel JARLAN and Gilles BOULET.

often costly, scarce and very localized. Methodologies presented in the previous chapters have shown how the electromagnetic variable measured by a space sensor could be turned into useful information. In particular, classification aims extracting data on land use. Inversion enables to assess key biophysical characteristics such as soil moisture, leaf area index or temperature. Thus, remote sensing gives access to a large amount of information on the time and space variability of surface characteristics.

However, besides uncertainties on the measurement itself, remote sensing only gives an overview of terrestrial surfaces. The first limitation is due to the physics of the interaction between the electromagnetic wave and the surface, which limits the penetration depth to a value close to the sensor's wavelength. In the visible part of the electromagnetic spectrum, sensors can only "see" leaves in the foreground and cannot observe undergrowth when canopy is dense, for example. In the microwave part (wavelengths $> 1\,\mathrm{cm}$), penetration depth is higher but data we may hope to inverse on soil moisture, for example, will only concern the first centimeters, yet agronomists are interested in the water content in the root zone. With the exception of geostationary satellites whose spatial resolution is low ($> 3\,\mathrm{km}$), the second limitation is due to the orbital characteristics of the satellite, which cause remote sensing images to provide only a snapshot of the terrestrial surfaces with a time step that does not allow to monitor the quick dynamics of land surfaces. For example, in the thermal infrared (TIR) domain, monthly temperature values are not enough to solve the multiscale evolution of the energy balance, because intra- and interdaily variabilities are extremely strong.

In the last two decades, deterministic models were developed to simulate the hydrological and ecological processes of continental surfaces. A model is a mathematical representation of biophysical processes that govern the functioning of land surfaces. A dynamic model simulates the evolution over time of the system it describes. Ideally, these processes should result from the fundamental principles of physics. In practice, even for mechanistic models, some processes that are not well known or too complex are set using empirical relationships. In reality, the high heterogeneity in both terrestrial surfaces and climate, as well as ill knowledge of initial conditions, add other uncertainty factors. Finally, surface models allow us to predict the main variables describing the functioning of land surfaces (temperature profile within the soil-vegetation continuum, space/time dynamics of vegetation and sub-surface water fluxes, etc.) in a continuous and spatialized, yet imperfect, manner.

The purpose of assimilating remote sensing data into surface models is to use the spatialized and repetitive snapshot view of the surface due to remote sensing, on the one hand, and the continuous prediction of dynamic models, on the other hand, in order then to obtain the best estimation of the state of continental surfaces, at every time and at every point. It brings together many approaches, from the complex system taking carefully into account all errors to the calibration of a small number of parameters. The advantage of data assimilation into a dynamic model, compared to satellite data inversion, is the dynamic link between observable variables and some variables not directly observable from space. This dynamic link allows us to propagate observed data toward non-observable variables, such as the root-zone soil moisture or evapotranspiration.

Sensor	Period	Layer	N	Satellite	Model	Assimilation
AMSR-E	Day	Surface	23	0.38 ± 0.02	0.43 ± 0.02	0.50 ± 0.02
		Root	22	-	0.40 ± 0.02	0.46 ± 0.02
AMSR-E	Month	Surface	12	0.41 ± 0.08	0.50 ± 0.09	0.57 ± 0.08
		Root	11	-	0.42 ± 0.10	0.54 ± 0.08
SMMR	Month	Surface	66	0.32 ± 0.03	0.36 ± 0.03	0.43 ± 0.03
		Root	33	-	0.32 ± 0.05	0.35 ± 0.05

Table 11.1. *Correlation coefficients of the comparison of surface soil moisture retrieved from AMSR and SMMR sensors and predictions of a land surface model to a set of in situ data before and after assimilation [REI 07]*

In geosciences, early developments in data assimilation into dynamical models have been initiated by the meteorology and oceanography communities [DAL 91]. While Maas [MAA 88] had laid the first theoretical foundations, it was not until Beven and Binley's [BEV 92] work in hydrology that real continental surface application of data assimilation emerged. Table 11.1 presents a very spectacular result (despite appearances) found in [REI 07]. the authors assimilated products of surface soil moisture derived from the observations of advanced microwave scanning radiometer for the Earth Observation (AMSR) sensor (onboard the Aqua satellite launched in 2002) and scanning multichannel microwave radiometer (SMMR) sensor (onboard the Nimbus-7 satellite launched in 1978) into a land surface model. Therefore, Table 11.1 shows the comparison between the moisture products of the two satellite acquisitions and the predictions of the model to a set of *in situ* data (the columns *satellite* and *model*, respectively), and the same with the joint use of both via data assimilation (*assimilation* column). Better correlation coefficients with *in situ* data are obtained if satellite products and the model are used jointly through data assimilation, rather than using the two separately. In addition, data assimilation of satellite products also improves the model performance on the root-zone soil moisture predictions although it is not directly observable from space. The purpose of this chapter is simply to provide the comprehension material of this result, which might seem just miraculous at first.

In this chapter, we have decided to discuss "data assimilation" for any technique aiming to combine models and observations (in the best possible way according to a statistic performance criteria or not) whether we are talking about real-time correction of variables of the model (data assimilation in the sense of meteorologists and oceanographers) or about parameter identification. These two approaches group together many different techniques and the main ones will be outlined from a theoretical and practical point of view. In section 11.2, we introduce the different elements of an assimilation system, including a few insights into the dynamic modeling of natural surfaces. Data assimilation and parameter identification, as well as their foundations, are presented from a theoretical point of view in section 11.3. Finally, more practical aspects of data assimilation are described in section 11.4.

11.2. Basic concepts in data assimilation

In this section, we present the different elements of a data assimilation system (Figure 11.1) and the main definitions of the terms used.

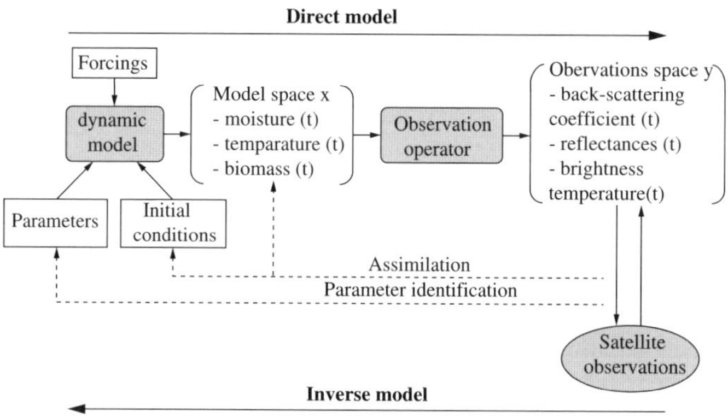

Figure 11.1. *Main elements of data assimilation*

11.2.1. *Elements of a data assimilation system*

11.2.1.1. *Dynamic modeling of natural land surfaces*

The Earth's surface represents a dynamic system[1]. Its mathematical representation is a deterministic[2] dynamic model that describes, over time, the evolution of the system's state. It is made up of algebraic or partial derivative equations enabling us to know the unique state of the system in the future, based on a set of initial conditions. These models are generally Markovian in the sense that the state at any moment is fully described from the state at the previous time interval and external "forcing" variables. For land surface models, the equation system relies on

1 That evolves throughout time.

2 That maps a given current condition to one single future stage.

energy, water and/or carbon balance equations. Practically, the resolution of this equation system, often nonlinear and potentially chaotic, is based on time discretization and numerical integration, with a time interval well below the typical time dimension of the process represented in order to avoid discrepancies due to nonlinearities.

The main equation of the dynamic model for the three budgets (energy, water and carbon) can be written under the generic form of a partial derivative equation like:

$$\text{capacity}\,(\mathbf{z})\,\frac{\partial \mathbf{z}}{\partial t} = \text{sources}(\mathbf{z}) - \text{sinks}(\mathbf{z}) \qquad [11.1]$$

where $\mathbf{z}$ is a vector (or a scalar). Figure 11.2 shows, in a (very) simplified manner, the three budgets solved by land surface models as well as the sources (in full lines) and sinks (in dotted lines) and the associated scalars. Thus, for the energy balance (Figure 11.3(b)), "capacity" is in this case thermal capacity; $\mathbf{z}$ refers to surface temperature or a soil temperature profile discretized at different depths, source is incoming radiation and sinks encompass all other dissipation terms (emitted radiation, latent and sensible heat fluxes, conductive flux in the ground, etc.).

In practice, surface models consider the vertical column made up of the soil and the vegetation canopy as the elementary unit, without explicitly considering interactions with neighboring units in order to limit both calculation time and complexity. This is obviously a rough approximation of reality. Moreover, they distinguish themselves in the mathematical representations adopted for the source and sink terms, and in the parameter values related to these models. The parameters involved in these mathematical representations[3] are either measured directly in the field and then spatialized using cartographic support (map of the ground texture, digital elevation model, etc.) or derived indirectly through calibration on field measurements (dynamic roughness length). Some parameters such as albedo[4] can be easily be measured at the local scale and on a larger scale due to remote sensing, and the relationship between the scales is linear. For some others, measures are more complex or even impossible at a local scale (aerodynamic resistance, stomatal conductance), and the multiscale relationship is nonlinear. Thus, it is very difficult to have representative values for an area covering several thousands of square kilometers. Their values are, therefore, calibrated on observations that are sensitive to these pseudo-constant parameters. These calibration approaches are grouped together under the term *"parameter identification"*.

---

3 These parameters will be considered to be constant on the time scale that is characteristic of the represented process: we then speak of "pseudo-constants".

4 The albedo, in natural surface modeling, has a value comprised between 0 and 1 and corresponds to the part of the incident solar energy reflected by the surface (see section 3.1.6).

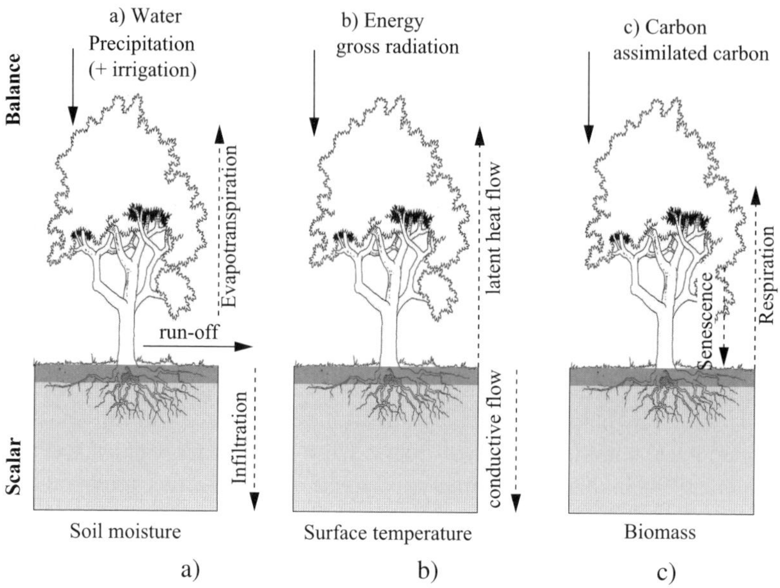

Figure 11.2. *a) Water balance, b) energy balance and c) carbon balance; the sink terms are identified (dotted line) and so are the source terms (full lines), which come up in the balance equation [11.1] as well as the scalar values considered*

The dynamic model is finally forced by external variables called *forcing variables* (see Figure 11.1), which are imposed onto the model. In the case of land surface models, the incoming solar radiation, the air temperature or the precipitations are forcing variables. For the water balance, the source term (Figure 11.2) is generally imposed onto the model (the water inputs are considered as independent from the state of the system). This is not the case for the carbon budget since the vegetation growth depends on the photosynthetic activity that depends itself on the present biomass quantity.

Simply suppressed, the state of natural terrestrial surfaces, at a given time, for the model (*model space*) (see Figure 11.1) is made up of a vector of variables defined on a space grid that can differ from the satellite data grid. For example, we can choose soil moisture and temperature in the soil at different depths, leaf area index, etc. These variables are simulated by the model from initial condition and are called *prognostic variables* or "state variables", because they suffice to characterize the system studied at time t. Some of these prognostic variables will be corrected through data assimilation, resulting in "analysis variables". Finally, *diagnostic variables* are those derived from *prognostic variables*. These variables are simulated by the model but no initial condition is required to assess them over time. For example, the sensible heat flux derived from the temperature is a diagnostic variable.

11.2.1.2. *The satellite measurements and the observation operator*

The biophysical characteristics of continental surfaces directly affect the spectral signature of surfaces at different wavelengths. In the visible and near-infrared domain, the sensor measures the solar energy reflected by the Earth's surface on a limited range of wavelength. This value, normalized by incident solar energy, is called "reflectance". Regarding surfaces covered by vegetation, chlorophyll pigments strongly absorb in the red while the leaf's internal structure reflects strongly in the near infrared. This contrasted behavior has led to use the *normalized difference vegetation index* (NDVI), given by equation [6.1], which can be directly linked to vegetation density (aerial biomass, leaf area index, etc.). In the TIR domain, sensors are radiometers measuring a brightness temperature[5], which is directly linked to the surface's temperature according to the Stefan–Boltzmann law[6]. This domain of wavelength seems, in particular, well adapted to monitor the plant's water stress (during water stress, the plant's temperature rises). Finally, the microwave domain is particularly sensitive to the surface's water content and, in particular to the moisture of the soil's first few centimeters[7]. For the microwave domain, the wave is very sensitive to the surface's dielectric constant that is dominated by the water content variations of the studied object. These three domains of wavelengths are, therefore, complementary because they allow us to have access to information on several key variables related to each main budget of the natural land surfaces.

However, the link between the electromagnetic magnitude measured by a sensor (reflectance, brightness temperature, etc.) and the prognostic variable of the model we wish to constrain through data assimilation (leaf area index, temperature, soil water content, etc.) is not direct. Furthermore, there can be a difference in scale between the size of the satellite pixel and that of the grid point of the surface model, resulting in spatial representativeness issues. The observation operator checks that the prognostic variable and satellite observation match. This observation operator will process the spatial representativeness of the satellite observation (by averaging the number of satellite pixels within a grid point of the model's grid, for example), and turn the electromagnetic magnitude measured by the sensor into a variable of the model. This last operation uses a model, the observation model, which can simply be

---

5 It corresponds to the temperature of the black body that would emit the same energy (i.e. it would have the same brightness) as the surface observed.

6 This law can be deduced from Planck's law, through equation [3.14].

7 For the wavelengths used by current sensors, the penetration depth does not go above several centimeters.

an empirical relationship between satellite measurements and field measurements of the prognostic variable based on regression (see [PRI 91] who carried out regression between the NDVI and the aerial biomass of Sahelian prairies), or a physically based model simulating the interaction of the electromagnetic wave with the surface's components. The most commonly used models are radiative transfer models as described in section 4.3.4. As shown in Figure 11.1, the surface's model outputs (soil moisture and temperature, biomass, etc.) are inputs of the observation model that predicts satellite measurements (reflectance, NDVI, brightness temperature, backscattering coefficient, etc.).

11.2.1.3. *The uncertainties*

Models, as well as observations, are uncertain. Practically, there are five types of uncertainties that may coexist within the modeling and observation system:

– those related to variables whose evolution is simulated through the dynamic model, and which depends de facto on uncertainties linked to initial conditions;

– those related to forcing variables that are often included in errors associated with prognostic variables or modeling;

– those related to parameters of the dynamic model that are difficult to quantify, especially in hydrology;

– those related to observations of state variables that are essentially linked to ill measurements of the instrument and representativeness errors;

– finally, those related to modeling dynamic model and observation model if any.

To take these uncertainties into account, we need information on errors. There is no reason why the error should be the same for each configuration of the system observation configuration, hydrological condition, etc. If we were capable of repeating the experiment, we could then get statistics on error distribution and, within the limits of a number of infinite achievements, its probability density function. This is not the case in general and, in practice, it is necessary to make assumptions on the form of this density. The most commonly used one is the Gaussian form. In this case, errors are fully documented according to their average and variance. In an algebraic formulation, they can be expressed as variance/covariance matrices. This is under this latter form that we will specify errors for all most popular data assimilation methods.

Determining these errors is the longest and the most delicate process in the implementation of a data assimilation system. It requires good knowledge of the system and if not, then close interaction between specialists in applied mathematics, who are in charge of implementing the method, and remote sensing and thematic specialists, would be needed.

11.2.2. *Notations and definitions*

The vector $\mathbf{x}$, of a size n, gathers all the "prognostic" variables allowing us to define at every point the "true" state[8] of the system that we study. Data assimilation, combining the predictions of the model with observations, will allow us to obtain an optimal estimate (in the statistic sense of the term) of this true state: the analysis variables, grouped together in the vector $\mathbf{x}^a$ (a stands for "analysis"). In some cases, we can have an approximate idea of the true state (the prediction of the model before using observations or a climatology). This *a priori* information is called the "background" and denoted as x_b.

The operator of the dynamic model will be denoted as $\mathcal{M}$. $\mathbf{x}_{i+1}$ predicted by the model will be given by:

$$\mathbf{x}_{i+1} = \mathcal{M}(\mathbf{x}_i, \mathbf{u}, \mathbf{p}) \qquad [11.2]$$

where $\mathbf{u}$ is a vector representing the set of forcing variables and $\mathbf{p}$ is the pseudo-constant vector or parameters. Several classic approaches outlined in the following section need a linearization of the nonlinear operator $\mathcal{M}$. The linearized version (the tangent linear model around the state $\mathbf{x}$ at the time I will be denoted by $\mathbf{M}_i$and $\mathbf{x}_{i+1}$) will be written as:

$$\mathbf{x}_{i+1} = \mathbf{M}_i(\mathbf{p})\, \mathbf{x}_i + \mathbf{F}_i(\mathbf{p})\mathbf{u} \qquad [11.3]$$

where $\mathbf{F}_i$ is the linear operator that connects state $\mathbf{x}$ to forcing.

The assimilation system is fed with a set of m observations denoted by vector $\mathbf{y}$. Its size is, in general, lower than that of the state vector of the model ($m < n$) because there are generally fewer observations than variables in the model. It may be composed of heterogeneous observations (for example satellite observation in different wavelength). The observation operator, which is also potentially nonlinear (for example complex radiative transfer model), connects the state of the system $\mathbf{x}$ to observation space $\mathbf{y}$. It is denoted by $\mathcal{H}$. The vector $\hat{\mathbf{y}} = \mathcal{H}(\mathbf{x})$ represents the projection into observation space of the estimated state $\mathbf{x}$. $\hat{\mathbf{y}}$ and $\mathbf{y}$ would coincide if these observations and the state vector were perfect (in the absence of modeling errors). Similarly, its linearized version around the state $\mathbf{x}$ at the instant i, $\mathbf{x}_i$ will be denoted by $\mathbf{H}_i$:

$$\hat{\mathbf{y}}_i = \mathbf{H}_i \mathbf{x}_i \qquad [11.4]$$

8 This "true" term is a little stretched because, for environmental systems, the "true" state is represented by an infinite number of variables or parameters. Furthermore, for the model, the so-called "true" is only a discretized representation of reality.

Errors are represented by probability density functions. For example, the background is separated from the true state of the system by an error vector denoted by $\mathcal{E}^b$:

$$\mathbf{x}^b = \mathbf{x}^t + \mathcal{E}^b \quad [11.5]$$

where $\mathcal{E}^b$ is called the background error. Analysis is also contaminated by errors and:

$$\mathbf{x}^a = \mathbf{x}^t + \mathcal{E}^a \quad [11.6]$$

where $\mathcal{E}^a$ is called the analysis error. Similarly, observations move away from the true state of the system:

$$\mathbf{y} = \mathcal{H}(\mathbf{x}^t) + \mathcal{E}^y \quad [11.7]$$

where $\mathcal{E}^y$ is the observation error. These errors include instrumental errors, representativeness errors as well as errors due to the observation model.

Finally, the dynamic model is also uncertain and equation [11.2] becomes:

$$\mathbf{x}_{i+1} = \mathcal{M}(\mathbf{x}_i, \mathbf{u}, \mathbf{p}) + \mathcal{E}^q \quad [11.8]$$

where $\mathcal{E}^q$ is the modeling error. Assimilation methods distinguish themselves through the consideration, or not, of modeling errors. We speak of a strong constraint if the model is supposed to be perfect (no modeling error, such as the 4D-VAR) and of a weak constraint if the model is not supposed to be perfect (such as the ensemble Kalman filter (EnKF)).

Classic linear approaches presented in the following part assume normal (Gaussian), null averages, and time-uncorrelated errors (white Gaussian noise of a null average):

$$\begin{aligned} \bar{\mathcal{E}^b} &= \langle \mathcal{E}^b \rangle = 0 \\ \bar{\mathcal{E}^y} &= \langle \mathcal{E}^y \rangle = 0 \\ \bar{\mathcal{E}^q} &= \langle \mathcal{E}^q \rangle = 0 \end{aligned} \quad [11.9]$$

These error distributions can, thus, be summed up by their variance/covariance matrices of background $\mathbf{B}$, observation $\mathbf{R}$ and modeling $\mathbf{Q}$ errors:

$$\begin{aligned} \mathbf{B} &= \langle (\mathcal{E}^b - \bar{\mathcal{E}^b})(\mathcal{E}^b - \bar{\mathcal{E}^b})' \rangle = \langle \mathcal{E}^b \mathcal{E}^{b'} \rangle \\ \mathbf{R} &= \langle (\mathcal{E}^y - \bar{\mathcal{E}^y})(\mathcal{E}^y - \bar{\mathcal{E}^y})' \rangle = \langle \mathcal{E}^y \mathcal{E}^{y'} \rangle \\ \mathbf{Q} &= \langle (\mathcal{E}^q - \bar{\mathcal{E}^q})(\mathcal{E}^q - \bar{\mathcal{E}^q})' \rangle = \langle \mathcal{E}^q \mathcal{E}^{q'} \rangle \end{aligned} \quad [11.10]$$

$\mathbf{B}$, $\mathbf{R}$ and $\mathbf{Q}$ are symmetrical and positive definite. The null average hypothesis means that there is no bias between the background, the observations and the true state of the system. A prerequisite to the implementation of a data assimilation system is to check for the absence of a bias and, where applicable, to use methods in order to eliminate this bias (see section 11.4).

Diagonal terms represent uncertainties we have on each observation for the matrix $\mathbf{R}$ (error variances of observations) and on each variables/parameters pair of the model that we will try to correct for $\mathbf{B}$ (error variances in the background). Non-diagonal terms are more subtle to interpret. They are proportionate to correlations between two observations for $\mathbf{R}$ and two variables of the background for $\mathbf{B}$. They represent the varying smoothness of the background (see section 11.4).

One of the essential diagnoses for data assimilation is the deviation between the observations and the projection of the state $\mathbf{x}$ into this space $\mathbf{y} - \mathcal{H}(\mathbf{x})$. When $\mathbf{x}$ is considered equal to the background $\mathbf{x}^b$, this vector is called the innovations vector. When considering the analysis state $\mathbf{x}^a$, this vector is called "residues" (meaning analysis residues). Finally, the vector $(\mathbf{x}^a - \mathbf{x}^b)$ is called "analysis increment". The statistic analysis of these vectors helps us diagnose the good or bad functioning of the assimilation system (see 11.4). These magnitudes are graphically shown in Figure 11.3 for the sequential and variational approaches.

11.2.3. *Data assimilation: an inverse problem*

Knowing the initial conditions, parameters and forcing values, the model in a direct mode allows us to predict the diagnostic and prognostic variables (Figure 11.1). As opposed to the direct mode, the data assimilation is an inverse problem that consists of using observations of one or several prognostic variables (or observations connected to the prognostic variables of the model through an observation operator) to correct the prognostic variables themselves or the inputs of the model (initial conditions, parameters and, less frequently, the forcing). Regardless of the mathematical methods used that we present in due detail hereafter, the following two main approaches, with contrasting philosophy, have been applied to land models of continental surfaces (see [MAA 88]):

– calibration or parameter identification: here the values of the parameters are adjusted, mainly in order to limit the systematic errors (bias) between model predictions and observations;

– the correction of prognostic variables: it may take the form of a correction of prognostic variables of the model, each time an observation is available (Kalman filter type sequential approach) or corrections of the initial conditions of this variable on a given time window by using all observations available on this window (variational approach of a 2D-, 3D- and 4D-VAR type).

The assimilation consists of merging observations and a dynamic model, as well as their associated uncertainties in order to obtain an estimate of the distribution of the true state of the system. Statistically speaking, we search the *a posteriori* distribution $p(\mathbf{x}|\mathbf{y})$ showing the distribution of the true state $\mathbf{x}$ given a series of uncertain observations $\mathbf{y}$. The Bayesian inference theory provides us with a statistical framework to rigorously formalize this problem and a good start for describing the numerous data assimilation methods. This theory allows us to express this *a posteriori* distribution without any hypothesis on the *a priori* distributions or the dynamic model by Bayes' law:

$$p(\mathbf{x}|\mathbf{y}) = \frac{p(\mathbf{y}|\mathbf{x})p(\mathbf{x})}{p(\mathbf{y})} \qquad [11.11]$$

where $p(\mathbf{x})$ describes the *a priori* distribution of the true state of the system; $p(\mathbf{y})$, that of observations, is a normalization factor so that the $p(\mathbf{y}|\mathbf{x})$ equals 1; and $p(\mathbf{y}|\mathbf{x})$ is the distribution that observations would take for the searched state $\mathbf{x}$, and which reflects measurements, modeling and representativeness errors. The maximum of this *a posteriori* distribution corresponds to the likeliest solution. In practice, the Bayesian inference is very difficult to implement for solving a concrete problem, except in the (very rare) cases of normal *a priori* distributions whose averages and variances are known, and of the linearity of dynamic and observation models. Apart from these hypotheses (hardly ever met in practice), the determination of $p(\mathbf{x})$ is literally impossible. However, certain approaches will try to represent this *a posteriori* distribution using discrete sampling with Monte Carlo type methods (particular filter, Kalman filter).

11.3. Different approaches

11.3.1. *Brief history and classification*

Historically, the first "analysis" was visually carried out by weather forecasters from the very few weather stations available, and using the previous analysis. Afterward, research was initiated by the need for a theoretical framework and, above all, the need for an automated process. The data assimilation issue was, thus, tackled in the past using the following two different approaches:

– The deterministic approach based on the theory of the optimal control and on the calculation of variations. This approach gave birth to variational assimilation techniques (3D-VAR and 4D-VAR) that have been traditionally used by meteorologists and oceanographers alike. The objective is to minimize a cost function so that predictions of the model are as close as possible to observations, the solution being searched around an *a priori* state. Minimizing this cost function in the case of a linear observation operator and normal distributions of observation and background errors leads to determining the maximum of distribution p. The best solution is the solution that is the likeliest.

– The statistic approach derived from the estimation theory or from the Bayesian inference, which translates into the *best linear unbiased estimate* (BLUE), a minimum variance estimator under the same hypotheses. In other words, rather than searching for complete *a posteriori* distributions, several "statistic" approaches are restricted to order 2 moments, thus allowing an algebraic formulation of the inverse problem and a relatively simple numerical resolution. The Kalman filter and its spin-offs, among others, are derived from the *best linear unbiased estimate*. This approach allows us to determine the optimal solution, defined as the one corresponding to the minimum variance as well as the error associated with this solution as opposed to the deterministic approach.

Bearing in mind that normality and linearity hypotheses, which are prerequisites to the optimality of the analyzed state, are hardly ever met in an actual problem, we briefly describe these two types of approaches because the resulting methods are very popular. Stochastic approaches allow us to bypass some of the hypotheses we began with, which leads most of the time to a longer calculation time; the latter, however, may be easily tolerable for a data assimilation issue into a land surface model that is the subject of this chapter. Some of them (ensemble Kalman and particle filters) are also mentioned.

There are several ways to classify data assimilation methods. Optimality may be a criterion, and methods would thus go into optimal methods and suboptimal methods, but this criterion is a little bit artificial since optimality conditions are hardly ever proven in a real problem. Bouttier and Courtier [BOU 99] introduced, in this respect, a more satisfying classification by speaking of intermittent assimilation, for which the state of the system is discontinuous each time an analysis is carried out, and, conversely, of continuous assimilation. In this introductory chapter, we will stay with the most commonly used classification that distinguishes between sequential methods (section 11.3.2), which correct the state of the system every time an observation is available (i.e. the Kalman filter and its variants), and variational methods[9] (section 11.3.3), which consist of adjusting observations to the model's predictions on a given time window (Figure 11.3). Finally, we will lay special emphasis on parameter identification in section 11.3.4, which could be part of variational methods because, from a technical point of view, the objective is also to adjust observations to the model's predictions on a given time window, except that it is the parameters of the model that are optimized, not the initial conditions.

9 The word variational generally refers to the technique used, based on the theory of variations and optimum control, but it is also used by extension to every inverse problem aiming to adjust model predictions to observations (see section 11.3.3).

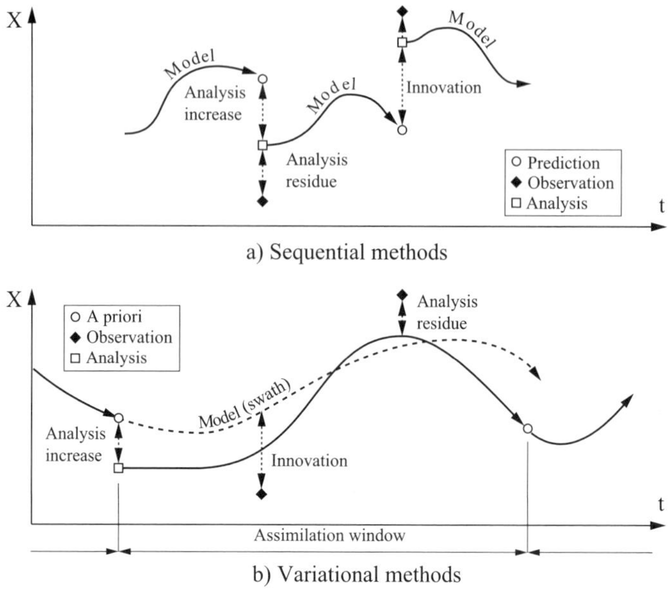

Figure 11.3. *Sequential and variational methods and definitions of some of the diagnoses in data assimilation. The case where the variable observed is also the variable analyzed (i.e. the observation operator is the identity matrix)*

11.3.2. *Sequential methods*

11.3.2.1. *The best linear unbiased estimate (BLUE) and the Kalman filter*

Let us suppose that a (certain) scalar quantity x has been measured at the same place and at the same date using two (uncertain) methods. Let y_1 and y_2 (respectively σ_1 and σ_2) be the values of x measured by the two methods (respectively the uncertainties associated with these two measures). We may consider that y_1 and y_2 are realizations of two random variables Y_1 and Y_2. If the measures are not biased, the equation can be written as:

$$Y_1 = x + E_1 \qquad \text{and} \qquad Y_2 = x + E_2 \qquad [11.12]$$

where E is the "observation noise". We can then associate two implementations of this noise e_1 and e_2, to y_1 and y_2, so that:

$$y_1 = x + e_1 \qquad \text{and} \qquad y_2 = x + e_2 \qquad [11.13]$$

Since the instruments are not biased, $\langle E_1 \rangle = \langle E_2 \rangle = 0$. Furthermore, $\langle E_1^2 \rangle = \sigma_1^2$ and $\langle E_2^2 \rangle = \sigma_2^2$. If, finally, the measurements are independent, $\langle E_1 E_2 \rangle = 0$.

We seek an intermediate (an estimator) value between the two measurements that considers their uncertainties. The estimator $\hat{x}$ is also a random variable, which can be written as a linear combination of the two measurements:

$$\hat{x} = a_1 y_1 + a_2 y_2 \qquad [11.14]$$

We want this estimator to be unbiased: $\langle \hat{x} \rangle = x$. Or, $\langle \hat{x} \rangle = (a_1 + a_2)x$ implies $a_1 + a_2 = 1$. Therefore, we can reformulate the estimator $\hat{x}$ by:

$$\hat{x} = y_1 + a_2(y_2 - y_1) \qquad [11.15]$$

The *best linear unbiased estimate* (BLUE) is given by the coefficients a_1 and a_2, which minimize the variance of the estimation error:

$$\hat{\sigma}^2 = \left\langle (\hat{x} - x)^2 \right\rangle \qquad [11.16]$$

This minimum is achieved by a choice of the coefficients a_1 and a_2 so that:

$$a_1 = \frac{\hat{\sigma}}{\sigma_1}^2 \quad \text{and} \quad a_2 = \frac{\hat{\sigma}}{\sigma_2}^2 \qquad \text{with} \quad \frac{1}{\hat{\sigma}^2} = \frac{1}{\sigma_1^2} + \frac{1}{\sigma_2^2} \qquad [11.17]$$

This estimator is at the origin of the Kalman filter [KAL 60]: when an observation is available, we readjust the variable of the state simulated by the model at the previous time step ("background") to set it at the level estimated to be intermediate ("analysis") between observation and simulation. In other words, in the case of a one-dimensional state and of a single observation, we want to obtain an estimate of the state x_i^a at the time i as a linear combination of the *a priori* state x_i^b at the time i right before the availability of the observation and the weighted difference between the observation y and this *a priori* estimate. The best solution is given by:

$$x_i^a = x_i^b + K_i(y - x_i^b) \qquad [11.18]$$

If we assimilate y and x_i^b to two realizations of the true state of the system x_i at the time i, we proceed analogously with the BLUE. K_i, the Kalman gain, is therefore equal to:

$$K_i = \frac{1}{1 + \frac{\sigma_y^2}{\sigma_b^2}} \qquad [11.19]$$

We note that K_i is lower than 1 and so analysis x_i^a is an intermediate value between the observation and the *a priori* estimate. Moreover, K_i tends toward 1 as uncertainty on x_i^b increases and the analysis becomes closer to the observation; and conversely, when the uncertainty is strong on observation, K_i tends toward zero and the analysis is close to the *a priori* estimate. Moreover, the analysis error is written as before:

$$\sigma_a^2 = \frac{1}{\frac{1}{\sigma_y^2} + \frac{1}{\sigma_b^2}} = \frac{\sigma_b^2 \sigma_y^2}{\sigma_b^2 + \sigma_y^2} = (1 - K_i)\sigma_b^2 \qquad [11.20]$$

To simplify, the analysis error is equal to the background error, reduced by a factor equal to 1 minus the optimal gain, which depends on the precision of the observations and of the background.

In the multidimensional general case, the space of the observations differs from the state of the system, the projection between these two spaces being ensured by the (linear) observation operator $\mathbf{H}$. Moreover, considering the general case of a state x with n components and an observation vector y, the analysis is expressed as (see also Figure 11.4):

$$\mathbf{x}_i^a = \mathbf{x}_i^b + \mathbf{K}_i(\mathbf{y} - \mathbf{H}\mathbf{x}_i^b) \qquad [11.21]$$

The background is derived from the analysis at the previous time step, if we consider a linear dynamic model:

$$\mathbf{x}_i^b = \mathbf{M}\mathbf{x}_{i-1}^a \qquad [11.22]$$

By minimizing the estimation error of the analyzed vector $\mathbf{x}_i^a$, as in the case of a single observation and a one-dimensional state, we obtain the expression of the best Kalman gain $\mathbf{K}_i$. In practice, the literal resolution of this minimization is more laborious because of the fact that state and observation vectors have several components, so variances σ_y^2 and σ_b^2 take the form of the covariance matrices $\mathbf{R}$ and $\mathbf{B}$, respectively, and the observation operator is introduced.

The different stages of this calculation will not be detailed but they can be found in most texts on data assimilation (see [GHI 91] among others). The resulting Kalman gain of this calculation is expressed as follows:

$$\mathbf{K}_i = \mathbf{B}_i\mathbf{H}^t(\mathbf{H}\mathbf{B}_i\mathbf{H}^t + \mathbf{R}_i)^{-1} \qquad [11.23]$$

where $\mathbf{H}^t$ represents the transposed matrix of $\mathbf{H}$. The Kalman gain is, therefore, equal to the background error in observation space $\mathbf{B}_i\mathbf{H}'$ multiplied by the inverse of the sum of the errors (background and observations). So the more significant the background error, the higher the optimum gain, and therefore the correction; conversely, the higher the observation errors, the less the Kalman gain. Similarly, the covariance matrix of analysis errors at the time i is expressed as follows:

$$\mathbf{A}_i = (\mathbf{I} - \mathbf{K}_i\mathbf{H})\mathbf{B}_i \qquad [11.24]$$

These two equations are simply the multidimensional analogues of equations [11.19] and [11.20].

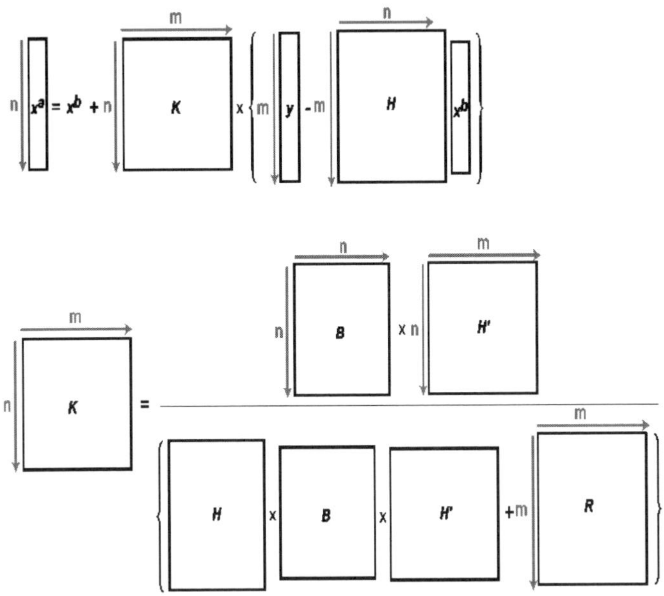

Figure 11.4. *Analysis equation of the gain with the size of the vectors and of the matrices involved; m is the number of observations and n is the size of the analysis vector ($m < n$)*

We have the following few remarks:

– The Kalman filter is sequential; observations are no longer used once they have been assimilated because the Kalman filter extracts all the useful information from these observations.

– In the extreme case where the background is very uncertain ($\mathbf{B}$ tends toward the infinite) and a very precise observation ($\mathbf{R}$ tends toward 0), we are in a particular case, often called "direct insertion" where the state analyzed in the space of the observations is simply replaced by the corresponding observation. This approach is still very much used for its simplicity and for demonstrative purposes as well.

– The matrix $\mathbf{B}_i$ is derived from the evolution of the covariance matrix of analysis error at the previous time step:

$$\mathbf{B}_i = \mathbf{M}\mathbf{A}_{i-1}\mathbf{M}^t + \mathbf{Q}_{i-1} \qquad [11.25]$$

where $\mathbf{Q}_{i-1}$ denotes the modeling errors. We will return to this in the practical section related to error specification. The propagation of the covariance matrix of background error in particular needs an approximate n^2 number of operations. The numerical cost of a Kalman filter can become very high should the number of variables be significant, limiting in particular its application in an operational context in meteorology, where less costly, variational approaches are preferred. For the assimilation of data in a surface model whose number of variables is well below an atmospheric model, the use of a Kalman filter is quite worth considering.

– The Kalman filter is applied in the case of a linear dynamic model as well as a linear observation operator. One variant has been developed for nonlinear cases[10]: the extended Kalman filter. The principle is to locally linearize the operator of the nonlinear dynamic model $\mathcal{M}$ around the background $\mathbf{x}_i^b$. This linearized operator $\mathbf{M}_i$ is the Jacobian matrix of the nonlinear operator. Its components are expressed as follows:

$$(\mathbf{M}_i)^{jk} = \frac{\partial \mathcal{M}^j(\mathbf{x})}{\partial \mathbf{x}^k}|_{\mathbf{x}=\mathbf{x}_i^b} \quad [11.26]$$

The different terms of the Jacobian are generally estimated literally starting from the equations of the model. The other solution is to calculate a numerical approximation through finite differences, and this solution consists of perturbing a reference run for each variable/parameter $\mathbf{x}^k$ of the state $\mathbf{x}$ with a perturbation $\delta\mathbf{x}^k$:

$$(\mathbf{M}_i)^{jk} = \frac{\mathcal{M}^j(\mathbf{x}) - \mathcal{M}^j(\mathbf{x} + \delta\mathbf{x}^k)}{\delta\mathbf{x}^k}|_{\mathbf{x}=\mathbf{x}_i^b} \quad [11.27]$$

In the case of meteorological models with a large number of variables (), this represents a colossal computational cost. This solution, based on the use of the tangent linear model, is satisfactory in the case where the nonlinearities are quite weak between the two analyses. The validity of this linear approximation, therefore, depends on the equations of the model but also on the time between two observations. If nonlinearities are strong, the use of the linearized model can lead to strong errors on the average and on the *a posteriori* covariance that can trigger the divergence of the filter. Indeed, even if the estimate of the state $\mathbf{x}_{i-1}$ at the time $i - 1$ is Gaussian, there is no reason for $\mathbf{x}_i$ to be Gaussian once it has gone through the nonlinear operator of the dynamic model $\mathcal{M}$. In this case, it is better to turn toward global approaches such as the Kalman filter and the particle filtering (see section 11.3.2.2).

– Optimum interpolation, which has been an assimilation method widely used in meteorology up until recently, is very close to a Kalman filter, except that the covariance matrix of background error is set *a priori*, thus skipping the very costly propagation stage (equation [11.25]).

– To reduce the cost of the propagation of the background error covariance matrix, approaches based on the Kalman filter have been proposed, seeking to project $\mathbf{B}$ on

10 The term "nonlinear" can be considered to be a little stretched if we speak of linearizing the system neighboring the current state in order to go back to the linear case. Therefore, we do not have a nonlinear approach *per se*, which is based on the initial equations of the model.

a reduced basis (empirical orthogonal functions or principal components). This is the case, for example, of the *singular evolutive extended kalman filter* (SEEK, [PHA 98]).

11.3.2.2. *The ensemble Kalman filter (EnKF) and particle filtering*

The EnKF [EVE 94] gives a spare solution to the costly equation for the propagation of the covariance matrix of background error of the Kalman filter. The relatively simple principle consists of propagating not only the background but also a set of perturbed states based on the background, each representing a potential trajectory of the system (Figure 11.5). The propagation of each member of the set comprising M members up to the time i where an observation is available takes the following form:

$$\mathbf{x}^b_{i,j} = \mathcal{M}(\mathbf{x}^b_{i-1,j}, \mathbf{u}) + \mathcal{E}^q \qquad [11.28]$$

where $\mathcal{E}^q$ is the modeling error.

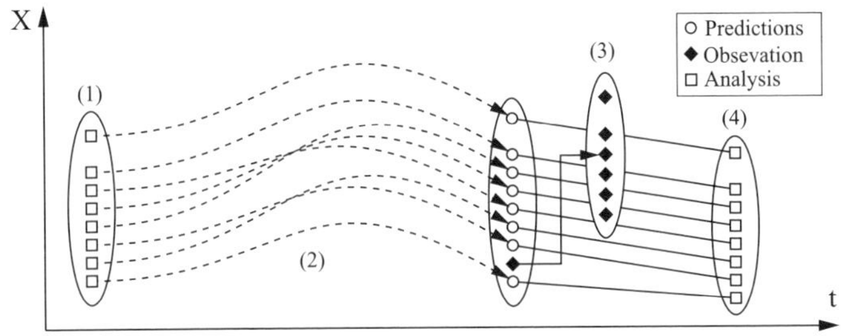

Figure 11.5. *Schematic principle of the ensemble Kalman filter. (1) The generation of the initial set of members (perturbation of an a priori estimation according to a normal distribution), (2) model-based propagation, (3) generating a set of observations and (4) analysis*

Therefore, the EnKF bypasses the linearization of the observation model and of the dynamic model by using the initial equations, which can be nonlinear. Analysis equations of the EnKF are identical to that of the Kalman filter (equations [11.23] and [11.24]) except that the estimate of the *a posteriori* state that is calculated for each member j of the set. Therefore, the covariance matrix of analysis error $\mathbf{A}_i$ is deduced from the following ensemble:

$$\mathbf{A}_i \approx \mathbf{A}^{En}_i = \frac{1}{M-1}\sum_{j=1}^{M}(\mathbf{x}^a_{j,i} - \bar{\mathbf{x}}^a_i)^t(\mathbf{x}^a_{j,i} - \bar{\mathbf{x}}^a_i) \qquad [11.29]$$

where $\bar{\mathbf{x}}^a_i$ is the average of the analyses ensemble.

The average of the ensemble is considered as the likeliest estimate of state x, and the ensemble spread is considered as an estimate of the second-order moment of the distribution of the model's potential trajectories. Furthermore, when the size of the set M tends toward the infinite, we show that the covariance matrix of analysis errors $\mathbf{A}_i^{En}$ estimated from the ensemble tends toward $\mathbf{A}_i$. The observations are also considered as random Gaussian variables. Therefore, a set of observations is formed at each time pace by perturbing them with a Gaussian noise with an average equal to the observation and a covariance $\mathbf{R}$:

$$\mathbf{y}_j = \mathbf{y} + \mathcal{E}_j \qquad [11.30]$$

with j ranging from 1 to M, the number of members of the ensemble.

Readers can refer to [EVE 03] where the theoretical basic elements of the approach are explained, as well as its application in geosciences and the details of its practical implementation. In addition, the application of the Ensemble Kalman Filter to the analysis of the soil moisture is illustrated at Figure 11.6.

We have the following few remarks:

– This filter is always based, just like the Kalman filter, on the hypothesis of normal distributions and an absence of bias between real state, estimated state and observations.

– Reichle *et al.* [REI 02] have shown that, in the case of surface soil moisture assimilation, EnKF gives similar results as the Kalman filter, with the added advantage of a better estimate of the covariance matrix of background error if the model presents strong nonlinearities.

– There is another variant of the Kalman filter close to the EnKF, that is the *unscented* Kalman filter [JUL 95]. Its difference to EnKF relates to the building of the ensemble. With the EnKF, sampling distribution of the backgrounds is carried out through a stochastic Monte Carlo type method, whereas with the *unscented* Kalman filter, members of the ensemble are chosen in a way that determines the Gaussian distribution of the background, from one part to the other part of the average. To a certain extent, this is a more rigorous version of the EnKF. Please refer to [WAN 01] for more details.

– As for the size of the ensemble, there are no exact methods but intuitively, it should be about the same order of magnitude as the number of degrees of freedom of the model. The problem comes from the fact that it is very difficult to assess this number of degrees of freedom. An ensemble containing less than 100 members was successfully applied to atmosphere models whose state vector size was around 10^6 [HOU 09]. For a land surface model whose state vector size and computing cost are much lower, a set of 100 members seems a good choice. Although Pauwels *et al.*

[PAU 07] show that 32 members are sufficient for the assimilation of leaf area index and soil moisture in a coupled vegetation-hydrology model, the analyses of [REI 02, MUN 07] in the case of data assimilation of surface soil moisture and of [BAT 12] for the assimilation of the surface temperature point also in favor of choosing 100 members.

Particle filtering [GOR 93] is different from EnKF in its basic elements, but it relies on a sampling of error distributions by a set of members, such as for the EnKF, which will hereon be called "particles". In relation to Kalman filters, we will no longer look for the "optimal" *a posteriori* discrete state (which we can only express algebraically for normal distributions) and that is definitely not optimal in several practical cases because of the tangent linear approximation for the EKF and of the limited size of the ensemble for the EnKF, but an approximation of the complete distribution p. It is a practical implementation of the Bayesian inference applied to the filtering. The general principle consists of propagating an ensemble of M particles since the initial state up to the time i when an observation is available. This propagation stage is identical to the EnKF except that the distribution of the initial ensemble of particles can be derived from non-normal *a priori* distributions and provides an estimate of $p(\mathbf{x})$ (Bayes' theorem, equation [11.11]). The two approaches differ at the stage of considering observations. Regarding particle filtering, a weight w_j will be assigned to each particle depending on its proximity with the observation but the particle is not modified, as opposed to the EnKF. In statistical terms, these weights, at the time i, are calculated as follows:

$$w_j = \frac{p(\mathbf{y}_i|\mathbf{x}_{i,j})}{\sum_{j=1}^{M} p(\mathbf{y}_i|\mathbf{x}_{i,j})} \qquad [11.31]$$

where $p(\mathbf{y}_i|\mathbf{x}_{i,j})$ is the probability density of the observation $\mathbf{y}_i$ considering the state $\mathbf{x}_{i,j}$. The denominator is a normalization factor for which the sum of the weights equals 1. In the Gaussian case, for example, this probability density is equal to:

$$A \exp\left[-1/2(\mathbf{y}_i - H(\mathbf{x}_{i,j}))^t \mathbf{R}^{-1} (\mathbf{y}_i - H(\mathbf{x}_{i,j}))\right].$$

The weight will therefore be conversely related to the deviation between prediction and observation. The "analyzed" state is then simply calculated as the weighted sum of all the particles. The process is then repeated until the next set of observations.

In practice, the fact that the particles are not modified poses a problem because if they move far from the observations, their weight will tend toward zero with nothing being able to bring them back to the observed reality. Their application to relatively simple systems [VAN 96] has shown that after several iterations, all the weight is concentrated into one particle. To overcome this problem, a resampling of the particles must be carried out by multiplying the particles proportionately to their weight, while eliminating those whose weight is null so as to preserve the same number as the previous ensemble. This technique is close to the principle of a genetic algorithm [BAE 93, GOL 94] where the most adapted individuals (i.e. the members of the ensemble with the highest weight) will become more numerous at the following iteration. This sampling is named sampling with resampling by significance. For more theoretical and practical detail, refer to [VAN 09].

11.3.3. *Variational assimilation*

A different and more intuitive viewpoint is to minimize the distance between a given trajectory of the model and a set of data. The objective of the inverse problem is, therefore, to find the value of the parameters and/or initial conditions of the analysis variables $\mathbf{x}_0$ of the dynamic model so that the predictions $\mathbf{y}_i = \mathbf{H}\mathbf{x}_i$ at the observation time i are closer to the m observations $\mathbf{y}_i$ on a given time window. Therefore, if we disregard the noise of the observations, in order to solve this problem, we must minimize the following cost function:

$$J = \left[\frac{1}{m} \sum_{i=1}^{m} ||\hat{\mathbf{y}}_i - \mathbf{y}_i||^2 \right]^{1/2} \qquad [11.32]$$

This function J gives a measure of the adequacy of the predictions to the observations in the norm L^2 and depends on the vector $\mathbf{x}_0$ and of the variables/parameters of the model with n components that we seek to estimate. The lower the cost, the better the predictions. In our case, the vector $\mathbf{y}_i$ corresponds to the satellite observations and the vector $\mathbf{x}$ corresponds to the variables/surface parameters (characteristic of the soil and the vegetation). In several cases, we may have an approximate idea of the relevant solution: the background $\mathbf{x}^b$. Besides the observations–predictions proximity criterion, we may wish that the solution is not too far from this *a priori* solution. Considering the error variance–covariance matrices, the cost function is written, thus, as:

$$J = \sum_{i=1}^{m} (\hat{\mathbf{y}}_i - \mathbf{y}_i)^t \mathbf{R}_i^{-1} (\hat{\mathbf{y}}_i - \mathbf{y}_i) + (\mathbf{x}^b - \mathbf{x})^t \mathbf{B}^{-1} (\mathbf{x}^b - \mathbf{x}) \qquad [11.33]$$

In the case of a linear observation operator and a linear dynamic model, we demonstrate that minimizing J leads to the BLUE. To achieve this result, we calculate the first derivative of J that must be null in $\mathbf{x} = \mathbf{x}^a$. The demonstration is given in [BOU 99]. We can also demonstrate that minimizing J is the same as maximizing the distribution of the possible states of the system, considering available observations $p(\mathbf{x}|\mathbf{y})$ since this cost function is nothing but subtracting the logarithm from this *a posteriori* distribution. These two approaches are, therefore, equivalent in the case of a linear problem with normal error distributions: the solution of a minimum variance given by the BLUE is equally the likeliest solution (i.e. the maximum of the *a posteriori* distribution). In practice, as for the Kalman filter, even if the normality and linearity hypothesis is verified, which is rarely the case, the algebraic solution of minimizing the cost function is still quite costly considering the size of the matrices that must be inverted, particularly in meteorology and oceanography.

The variational approaches used in meteorology, therefore, also rely on the linearization of the observation operator and of the dynamic model. Moreover, the minimization of the cost function is implemented using gradient descent methods. These so-called "local" approaches start from an initial "first guess" point within the search space. They proceed via successive iteration where the following point is sought in the direction of the steepest slope. Therefore, they need the cost function and its gradient to be assessed at each point tested. The evaluation of the gradient of the cost function in relation to the n variables/parameters of the state vector is an extremely costly operation. This evaluation must also be carried out as many times as there are potential solutions tested by the gradient descent method. The 3D-VAR approach considers that all the observations available on the assimilation window have been acquired at the same time i. Stated differently, there is no summation on the left member of the cost J. We clearly note the similitude of the 3D-VAR with the Kalman filter. However, there are several significant points where they differ:

– The solution is not sought algebraically but through the minimization of the cost J.

– There is no propagation equation of the covariance matrix of background error.

– As a result, the model is considered to be perfect. This can be demonstrated that for a null model error in a normal, linear case, the Kalman Filter leads to an analyzed value that is identical to the value obtained through the 3D-VAR (see especially [KAL 03]).

The 4D-VAR approach (three space dimensions + one time dimension) also aims to minimize the cost function, as described in equation [11.25]. The introduction of the time variable implies the propagation of the state $\mathbf{x}$ up to a date i where an observation is available. The estimation of the term of cost that quantifies the deviation from the data requires the integration of the model m times (dimension of the observation vector). Besides the introduction of the time variable, the minimization is not done with the help of a minimization technique such as gradient descent but relies on a more computationally efficient approach based on the use of the adjoint model of the linear tangent one. The adjoint model is a backward integration of the model that propagates the state $\mathbf{x}$ "backwards" from $i + 1$ to i. In the case of a linear operator $\mathbf{M}_i$, the adjoint model is simply its transposed $\mathbf{M}_i^t$. To save calculation time, in weather forecast centers the code of tangent linear and adjoint models is developed literally from equations of the initial dynamic model. However, we can still use a numerical approximation by finite differences. For more detail, the reader can refer to [KAL 03] and [BOU 99].

We will not detail the finer points of this approach, which has mainly been developed to reduce the numerical cost of minimization. In particular, for the 4D-VAR method, the adjoint model used is the model related to the dynamic tangent linear model, and the weak nonlinearity of the problem to be solved is a prerequisite to guarantee convergence to the optimal solution (the most cost-minimizing one). If not, the search space is non-convex and the approach may be trapped by a local minimum. Moreover, designing a tangent linear model and its adjoint model is durable and labor-intensive. If the model evolves, the process starts anew. Most assimilation problems of data applied to continental surfaces can be solved using direct methods[11], which are much easier to implement and less restrictive in terms of starting hypothesis, as we will discuss in the following sections.

We have the following few remarks:

– Analyses are carried out on successive windows with no overlapping, so as not to use the same observation several times, which would give more weight to the observations than they have in reality at the expenses of the model.

– The 4D-VAR assumes the hypothesis that the model is perfect. The method of representers developed by Bennett and Thornburn [BEN 92] increases the function cost of an additional term that allows us to consider the modeling error. This very elegant approach based on a resolution in the space of the observations is detailed in [BEN 02].

– The cost of the EnKF is directly connected to the size of the ensemble. The size is generally comprised between 10 and 100 members that, in relation to a 4D-VAR, represents an increase of the cost between 10 and 100 times. This remains, however, reasonable compared to the extended Kalman filter that needs n additional operations (the size of the state vector $\mathbf{x}$) for the propagation of the covariance matrix of background error. Several teams work actively in the development of hybrid approaches [HAM 00] coupling the variational approaches of type 3 and 4D-VAR that do not allow the propagation of the background error covariance matrix (which is constant), yet remain much more efficient than the EnKF for the estimation of the analysis state, to a Kalman ensemble for the error matrix propagation.

11 Methods that do not need the tangent linear model, and which are based on the use of the dynamic model/observation model couple in direct mode (see Figure 11.1).

11.3.4. *Parameter identification*

Poor parameterization manifests itself through systematic errors (or bias) between predictions of the model and observations. For the data assimilation specialists, the identification of parameters is, therefore, a preliminary stage of data assimilation, simply to eliminate this bias. For the modeling specialist, the identification of the parameters is an important stage to better apprehend the functioning of its model. Here, we will stick to the data assimilation specialist's point of view.

Identification relies on the minimization of a cost function (i.e. a measurement of the capacity of the model to reproduce the variables observed for a given set of parameters). Just like for the variational approach, this cost can be a measurement of the distance to the observations, potentially increased with a measurement of the distance to an *a priori* if we have an estimate of the value of the searched parameters. However, in opposition to data assimilation, the temporal window will be much longer because the searched parameters must allow the model to reproduce the behavior observed in a variety of situations: humid periods and dry periods for a hydrological model, several growth seasons for a vegetation model, etc. So we need a series of observations covering this wide range of situations.

From a purely technical point of view, minimizing this cost is essentially an optimization problem. Optimization is a separate field beyond the scope of this chapter. Most methods presented so far rely on the literal or numerical linearization of the dynamic model, and need a starting point in the search (the *a priori* value). These are the so-called "local methods". Now the cost function, being a paraboloid in the linear case, becomes a complex function that can present several local minima in the nonlinear case (equations of the dynamic surface models) and the non-normal case. Local optimization techniques such as the gradient descent method may be trapped in a local minimum. We would like to underline the existence of stochastic global approaches (Monte Carlo, genetic algorithms, ants colonies, etc.)[12] that can represent interesting alternatives. The double advantage of these approaches is:

1) enabling us to get a sketch of the set of acceptable solutions[13];

2) solving the inverse problem through a direct approach and therefore bypassing the calculation of the cost gradient.

Optimization is traditionally carried out *a posteriori* on historical data. We can also refine an estimate of the model's parameters. To do so, it is possible to deal with parameters as state variables (we speak of an increase in the state vector) by

12 In opposition to the local methods such as gradient descent techniques that need a starting point in the research.

13 That simply passes through the error bars of the observations.

giving them a dynamic equation (for example $\mathbf{x}(t + dt) = \mathbf{x}(t) + \mathbf{error}$ or even a decorrelation in time).

11.4. Assimilation into land surface models

With the recent launch of numerous sensors dedicated to Earth observation, data assimilation in surface models is an extremely active research subject. We have selected several key papers here from the abundant literature related to assimilation of satellite products: the *leaf area index* (LAI), the vegetation index NDVI, the soil moisture and the radiative temperature.

11.4.1. *Soil moisture*

Soil moisture is a key variable of the terrestrial hydrological cycle because it partly conditions the processes of sensible and latent heat exchange at the ground-vegetation-atmosphere interface [SHU 82]. Beljaars *et al.* [BEL 96] and Koster and Suarez [KOS 02] have, furthermore, shown their interest for weather and climate forecasts. Nevertheless, it was only at the beginning of the 1990s that the first attempts to analyze soil moisture in the land surface models were implemented. In particular, Mahfouf and Noilhan [MAH 91] have come up with the idea of relying on indirect observations of soil moisture: the temperature and the air humidity at 2 m, which are measured by the meteorological station of the synoptic network.

The recent launch of the microwave radiometers AMSR (in 2002 on the Aqua satellite), soil moisture and ocean salinity (SMOS) in 2009) and the associated distribution of operational products has initiated the development of the assimilation of soil moisture observations in surface models. These two sensors offer a measurement of soil moisture in the first 5 cm of the soil. Superficial and root zone moisture are physically connected through water diffusion processes in the ground. On this basis, several authors have proposed to analyze the water content of the root area, starting from an observation of the moisture of the superficial layer.

The first works introducing the observation of surface soil moisture were carried out by Jackson *et al.* [JAC 81] with the help of *in situ* observations and a simple method of direct insertion. Prévot *et al.* replaced these *in situ* observations by measures derived from the backscattering coefficient measured by a scatterometer (a radar instrument) installed on the field. Milly [MIL 86] also with *in situ* observations followed by Entekhabi *et al.* [ENT 94] with synthetic data of brightness temperature were then the first ones to propose an optimal solution to this problem, based on a Kalman filter. The demonstration of the potentialities of this approach for the correction of the soil moisture of the root zone in a *soil vegetation atmosphere transfer* (SVAT) model was provided by [CAL 98] and [FRA 03]. [WAL 01] proved the superiority of the Kalman filter over the simple approach of direct insertion that was still very popular at this time. [REI 01] describe a 4D-VAR applied to the assimilation of microwave brightness temperatures before the launch of the AMSR sensor. The study by [MUN 07], on an instrumented site, compares several assimilation approaches. Figure 11.6 displays the time series of the root-zone soil moisture simulated by the ISBA land surface model without data assimilation (open-loop) and with data assimilation of surface soil moisture observations (analysis) at a site located in Southwest of France. *In situ* observations have been superimposed for comparison purpose. Analysis are closer to the *in situ* data than the open-loop, in particular, during summer 2001 and, to a lesser extent, summers 2003 and 2004. Since the beginning of the 2000s and the arrival of the first global data, the studies based on real remote-sensing data are numerous [DRU 09].

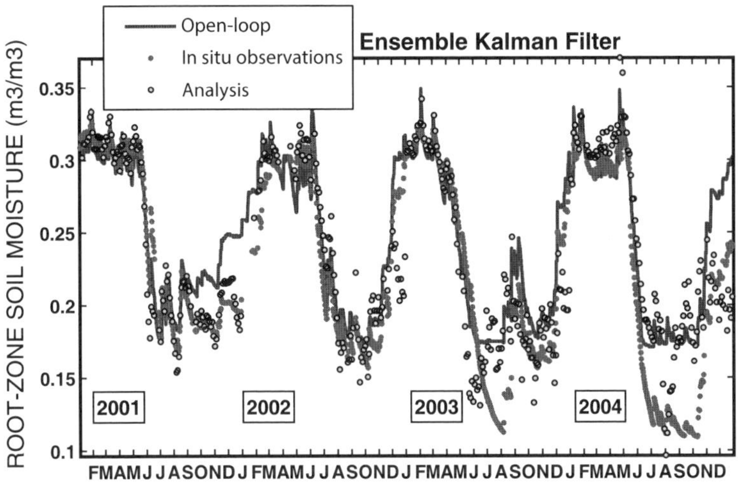

Figure 11.6. *Time serie of root-zone soil moisture simulated by the ISBA land surface model before (open-loop) and after data assimilation (analysis) of surface soil moisture observations over an experimental site located in the Southwest of France (2001–2004); from [MUN 07]*

11.4.2. *The surface temperature*

Since evapotranspiraton is the most thermodynamically efficient means for dissipating the energy received as radiation (there is no other form of transfer -sensible heat, ground conduction-able to evacuate so much energy during water stress), the surface under stress will warm up. There is therefore a relationship between surface temperature and soil moisture when humidity is below the threshold that triggers water stress, which is in fact a remote sensing observable particularly interesting in data assimilation. The radiative surface temperature is not really a state variable, but it is controlled by the ground surface temperature, which is generally described according to a dynamic model in most SVAT models. For the double-source models (one source of evaporation and one source of transpiration), this temperature combines the effects of reducing soil evaporation because of lower surface soil moisture, and of plant water stress because of lower soil moisture of the root area. This observable is, therefore, connected to the whole soil moisture profile. The [OTT 94] paper is one of the very first articles on the topic. Caparrini *et al.* [CAP 04] assimilated, based on a 4D-VAR, surface temperature in a model of thermal inertia of the surface, in order to find out daily evaporation values. Considering the difficulty of specifying errors but also of thermally documenting radiative transfer models, we resort more and more often to the inversion of evapotranspiration or of the level of water stress, starting from the temperature observed mostly through the resolution of the energy balance, and assimilate this product in a model that can then be a hydrological model of a daily time step.

11.4.3. *The vegetation*

Despite the first works being very old and aiming to assimilate products on directly observed vegetation (reflectance, vegetation index) or estimated vegetation based on satellite data (leaf area index, part of the incident radiation intercepted by the plants) in surface models [MAA 88], less progress has been made than with the assimilation of soil moisture products, which have been strongly stimulated through the significant role played by soil moisture on weather forecast. There are, however, a few works on the subject but for the majority, these studies only regard an experimental site or a small region. The first works presented for demonstration purposes aimed at identifying several parameters of crop models or natural vegetation models [CLE 96, DUC 08, GUE 98] or the global vegetation models for monitoring the carbon fluxes [KER 95, KNO 01]. By contrast, only a few studies are related to correction in real time of crop models using data assimilation approaches [PAU 07, DEM 07]. Curnel *et al.* [CUR 11] compared parameters identification and data assimilation with the help of an EnKF, and showed the superiority of the identification for grain yields forecasts. Barbu *et al.* [BAR 11] combined soil moisture and leaf area index products with the help of an extended Kalman filter in the Météo-France ISBA-A-gs model[14] and shows a moderate improvement of the predicted carbon fluxes (around 5%). Finally, the recent works of Boussetta *et al.* [BOU 13] have shown for the first time a potential impact of leaf area index data

14 Interaction-soil-biosphere-atmosphere (ISBA) is a model solving the energy and water budgets of land surfaces. It is used to represent the exhanges of matter and energy at the soilvegetation-atmosphere interface in the atmospheric and hydrological models at Météo-France.

assimilation on short-term weather forecasts, which leads the way for the operational assimilation of leaf area index products in weather centers.

11.5. Data assimilation – in practice

11.5.1. *Overdetermined problem, underdetermined problem and ill-posed problem*

The solution to the inverse problem of data assimilation will always be a compromise between the information we wish to obtain and what can really be obtained through a series of observations. This concept can simply be illustrated by comparing the length of data vector m and that of the unknowns vector n. The problem is called "underdetermined" or "overdetermined" if it has less or more, respectively, data than degrees of freedom. Let us take the example of data assimilation in a SVAT model, whose prognostic variable vector consists of temperatures and soil moisture at l different depths in the ground (i.e. $2l$ prognostic variables). Observations that might be accessed from space in order to analyze these variables are at best surface soil moisture and surface temperature. At one single point, we therefore have two observations for $2l$ variables to analyze. The problem as a whole is, therefore, underdetermined. If we now consider only the analysis of the initial condition of surface soil moisture within an assimilation window of 10 days with an observation every 3 days, we have three observations available for one variable to be analyzed. This specific problem, in the real case of an imperfect model, is overdetermined. In this case, the problem cannot be resolved in the sense that there is no exact solution (that should lead to a null cost), and we will try to reduce cost to a minimum. In practice, problems are therefore overdetermined and underdetermined at the same time. We say that they are ill posed. Determining that a problem is ill posed is not just about comparing the size of the data vector and that of the unknowns vector. The nature of data and the error level may contain enough information for constraining certain parameters/variables and not others. Particularly, if data are very noisy, it is only very slightly sensitive to the prognostic variable or to the parameters that we seek to constrain, and/or if the time and space data sampling is well below the typical time dimension of the process, then the problem is also ill posed and the state analyzed will be, unsurprisingly, of bad quality. This will translate into an optimal solution that will present no improvement compared to "simulation without assimilation" (i.e. the *open loop*).

11.5.2. *The adjustment criterion*

The majority of the classical assimilation methods described above are tightly connected to a very particular criterion, that is distance to data in norm L^2. The L^2 norm provides non-biased estimators of the parameters/variables that have a minimal variance if the residues are identically distributed and independent from one another.

However, by generalizing equation [11.32], deviation from data can be calculated following a norm L^q as follows:

$$J = \left[\frac{1}{m} \sum_{i=1}^{m} ||\hat{\mathbf{y}}_i - \mathbf{y}_i||^q \right]^{1/q} \qquad [11.34]$$

In some cases, it can be interesting not to cling on to a norm L^2: the higher the norm (i.e. the q value in equation [11.34]), the more it will give weight to significant errors. If observations are precise, we will therefore prefer a higher norm that will make the observations–predictions deviations more significant. This free choosing of the norm will only be considered in the case of an approach aiming at minimizing a cost function, such as equation [11.32] or [11.34]. Furthermore, the solution obtained will no longer be optimal in the sense of the least squares method.

11.5.3. *The analysis or control vector: observability and equifinality*

The control or analysis vector is made up of variables that will be corrected by data assimilation. Observability translates the intensity of the connection between observations and variables/parameters of the model to be constrained. Observability may come up, for example, in the case of a spatial resolution difference: is the observation representative of the grid point of the model? Staying with the example of soil moisture, remote sensing soil moisture products have a resolution of several tens of kilometers approximately 40 km for the SMOS satellite. It seems obvious this average information on the satellite pixel will not be able to constrain the small-scale processes of a hydrological model discretizing space into a kilometer resolution grid. conversely, a local measurement of surface soil moisture will not be representative enough for constraining the average dynamic of a grid point of several hundreds of kilometers, predicted by a climate model. Therefore, it is necessary that prior to assimilation, we carry out studies in order to ensure that variables/parameters we try to correct are actually observable. To this end, there are two solutions that can be considered:

– Implementing a sensitivity analysis: the simplest method consists of varying on a realistic range a single parameter or variable to be constrained, while keeping the others constant, and in renewing the operation for all components of the control vector. The advantage of this so-called "local approach" is that it is very simple, but it can give a wrong image of the model's sensitivity if the model is nonlinear. In this case, the sensitivity will only correspond to a particular configuration of the model and nothing guarantees that a different configuration (with new parameter/variable values) would lead to the same results. A more satisfying alternative is to turn toward so-called "global methods". Generally speaking, these approaches rely on a random sampling of a number of control vectors in a realistic space (i.e. the minimum and maximum values of the parameters). Each of these potential parameter vectors is then assessed through a target criterion, which is often the distance to observation (or to a reference simulation) whose information content is to be analyzed. Finally, solutions are dispatched within a subset of solutions called "acceptable" (i.e. whose target criterion is low, synonymous with the proximity to the observations or to the reference simulation); the remaining solutions represent the subset of "non-acceptable" solutions. Sensitivity of the parameters and/or variables is measured by quantifying the deviation from the initial uniform distribution (used for random sampling) of parameter distributions within the acceptable solutions subset. Several methods are based on this general principle [DEM 07].

– A twin experiment: in data assimilation, a twin experiment is an almost indispensable prelude before applying it to real data. It consists of creating a synthetic observation set starting from the dynamic model, perturbing this data set for simulating an observation error, and assimilating these data into the dynamic model in order to analyze the constraint brought upon the control vector. Going through a synthetic data set is very easy as it allows us to test sensitivity of the observations to different noise levels, time step, etc. [PEL 04].

11.5.4. *Algorithmic parameters*

11.5.4.1. *Bias reduction*

Assimilation methods traditionally used for constraining models are supposed to correct a random error, not a bias. Prior to implementing an assimilation technique, it is worth eliminating it or reducing it as much as possible. A bias between predictions and observations often reveals the poor parameterization of the dynamic model, the observation model or both. In this case, parameter identification methods are often recommended. In practice, significant biases between satellite products or observations and model predictions still exist, despite attention being paid to parameter calibration (see [REI 04a] for remote sensing soil moisture products). To reduce bias, several works propose adjusting the cumulative distribution of satellite products to the climatology of the dynamic model (i.e. histogram matching). These techniques rely on varyingly complex adjustment methods of the moments (often the average and the standard deviation) of the two distributions. This is equivalent to no longer constraining the absolute value of the predicted variable by the dynamic model but only its relative variations (or anomalies). Numerous works were related to histogram matching of satellite soil moisture products [DRU 05]. This approach was also applied to the adjustment of leaf area index products [JAR 08]. These adjustment approaches are, however, difficult to implement as they assume:

1) that sound estimates of the statistics of the two distributions, which means long enough time series, are available; this is not always the case, especially for soil moisture products measured by sensors whose satellites have been launched recently (SMOS, AMSR);

2) to adopt a strategy that could locally adjust statistics showing high spatiotemporal variability. In the latter case, we often substitute spatial variability with temporal variability, that is the ergodicity principle. Reichle *et al.* [REI 04b] have proposed a local adjustment strategy. Please also refer to the approach based on the use of copula functions [LER 13] or the variational bias correction method [DEE 05], which aims at automatically adjusting this bias jointly with the analysis procedure by increasing the cost function.

11.5.4.2. *Error estimate*

Covariance matrices of observation error $\mathbf{R}$, of the model error $\mathbf{Q}$ and of the background error $\mathbf{B}$ will gather together all information on the system's errors. Their estimation is an essential phase in the implementation of a data assimilation system (see, in particular, [REI 03, CRO 06] for the soil moisture product assimilation), as they will determine the relative weight given to observations and to the background. In areas with a high density of observations, the non-diagonal terms of the covariance matrix will stop the analyzed field from adjusting to all observations that, by nature, have a limited representativeness, and thus present a variability compatible with the correlation length that is typical of the observed process. Unfortunately, we will see that there is no simple and generic approach for error assessment. The latter will appeal to the common sense of the modeler and approximations cannot be avoided.

11.5.4.3. *Observation error*

For the observation error matrix $\mathbf{R}$, diagonal elements of the inverse of the covariance matrix will be the inverse of variances of data error. Thus, data with a strong error will have a low contribution to the cost, whereas precise data will have a big contribution. If an observation vector contains data of different kinds (soil moisture and leaf area index for example, or measures carried out by two different sensors), then the error observation matrix will be structured into blocks. Blocks outside the diagonal are null if we suppose that errors between two types of instruments or products are not correlated. In the best scenario, the provider of the observation documents the relevant uncertainties (i.e. the diagonal terms of matrix $\mathbf{R}$). This error can, thus, be set to a certain percentage of the climatology (the average of the observations over a long period of time). We can also assess the variance of the observation acquired on a target that we know is stable, and consider that it corresponds to the observation error by applying the ergodicity principle. Finally, if two instruments A and B measure the same magnitude on the same target and we know the error on one of the two, we can estimate the error for the other instrument ($var(A - B) = var(A) + var(B)$). Scipal *et al.* [SCI 08] followed by Crow and van den Berg [CRO 10] have applied this type of approach to assess observation errors due to several co-localized soil moisture satellite products. Gu *et al.* [GU 06] have proposed a method to assess observation errors of leaf area index products of the MODIS sensor. Given the difficulties in estimating non-diagonal terms, we often consider them to be null.

11.5.4.4. *Background error*

With regard to background error, there are several, more satisfactory assessment methods. We can simply rely on the climatology of the model that we compare to an observed climatology or to model predictions [JAR 08]. Hollingsworth and Lonnberg [HOL 86] have also proposed quite an attractive approach, as it is more rigorous than previous empiricism, but it supposes a strong data density. It is based on the often obvious idea of comparing background to observations in order to estimate this background error. In practice, we calculate the histogram of deviations to the background ($\mathbf{y} - \mathbf{H}\mathbf{x}^b$) depending on the distance to observations. Let us imagine a spatial grid made of several grid points where, at each point, we have background $\mathbf{H}\mathbf{x}_i^b$ and observation $\mathbf{y}_i$ handy. The histogram of deviations between background and the corresponding observation calculated on the entire field gives us the means of observation error and background error $(var(\mathbf{y} - \mathbf{H}\mathbf{x}^b) = var(\mathbf{y}) + var(\mathbf{H}\mathbf{x}^b))$. If the calculation is now carried out between a background $\mathbf{H}\mathbf{x}_i^b$ and an observation $\mathbf{y}_j$ issued from a different grid point situated at a certain distance, we obtain an estimate of the background error if the observation errors are not correlated (in the opposite case, it is impossible to separate the two errors). A demonstration is given in [BOU 99].

The covariance matrix of background error $\mathbf{B}$ has to meet a certain number of criteria. It should be positive definite and therefore inversible; non-diagonal terms should tend toward zero beyond a certain distance if observations only have a local influence. The numerical approach presented above may not meet these conditions as a whole. In that case, we can approximate these errors by using a parameterized covariance function (the choice is however very wide, see [MEI 81] for several examples suitable for background error prescription) that will decrease depending on the distance between two observations, according to a correlation length:

$$\mathbf{B}(i,j) \cong \sigma_i \sigma_j \exp\left[-\frac{d(i,j)}{d_0}\right] \qquad [11.35]$$

Correlation length d_0 may be considered as a constant (stationary error covariance) or it may vary with time (non-stationary error covariance). In [REI 01], for example, their error correlations are determined by using a covariance function and by setting time and space correlation lengths, according to their knowledge of the system.

Other approaches that are more complex to implement, such as the *national meteorological center* (NMC) method [PAR 92], exist. Although without theoretical foundation, this approach is still very much used within national meteorological centers. It consists of estimating the background error on the basis of the dispersion of a set of predictions taken at the same time, but launched at different initial times. Finally, a very good review of the ensemble approaches for estimating background error is carried out by Buehner [BUE 05].

11.5.4.5. *Modeling error*

If for a Kalman filter and its derivatives, the background error matrix is automatically updated, the main difficulty resides in the modeling error assessment. If observations of the variables to be analyzed are available on the same spatial scale and all over the field that the model must be used on, the comparison between predictions of the model and measured values can give an idea[15] of modeling errors. The only scale where this configuration is possible is the scale of the *in situ* station, where soil moisture and temperature sensors will be installed exactly at the same depth as the discretization of the ground in the model. In this case, the analysis of residues and their correlations (from one layer to another) can help document the modeling error matrix. The second option, which is more widespread at the moment, is to perturb either:

– 1) key parameters of the model that we know have a strong sensitivity to the variable to be analyzed (i.e. saturated hydraulic conductivity for soil moisture);

– 2) the forcing variables (i.e. precipitations);

– 3) both.

To determine the amplitudes and the distribution of these perturbations, we should, again, turn to the knowledge of the system studied by the modeler. We will speak, for example, of perturbing the forcings according to the error statistics of this input data (instrumental and representation errors). Examples of this approach abound in specialized literature ([REI 02, PAU 07, BAT 12] among others). Finally, the last option, which can be combined with the previous one, consists of multiplying the covariance matrix of background error by a factor higher than 1, called the "inflation factor" [AND 99]. The use of an inflation factor is also possible in the case of a dissipating model in order to avoid a crash of the ensemble [EVE 03]. The specification of these errors remains subject to approximations and they are often set to *ad hoc* values because of this. Adaptive filtering is one of the methods that has gained quite some acclaim for overcoming this subjectivity.

11.5.4.6. *Assessment of the quality of the analysis*

Generally speaking, the tools we have at our disposal for the assessment are mainly the innovation vector, analysis residues and the increments, as well as the error covariance matrix of the analysis. The analysis of temporal and spatial evolutions of these quantities and their statistical characteristics (average, standard

15 However, it cannot document it completely if the modeling error is not stationary. In this hypothesis, which is very likely in practice, we should have various observations of this same variable at the same instant in order to validate its distribution or observations distributed in time with no evolution of the system.

deviation) will provide us with information on the smooth operation of the assimilation system, and will allow us to detect a potentially defective element: existence of a bias in the observations or in the model, non-linearities, poor error prescriptions, etc.

11.5.4.7. *Validation with additional observations*

The most natural approach is to validate predictions of the model after the analysis, with observations not used in the analysis. These may be observations of the same variable as the one that is analyzed, but coming from a different measuring instrument: another satellite, for example, or *in situ* data (as in the study by [REI 07] described in section 11.1). There is also an approach consisting, in areas where data density is significant, of taking a certain number of observations out of the data set to assimilate, and then comparing predictions of the model after the analysis to these observations (a sort of *leave-n-out* type of validation often used in statistics). Finally, the last approach consists of validating the system with the observations of the variables that cannot be directly observed from the satellite: in the case of assimilating surface soil moisture to analyze root-zone soil moisture, we can check that the model's predictions of evapotranspiration are more in line with *in situ* observations than before data assimilation.

11.5.4.8. *Distance to the data: innovations and residues*

Another quite obvious approach is to check if predictions in the observation space after the analysis are closer to the observations than before: by reusing already defined terms, the residues mean $(\mathbf{y} - \mathbf{H}\mathbf{x}^a)$ is lower than that of innovations $(\mathbf{y} - \mathbf{H}\mathbf{x}^b)$. In this case, we will say that assimilation has drawn the model toward observations. This is obviously a prerequisite for the good functioning of the algorithm because most assimilation algorithms try to minimize a criterion measuring the distance to the data. However, good functioning is not and should not be based on this criterion, nor should either the performance of an assimilation system be quantified on this quantity only. Observations are uncertain and it is not particularly recommended that the optimal solution be the closest to observations. This is a significant criterion but one that will, at best, enable us to highlight an implementation problem, i.e. an error in the code.

Taken separately, statistics (the average and the standard deviation) related to residues and innovation vectors and their evolutions over time are essential diagnoses in data assimilation. If the analysis is optimal, they must be particularly similar to Gaussian white noise. If the model is linear and error statistics verify the normality hypotheses, the error covariance of the innovation term is simply $\mathbf{HBH}^t + \mathbf{R}$ (the sum of the error covariance of the model's predictions in the space of observations and the error covariance of the observations; see [DES 05]). Therefore, it is possible to normalize innovations by this value calculated *a posteriori*, and to compare the distribution of the innovations with the normal distribution $\mathcal{N}(0, 1)$ (see [REI 02]). A deviation from the normal allows us to detect a non-optimal solution that can come from the nonlinearity of the equations of the model, of an inadequate prescription error. The review of the temporal evolution of the average of the innovations vector will, moreover, allow us to highlight the presence of a possible bias or to detect a sudden malfunction of the instruments providing the assimilated observations.

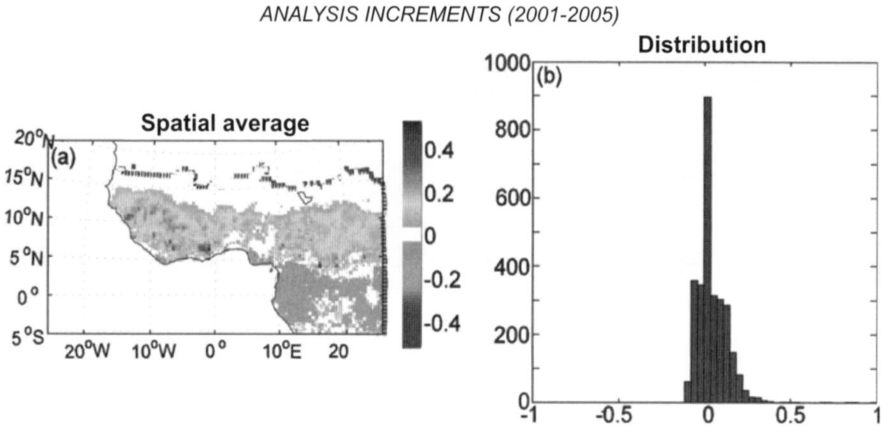

Figure 11.7. *Map and distribution of the analysis increments on West Africa for the assimilation system of the MODIS foliar surface products in the surface model of the European Center for Medium-range Weather Forecast [JAR 08]*

11.5.4.9. *The analysis increments*

Analysis increments $(\mathbf{x}^a - \mathbf{x}^b)$ are also a good diagnosis of a data assimilation system. Analysis techniques are identical to those listed above: evaluation of the average, the standard deviation and their evolution over time. Again, the distribution of the increments has to have a null average. If not, it is also the sign of a bias between observations and model. Figure 11.7 shows the map and distribution of increments for the assimilation system of MODIS leaf area index products in the land surface model of the European center. This figure shows, in particular, that increments are significantly positive in savannah areas and, to a lesser extent, Sahelian areas. This map has allowed us to identify a problem in the vegetation growth model and, particularly, a systematic delay of predicted leaf area index compared to observations, and thus to modify the parameters of the model.

11.6. Perspectives

Data assimilation in surface models has made considerable achievements in the last few years. The research currently in progress mainly deals with the following:

– Calibrating parameters together with data assimilation assimilation/inversion. By increasing the state vector with parameters processed as pseudo-constant state variables (i.e. which evolve slowly), we can adjust "on line" certain parameters once they have been estimated over a long calibration period. Thus, Pellenq *et al.* [PEL 03] have managed, for example, to find the deepening of the root extraction area in the middle of an agricultural season by assimilating soil moisture (see also [VRU 06] for an application to hydrometeorological predictions).

– Adaptive filtering: the principle of adaptive filtering relies on the statistics of the diagnoses described. So, in order to avoid the slightly *ad hoc* character of their estimation, modeling and observation errors are adjusted in real time if statistics of the innovation or residue vector do not follow the expected distribution [DES 05]; see [REI 08, CRO 08] for an application to soil moisture assimilation.

– Soft data assimilation: the large error on remote sensing observations that could result from a dense cloud cover or from unfavorable observation configurations can cause the assimilation to have a weak constraint on the trajectory of the model. However, certain hydrological events, for example, which are clearly observed by satellite, even without being quantitatively used may have a strong impact on the trajectory of the model. Thus, knowing if a surface is stressed, if snow has fallen or if crops have emerged is often more important as a piece of information (i.e. soft data) than the quantitative (and quite uncertain) monitoring on snow thaw, evolution of the aerial biomass or water stress. We then speak of em soft data assimilation.

Bibliography

[ADA 09] ADAM N., PARIZZI A., EINEDER M., *et al.*, "Practical persistent scatterer processing validation in the course of the Terrafirma project", *Journal of Applied Geophysics*, vol. 69, pp. 59–65, 2009.

[AIA 02] AIAZZI B., ALPARONE L., BARONTI S., *et al.*, "Context-driven fusion of high spatial and spectral resolution images based on oversampled multiresolution analysis", *IEEE Transactions on Geosciences and Remote Sensing*, vol. 40, no. 10, pp. 2300–2312, 2002.

[ALL 00] ALLWEIN E.L., SCHAPIRE R.E., SINGER Y., "Reducing multiclass to binary: a unifying approach for margin classifiers", *Proceedings 17th International Conference on Machine Learning*, Morgan Kaufmann, San Francisco, CA, pp. 9–16, 2000.

[ALP 07] ALPARONE L., WALD L., CHANUSSOT J., *et al.*, "Comparison of pansharpening algorithms: outcome of the 2006 GGRS-S data fusion contest", *IEEE Transactions on Geosciences and Remote Sensing*, vol. 45, no. 10, pp. 3012–3021, 2007.

[AND 99] ANDERSON J., ANDERSON S., "A Monte-Carlo implementation of the non-linear filtering problem to produce ensemble assimilations and forecasts", *Monthly Weather Review*, vol. 127, pp. 2741–2758, 1999.

[BAE 93] BAECK T., SCHWEFEL H., "An overview of evolutionnary computation for parameter optimization", *Evolutionary Computation*, vol. 1, pp. 1–23, 1993.

[BAI 97] BAILLARD C., Analyse d'images aériennes stéréoscopiques pour la restitution 3D des milieux urbains, PhD Thesis, École nationale supérieure des télécommunications, 1997.

[BAR 11] BARBU A.L., CALVET J.C., MAHFOUF J.F., *et al.*, "Assimilation of soil wetness index and leaf area index into the ISBA-A-gs land surface model: grassland case study", *Biogeosciences*, vol. 8, no. 7, pp. 1971–1986, 2011.

[BAT 12] BATENI S.M., ENTEKHABI D., "Surface heat flux estimation with the ensemble Kalman smoother. Joint estimation of state and parameters", *Water Resources Research*, vol. 48, pp. 1–16, 2012.

[BEA 99] BEAUDUCEL F., BRIOLE P., FROGER J., "Volcano wide fringes in ERS SAR inteferograms of Etna: deformation or tropospheric effects?", *Journal of Geophysysical Research*, vol. 105, no. 11, pp. 1145–1150, 1999.

[BEL 96] BELJAARS A., VITERBO P., MILLER M., *et al.*, "The anomalous rainfall over the United States during July 1993: Sensitivity to land surface parameterization and soil moisture anomalies", *Monthly Weather Review*, vol. 124, pp. 362–383, 1996.

[BEN 92] BENNETT A., THORNBURN M., "The generalized inverse of a non-linear quasi-geostrophic ocean circulation model", *Journal of Physical Oceanography*, vol. 3, pp. 213–230, 1992.

[BEN 02] BENNETT A.F., *Inverse Modeling of the Ocean and Atmosphere*, Cambridge Press, Cambridge, 2002.

[BEN 03] BENEDIKTSSON J.A., PESARESI M., ARNASON K., "Classification and feature extraction for remote sensing images from urban areas based on morphological transformations", *IEEE Transactions on Geosciences and Remote Sensing*, vol. 41, no. 9, pp. 1940–1949, 2003.

[BER 02] BERARDINO P., FORNARO G., LANARI R., *et al.*, "A new algorithm for surface deformation monitoring based on small baseline differential SAR interferograms", *IEEE Transactions on Geoscience and Remote Sensing*, vol. 40, no. 11, pp. 2375–2383, 2002.

[BER 06] BERTHIER E., VADON H., BARATOUS D., *et al.*, "Surface motion of mountain glaciers derived from satellite optical imagery", *Remote Sensing of Environment*, vol. 95, pp. 14–28, 2006.

[BEV 92] BEVEN K., BINLEY A., "The future of distributed models: model calibration and uncertainty prediction", *Hydrological Processes*, vol. 6, pp. 279–298, 1992.

[BIO 09] BIOUCAS-DIAS J., "Variable splitting augmented Lagrangian approach to linear spectral unmixing", *IEEE Workshop on Hyperspectral Image and Signal Processing*, 2009.

[BIO 10] BIOUCAS-DIAS J. M., PLAZA A., "Hyperspectral unmixing: geometrical, statistical and sparse regression-based approaches", *SPIE Conference in Image and Signal Processing and Remote Sensing*, Toulouse, France, 2010.

[BJO 97] BJORNSSON H., SILVIA A.V., A manual for EOF and SVD analyses of climate data Report no. 97-1, CCGCR Report McGill University, Montreal, Quebec, 1997.

[BLO 03] BLOCH I., *Fusion d'informations en traitement du signal et des images*, Hermes–Lavoisier, Paris, 2003.

[BOR 70] BORN M., WOLF E., *Principles of Optics*, Pergamon Press, Oxford, 1970.

[BOR 94] BOREL C.C., GERSTL S.A., "Nonlinear spectral mixing models for vegetative and soils surface", *Remote Sensing of Environment*, vol. 47, no. 2, pp. 403–416, 1994.

[BOU 99] BOUTTIER F., COURTIER P., *Meteorological training course lecture series, European Centre for Medium Range Weather Forecasts*, Reading, vol. 59, 1999.

[BOU 13] BOUSSETTA S., BALSAMO G., BELJAARS A., *et al.*, "Impact of a satellite-derived leaf area index monthly climatology in a global numerical weather prediction model", *International Journal of Remote Sensing*, vol. 34, pp. 37–41, 2013.

[BOV 07] BOVOLO F., BRUZZONE L., "A theoretical framework for unsupervised change detection based on change vector analysis in polar domain", *IEEE Transactions on Geoscience and Remote Sensing*, vol. 45, no. 1, pp. 218–236, 2007.

[BOY 01] BOYKOV Y., VEKSLER O., ZABIH R., "Fast approximate energy minimization via graph cuts", *IEEE Transactions on Pattern Analysis and Machine Intelligence*, vol. 26, no. 2, pp. 147–159, 2001.

[BRO 09] BROADWATER J.B., BANERJEE A., BURLINA P., "Kernel methods for unmixing hyperspectral imagery", in CAMPS-VALLS G., BRUZZONE L. (eds.), *Kernel Methods for Remote Sensing Data Analysis*, John Wiley & Sons, New York, 2009.

[BRU 02] BRUZZONE L., MELGANI F., "Support vector machines for classification of hyperspectral remote-sensing images", *IEEE International Geoscience and Remote Sensing Symposium, IGARSS*, vol. 1, pp. 506–508, June 2002.

[BUA 07] BUADES A., COLL B., MOREL J.M., "Nonlocal image and movie denoising", *International Journal of Computer Vision*, vol. 10, pp. 101–117, 2007.

[BUE 05] BUEHNER M., "Ensemble-derived stationary and flow-dependent background-error covariances. Evaluation in a quasi-operational NWP setting", *Quarterly Journal of the Royal Meteorological Society*, vol. 131, pp. 1013–1043, 2005.

[BUR 83] BURT P.J., ADELSON E.H., "The Laplacian pyramid as a compact image code", *IEEE Transactions on Communications*, vol. 31, no. 4, pp. 532–540, 1983.

[CAL 98] CALVET J., NOILHAN J., BESSEMOULIN P., "Retrieving the root-zone soil moisture from surface soil moisture or temperature estimates. A feasibility study based on field measurements", *Journal of Applied Meteorology*, vol. 37, pp. 371–386, 1998.

[CAL 01] CALOZ R., COLLET C., *Précis de télédétection*, Presses de l'Université du Québec, Sainte-Foy, Quebec, 2001.

[CAM 06] CAMPS-VALLS G., GÓMEZ-CHOVA L., MUÑOZ-MARÍ J., *et al.*, "Composite kernels for hyperspectral image classification", *IEEE Geoscience and Remote Sensing Letters*, vol. 3, no. 1, pp. 93–97, 2006.

[CAN 86] CANNY J., "A computational approach to edge detection", *IEEE Transactions on Pattern Analysis and Machine Intelligence*, vol. 8, no. 6, pp. 679–698, November 1986.

[CAP 03] CAPDEROU M., *Satellites: orbites et missions*, Springer, Berlin, 2003.

[CAP 04] CAPARRINI F., CASTELLI F., ENTEKHABI D., "Variational estimation of soil and vegetation turbulent transfer and heat flux parameters from sequences of multisensor imagery", *Water Resources Research*, vol. 40, 2004.

[CAR 90] CARPER W., LILLESAND T.M., KIEFER R., "The use of intensity-hue-saturation transformations for merging SPOT panchromatic and multispectral image data", *Photogrammetric Engineering and Remote Sensing*, vol. 56, no. 4, pp 459–467, 1990.

[CAR 96] CARNEC C., MASSONNET D., KING C., "Two examples of the use of SAR interferometry on displacement fields of small spatial extent", *Geophysical Research Letters*, vol. 23, no. 24, pp. 3579–3582, 1996.

[CAR 02] CARDOSO J., "Analyse en composantes indépendantes", *Proceedings of XXXIV Journées de Statistique*, Brussels, Belgium, 2002.

[CAS 11] CASU F., MANCONI A., PEPE A. *et al.*, "Deformation time-series generation in areas characterized by large displacement dynamics: the SAR amplitude pixel-offset SBAS technique", *IEEE Transactions on Geoscience and Remote Sensing*, vol. 49, no. 99, pp. 1–12, 2011.

[CAV 07] CAVALIÉ O., DOIN M.P., LASSERRE C., *et al.*, "Ground motion measurement in the lake Mead area, Nevada, by differential synthetic aperture radar interferometry time series analysis probing the lithosphere rheological structure", *Journal of Geophysical Research*, vol. 112, no. B03403, 2007.

[CHA 60] CHANDRASEKHAR S., *Radiative Transfer*, Dover, New York, 1960.

[CHA 91] CHAVEZ JR. P., SIDES S., ANDRESON J., "Comparison of three different methods to merge multiresolution and multispectral data Landsat TM and SPOT panchromatic", *Photogrammetric Engineering and Remote Sensing*, vol. 57, no. 3, pp. 295–303, 1991.

[CHA 03] CHAMBON S., CROUZIL A., "Mesures de corrélation robustes aux occultations", *ORASIS 2003–Journées jeunes chercheurs en vision par ordinateur*, Gérardmer, France, May 2003.

[CHA 04a] CHAABANE F., Suivi multi-temporel en interférométrie radar et prise en compte des effets atmosphériques, PhD Thesis, University of Paris XI, 2004.

[CHA 04b] CHAMBOLLE A., "An algorithm for total variation minimization and applications", *Journal of Mathematical Imaging and Vision*, vol. 20, pp. 89–97, 2004.

[CHA 04c] CHANG C.I., DU Q., "Estimation of number of spectrally distinct signal sources in hyperspectral imagery", *IEEE Transactions on Geosciences and Remote Sensing*, vol. 42, no. 3, pp. 608–619, 2004.

[CHA 06] CHANG C.I., JI B., "Weighted abundance-constrained linear spectral mixture analysis", *IEEE Transactions on Geoscience and Remote Sensing*, vol. 44, no. 2, pp. 378–388, February 2006.

[CHA 13] CHANDER G., HEWISON T., FOX N., *et al.*, "Overview of Intercalibration of satellite instruments", *IEEE Transactions on Geoscience and Remote Sensing*, vol. 51, no. 3, pp. 1056–1080, 2013.

[CHE 95] CHENG Y., "Mean shift, mode seeking, and clustering", *IEEE Transactions on Pattern Analysis and Machine Intelligence*, vol. 17, no. 8, pp. 790–799, August 1995.

[CHE 00] CHEN C., ZEBKER H., "Network approaches to two-dimensional phase unwrapping: intractability and two new algorithmes", *Journal of the Optical Society of America A*, vol. 17, pp. 401–414, 2000.

[CHE 02] CHEN C., ZEBKER H., "Phase unwrapping for large SAR interferograms: Statistical segmentation and generalized network models", *IEEE Transaction on Geoscience and Remote Sensing*, vol. 40, no. 8, pp. 1709–1719, 2002.

[CHE 11] CHEN J., CHEN X., CUI X. *et al.*, "Change vector analysis in posterior probability space: A new method for land cover change detection", *IEEE Geoscience and Remote Sensing Letters*, vol. 8, no. 2, pp. 317–321, March 2011.

[CLE 96] CLEVERS J.G.P.W., VAN LEEUWEN H.J.C., "Combined use of optical and microwave remote sensing data for crop growth monitoring", *Remote Sensing of Environment*, vol. 56, pp. 42–51, 1996.

[CLO 97] CLOUDE S., POTTIER E., "An entropy based classification scheme for land applications of polarimetric SAR", *IEEE Transactions of Geoscience and Remote Sensing*, vol. 35, no. 5, pp. 66–78, 1997.

[CLO 98] CLOUDE S., PAPATHANASSIOU K., "Polarimetric SAR interferometry", *IEEE Transactions on Geoscience and Remote Sensing*, vol. 36, no. 5, pp. 1551–1565, 1998.

[COL 06] COLESANTI C., WASOWSKI J., "Investigating landslides with space-borne synthetic aperture radar (SAR) interferometry", *Engineering Geology*, vol. 88, nos. 3–4, pp. 173–199, 2006.

[COM 94] COMON P., "Independent component analysis, a new concept?", *Signal Processing*, vol. 36, 3, pp. 287–314, 1994.

[COM 02] COMANICU D., MEER P., "Mean shift: a robust approach toward feature space analysis", *IEEE Transactions on Pattern Analysis and Machine Intelligence*, vol. 24, pp. 603–619, May 2002.

[COP 04] COPPIN P., JONCKHEERE I., NACKAERTS K., *et al.*, "Digital change detection methods in ecosystem monitoring: a review", *International Journal of Remote Sensing*, vol. 25, no. 9, pp. 1565–1596, May 2004.

[COV 65] COVER T.M., "Geometrical and statistical properties of systems of linear inequalities with applications in pattern recognition", *IEEE Transactions on Electronics and Computers*, vol. 14, no. 3, pp. 326–334, 1965.

[CRO 06] CROW W.T., LOON E.V., "Impact of incorrect model error assumptions on the sequential assimilation of remotely sensed surface soil moisture", *Journal of Hydrometeorology*, vol. 7, pp. 421–432, 2006.

[CRO 08] CROW W.T., REICHLE R.H., "Comparison of adaptive filtering techniques for land surface data assimilation", *Water Resources Research*, vol. 44, pp. 1–12, 2008.

[CRO 10] CROW W., VAN DEN BERG M., "An improved approach for estimating observation and model error parameters in soil moisture data assimilation", *Water Resources Research*, vol. 46, pp. 1–12, 2010.

[CUR 11] CURNEL Y., DE WIT A.J., DUVEILLER G., *et al.*, "Potential performances of remotely sensed LAI assimilation in WOFOST model based on an OSS experiment", *Agricultural and Forest Meteorology*, vol. 151, no. 12, pp. 1843–1855, 2011.

[DAL 91] DALEY R., *Atmospheric Data Analysis*, Cambridge University Press, Cambridge, 1991.

[DAL 04] VAN DALFSEN E.D.Z., PEDERSEN R., SIGMUNDSSON F., "Satellite radar interferometry 1993-1999 suggests deep accumulation of magma near the crust-mantle boundary at the Krafla volcanic system, Iceland", *Geophysical Research Letters*, vol. 49, no. 4, pp 1441–1452, 2004.

[DAL 10] DALLA MURA M., BENEDIKTSSON J.A., WASKE B., *et al.*, "Morphological attribute profiles for the analysis of very high resolution images", *IEEE Transactions on Geoscience and Remote Sensing*, vol. 48, no. 10, pp. 3747–3762, 2010.

[DAR 06] DARBON J., SIGELLE M., "Image restoration with discrete constrained total variation part I: fast and exact optimization – Part II: levelable functions, convex priors and non-convex cases", *Journal of Mathematical Imaging and Vision*, vol. 26, no. 3, pp. 261–276, 2006.

[DEB 06] DEBBA P., CARRANZA E., VAN DER MEER F., *et al.*, "Abundance estimation of spectrally similar minerals by using derivative spectra in simulated annealing", *IEEE Transactions on Geoscience and Remote Sensing*, vol. 44, no. 12, pp. 3649–3658, 2006.

[DEE 05] DEE D.P., "Bias and data assimilation", *Quarterly Journal of the Royal Meteorological Society*, vol. 131, pp. 3323–3343, 2005.

[DEL 98] DELACOURT C., BRIOLE P., ACHACHE J., "Tropospheric corrections of SAR interferograms with strong topography application to Etna", *Geophysical Research Letters*, vol. 25, pp. 2849–2852, 1998.

[DEL 09] DELEDALLE C., DENIS L., TUPIN F., "Iterative weighted maximum likelihood denoising with probabilistic patch-based weights", *IEEE Transactions on Image Processing*, vol. 18, no. 12, December 2009.

[DEL 10a] DELEDALLE C., NICOLAS J.M., TUPIN F., *et al.*, "Glacier monitoring correlation versus texture tracking", *IEEE International Geoscience and Remote Sensing Symposium (IGARSS)*, pp. 513–516, 2010.

[DEL 10b] DELEDALLE C., TUPIN F., DENIS L., "Polarimetric SAR estimation based on non-local means", *IEEE International Geoscience and Remote Sensing Symposium (IGARSS)*, pp. 2515–2518, 2010.

[DEL 11a] DELEDALLE C., Débruitage d'images au-delà du bruit additif gaussien. Estimateurs à patchs et leur application à l'imagerie SAR, PhD Thesis, Telecom ParisTech, 2011.

[DEL 11b] DELEDALLE C., DENIS L., TUPIN F., "NL-InSAR nonlocal interferogram estimation", *IEEE Transactions on Geoscience and Remote Sensing*, vol. PP, no. 99, March 2011.

[DEL 12] DELEDALLE C., DENIS L., TUPIN F., "How to compare noisy patches? Patch similarity beyond Gaussian noise", *International Journal of Computer Vision*, vol. 99, no. 1, pp. 86–102, February 2012.

[DEM 07] DEMARTY J., CHEVALLIER F., FRIEND A.D., *et al.*, "Assimilation of global MODIS leaf area index retrievals within a terrestrial biosphere model", *Geophysical Research Letters*, vol. 34, pp. 1–19, 2007.

[DES 00] DESOLNEUX A., MOISAN L., MOREL J.M., "Meaningful alignments", *International Journal of Computer Vision*, vol. 40, no. 1, pp. 7–23, 2000.

[DES 05] DESROZIER G., BERRE L., CHAPNIK B., *et al.*, "Diagnosis of observation, background and analysis-error statistics in observation space", *Quarterly Journal of the Royal Meteorological Society*, vol. 131, pp. 3385–3396, 2005.

[DES 11] DESCOMBES X., *Applications de la géométrie stochastique à l'analyse d'images*, Hermès–Lavoisier, Paris, 2011.

[DEV 97] DEVERNAY F., Vision stéréoscopique et propriétés différentielles des surfaces, PhD thesis, École Polytechnique, February 1997.

[DIE 02] DIERKING W., SKRIVER H., "Change detection for thematic mapping by means of airborne multitemporal polarimetric SAR imagery", *IEEE Transactions on Geoscience and Remote Sensing*, vol. 40, no. 3, pp. 618–636, 2002.

[DOI 09] DOIN M.P., LASSERRE C., PELTZER G., *et al.*, "Corrections of stratified tropospheric delays in SAR interferometry validation with global atmospheric models", *Journal of Applied Geophysics*, vol. 69, no. 1, pp. 35–50, 2009.

[DOI 11] DOIN M.P., LODGE F., GUILLASO S., *et al.*, "Presentation of the small baseline NSBAS processing chain on a case example: the Etna deformation monitoring from 2003 to 2010 using ENVISAT data", ESASP, SP-697, 2011.

[DOR 11] DORON M., BÉLANGER S., DOXARAN D., *et al.*, "Spectral variations in the near-infrared ocean reflectance", *Remote Sensing of Environment*, vol. 115, pp. 1617–1631, 2011.

[DRA 01] DRAGERT H., WANG K., JAMES T., "A silent slip event on the deeper Cascadia subduction interface", *Science*, vol. 292, pp. 1525-1528, 2001.

[DRU 05] DRUSCH M., WOOD E.F., GAO H., "Observation operators for the direct assimilation of TRMM microwave imager retrieved soil moisture", *Geophysical Research Letters*, vol. 32, pp. 32–35, 2005.

[DRU 09] DRUSCH M., SCIPAL K., DE ROSNAY P., *et al.*, "Towards a Kalman filter based soil moisture analysis system for the operational ECMWF integrated forecast system", *Geophysical Research Letters*, vol. 36, no. 10, pp. 1–6, 2009.

[DUB 92] DUBOIS D., PRADE H., "Combination of fuzzy information in the framework of possibility theory", in ABIDI M. A., GONZALEZ R.C. (eds), *Data Fusion in Robotics and Machine Intelligence*, Chapter 12, Academic Press, Salt Lake City, UT, pp. 481–505, 1992.

[DUB 04] DUBOIS D., FOULLOY L., MAURIS G., *et al.*, "Probability-possibility transformations, triangular fuzzy sets and probabilistic inequalities", *International Journal on Reliable Computing*, vol. 10, no. 4, pp. 273–297, 2004.

[DUB 07] DUBOIS D., PRADE H., "Possibility theory", *Scholarpedia*, vol. 2, no. 10, 2007.

[DUC 08] DUCHEMIN B., MAISONGRANDE P., BOULET G., *et al.*, "A simple algorithm for yield estimates evaluation for semiarid irrigated winter wheat monitored with green leaf area index", *Environmental Modelling and Software*, vol. 23, pp. 876–892, 2008.

[DUD 72] DUDA R.D., HART P.E., "Use of the Hough transformation to detect lines and curves in pictures", *Communications of the ACM*, vol. 15, no. 1, pp. 11–15, 1972.

[DUR 85] DURDEN S., VESECKY J., "A physical radar cross-section model for a wind driven sea with swell", *IEEE Journal of Oceanic Engineering*, vol. 10, no. 4, pp. 445–451, 1985.

[ENT 94] ENTEKHABI D., NAKAMURA H., NJOKU E.G., "Solving the inverse problem for soil moisture and temperature profiles by sequential assimilation of multifrequency remotely sensed observations", *IEEE Transactions on Geoscience and Remote Sensing*, vol. 32, pp. 438–448, 1994.

[ERT 09] ERTEN E., REIGBER A., HELLWICH O., *et al.*, "Glacier velocity monitoring by maximum likelihood texture tracking", *IEEE Transactions on Geoscience and Remote Sensing*, vol. 47, no. 2, pp. 394–405, 2009.

[ERT 13] ERTEN E., "Glacier velocity estimation by means of a polarimetric similarity measure", *IEEE Transactions on Geoscience and Remote Sensing*, vol. 51, no. 6, pp. 3319–3327, 2013.

[ESA 07] EUROPEAN SPACE AGENCY, AATSR product handbook, Report no. 2.2, 2007.

[EVE 94] EVENSEN G., "Sequential data assimilation with a nonlinear quasi-geostrophic model using monte carlo methods to forecast error statistics", *Journal of Geophysical Research*, vol. 99, pp. 43–162, 1994.

[EVE 03] EVENSEN G., "The ensemble Kalman filter. theoretical formulation and practical implementation", *Oceans Dynamics*, vol. 53, pp. 343–367, 2003.

[FAL 11] FALLOURD R., HARANT O., TROUVÉ E., *et al.*, "Monitoring temperate glacier displacement by multi-temporal terraSAR-X images and continuous GPS measurements", *IEEE Journal of Selected Topics in Applied Earth Observations and Remote Sensing*, vol. 4, no. 2, pp. 372–386, 2011.

[FAL 12] FALLOURD R., Suivi des glaciers alpins par combinaison d'informations hétérogènes: images SAR Haute Résolution et mesures terrain, PhD Thesis, University of Grenoble, 2012.

[FAU 93] FAUGERAS O., HOTZ B., MATHIEU H., *et al.*, Real time correlation based stereo: algorithm implementations and applications, Report no. RR-2013, INRIA, August 1993.

[FAU 06] FAUVEL M., CHANUSSOT J., BENEDIKTSSON J., "Decision fusion for the classification of urban remote sensing images", *IEEE Transactions on Geoscience and Remote Sensing*, vol. 44, no. 10, pp. 2828–2838, 2006.

[FAU 08] FAUVEL M., BENEDIKTSSON J.A., CHANUSSOT J., *et al.*, "Spectral and spatial classification of hyperspectral data using SVMs and morphological profiles", *IEEE Transactions on Geoscience and Remote Sensing*, vol. 46, no. 11, pp. 3804–3814, 2008.

[FAU 13] FAUVEL M., TARABALKA Y., BENEDIKTSSON J.A., *et al.*, "Advances in Spectral-spatial classification of hyperspectral images", *Proceedings of the IEEE*, vol. 101, no. 3, pp. 652–675, 2013.

[FEI 09] FEIGL K.L., THURBER C.H., "A method for modelling radar interferograms without phase unwrapping: application to the M 5 Fawnskin, California earthquake of 1992 December 4", *Geophysical Journal International*, vol. 176, no. 2, pp. 491–504, 2009.

[FER 01] FERRETTI A., PRATI C., ROCCA F., "Permanent scatterer in SAR interferometry", *IEEE Transactions on Geoscience and Remote Sensing*, vol. 39, no. 1, pp. 8–20, 2001.

[FJO 97] FJORTOFT R., MARTHON P., LOPES A., *et al.*, "Multi-edge detection in SAR images", *International Conference on Acoustics, Speech, and Signal Processing (ICASSP'97)*, 1997.

[FOR 12] FORNARO G., PAUCIULLO A., REALE D., *et al.*, "SAR-tomography an advanced tool for 4D spaceborne radar scanning with application to imaging and monitoring of cities and single buildings", *IEEE Geoscience and Remote Sensing Society Newsletter*, pp. 10–18, 2012.

[FRA 99] FRADKIN M., ROUX M., MAÎTRE H., *et al.*, "Surface reconstruction from multiple aerial images in dense urban areas", *IEEE Conference on Computer Vision and Pattern Recognition*, vol. 1, Fort Collins, CO, pp. 262–267, June 1999.

[FRA 03] FRANÇOIS C., QUESNEY A., OTTLÉ C., "Sequential assimilation of ERS-1 SAR data into a coupled land surface–hydrological model using an extended Kalman filter", *Journal of Hydrometeorology*, vol. 4, no. 2, pp. 473–488, 2003.

[FRE 92] FREEMAN A., "SAR calibration: an overview", *IEEE Transactions on Geoscience and Remote Sensing*, vol. 30, pp. 1107–1121, 1992.

[FRU 96] FRUNEAU B., ACHACHE J., DELACOURT C., "Observation and modelling of the saint-etienne-de-tinée landslide using SAR interferometry", *Tectonophysics*, vol. 265, pp. 181–190, 1996.

[FUK 75] FUKUNAGA K., HOSTETLER L.D., "The estimation of the gradient of a density function, with applications in pattern recognition", *IEEE Transactions on Information Theory*, vol. 21, no. 1, pp. 320–340, January 1975.

[FUN 92] FUNG A., LI Z., CHEN K., "Backscattering from randomly rough dielectric surfaces", *IEEE Transactions on Geoscience and Remote Sensing*, vol. 30, no. 6, pp. 256–369, November 1992.

[GAL 03] GALLAND F., BERTAUX N., RÉFRÉGIER P., "Minimun description length synthetic aperture radar image segmentation", *IEEE Transactions on Image Processing*, vol. 12, no. 9, pp. 995–1006, September 2003.

[GAT 94] GATELLI F., GUARNIERI A.M., PARIZZI F., *et al.*, "The wavenumber shift in SAR interferometry", *IEEE Transactions on Geoscience and Remote Sensing*, vol. 32, no. 4, pp. 855–865, July 1994.

[GEM 84] GEMAN S., GEMAN D., "Stochastic relaxation, Gibbs distribution, and the Bayesian restoration of images", *IEEE Transactions on Pattern Analysis and Machine Intelligence*, vol. 6, no. 6, pp. 721–741, November 1984.

[GHI 91] GHIL M., MALANOTTE-RIZOLLI P., "Data assimilation in meteorology and oceanography", *Advance in Geophysics*, vol. 33, pp. 141–266, 1991.

[GHI 98] GHIGLIA D.C., PRITT M.D., *Two-Dimensional Phase Unwrapping*, John Wiley & Sons, New York, 1998.

[GIT 02] GITELSON A.A., ZUR Y., CHIVKUNOVA O.B., *et al.*, "Assessing carotenoid content in plant leaves with reflectance spectroscopy", *Photochemistry and Photobiology*, vol. 75, pp. 272–281, 2002.

[GIT 11] GITELSON A.A., GAO B., LI R., BERDNIKOV S., *et al.*, "Estimation of chlorophyll-a concentration in productive turbid waters using a hyperspectral imager for the coastal ocean; the Azov Sea case study", *Environmental Research Letters*, vol. 6, no. 2, 2011.

[GOL 88] GOLDSTEIN R.M., ZEBKER H.A., WERNER C.L., "Satellite radar interferometry: two-dimensional phase unwrapping", *Radio Science*, vol. 23, pp. 713–720, 1988.

[GOL 93] GOLDSTEIN R., ENGLEHARDT H., KAMB B., *et al.*, "Satellite radar interferometry for monitoring ice sheet motion: application to an Antarctic ice stream", *Science*, vol. 262, no. 1, pp. 525–530, 1993.

[GOL 94] GOLDBERG D.E., CORRUBLE V., GANASCIA J.G., *et al.*, *Algorithmes génétiques exploration, optimisation et apprentissage automatique*, Addison-Wesley, Paris, 1994.

[GOL 98] GOLDSTEIN R., WERNER C., "Radar interferogram filtering for geophysical applications", *Geophysical Research Letters*, vol. 25, no. 21, pp. 4035–4038, 1998.

[GOM 07] GOMEZ-CHOVA L., CAMPS-VALLS G., CALPE J., GUANTER L., *et al.*, "Cloud screening algorithm for ENVISAT/MERIS multispectral images", *IEEE Transactions on Geosciences and Remote Sensing*, vol. 45, no. 12, pp. 4105–4118, 2007.

[GOO 75] GOODMAN J., "Statistical properties of laser speckle patterns", in DAINTY J.C., *et al.* (eds) , *Laser Speckle and Related Phenomena*, vol. 9, Springer Verlag, Heidelberg, pp. 9–75, 1975.

[GOO 76] GOODMAN J., "Some fundamentals properties of speckle", *Journal Optical Society of America*, vol. 66, no. 11, pp. 1145–1150, 1976.

[GOR 93] GORDON N., SALMOND D., SMITH A., "Novel approach to non-linear/non-gaussian Bayesian state estimation", *IEEE Proceedings Radar and Signal Processing*, vol. 140, no. 2, pp. 107–113, 1993.

[GOU 10] GOURMELEN N., AMELUNG F., LANARI R., "Interferometric synthetic aperture radar-GPS integration: interseismic strain accumulation across the Hunter Mountain fault in the eastern California shear zone", *Journal of Geophysical Research*, vol. 115, 2010.

[GRA 01] GRAY A., SHORT N., MATTAR K., *et al.*, "Velocities and flux of the Filchner ice shelf and its tributaries determined from speckle tracking interferometry", *Canadian Journal of Remote Sensing*, vol. 27, no. 3, pp. 193–206, 2001.

[GRA 09] GRANDIN R., SOCQUET A., BINET R., *et al.*, "September 2005 Manda Hararo-Dabbahu rifting event, Afar (Ethiopia): constraints provided by geodetic data", *Journal of Geophysical Research*, vol. 114, no. B08404, pp. 1–20, 2009.

[GU 06] GU Y., BÉLAIR S., MAHFOUF J., *et al.*, "Optimal interpolation analysis of leaf area index using MODIS data", *Remote Sensing of Environment*, vol. 104, pp. 283–296, 2006.

[GUE 98] GUERIF M., DUKE C., "Calibration of the SUCROS emergence and early growth module for sugar beet using optical remote sensing data assimilation", *European Journal of Agronomy*, vol. 9, pp. 127–136, 1998.

[GUE 08] GUEGUEN L., DATCU M., "A similarity metric for retrieval of compressed objects: application for mining satellite image time series", *IEEE Transactions on Knowledge and Data Engineering*, vol. 20, pp. 562–575, 2008.

[GUY 99] GUYOT G., *Climatologie de l'environnement*, Dunod, Paris, 1999.

[HAG 08] HAGOLLE O., DEDIEU G., MOUGENOT B., *et al.*, "Correction of aerosol effects on multi-temporal images acquired with constant viewing angles: application to Formosat-2 images", *Remote Sensing of Environment*, vol. 112, no. 4, pp. 1689–1701, April 2008.

[HAI 98] HAIR J.F., ANDERSON R.E., TATHAM R.L., *et al.*, *Multivariate Data Analysis*, Prentice Hall, Upper Saddle River, 1998.

[HAL 85] HALLIKAINEN M., ULABY F., DOBSON M., *et al.*, "Microwave dielectric behaviour of wet soil. Part I: empirical models and experimental observations", *IEEE Transactions on Geoscience and Remote Sensing*, vol. 23, pp. 25–34, 1985.

[HAM 00] HAMILL T., SNYDER C., "A hybrid ensemble Kalman filter. 3D variational analysis scheme", *Monthly Weather Review*, vol. 128, pp. 2905–2919, 2000.

[HAN 01] HANSSEN R.F., Radar interferometry – Data–Interpretation and Error Analysis, (Remote Sensing and Digital Image Processing), vol. 2, Kluwer Academic Publishers, Dordrecht, the Netherlands, 2001.

[HAR 73] HARALICK R., SHANMUGAM K., DINSTEIN I., "Textural features for image classification", *IEEE Transactions on Systems, Man and Cybernetics*, vol. 3, no. 6, pp. 610–621, 1973.

[HAR 11] HARANT O., BOMBRUN L., VASILE G., *et al.*, "Displacement estimation by maximum-likelihood texture tracking", *IEEE Journal of Selected Topics in Signal Processing*, vol. 5, no. 3, pp. 398–407, 2011.

[HAS 80] HASSELMANN D., DUNCKEL M., EWING J., "Directional wave spectra observed during JONSWAP (Joint North Sea Wave Project)", *Journal of Physical Oceanography*, vol. 10, pp. 1264–1280, 1980.

[HER 04] HERNÁNDEZ C., Stereo and silhouette fusion for 3D object modeling from uncalibrated images under circular motion, PhD Thesis, École Nationale Supérieure des Télécommunications, May 2004.

[HOL 86] HOLLINGSWORTH B.A., LONNBERG P., "The statistical structure of short-range forecast errors as determined from radiosonde data. Part I: the wind field", *Tellus*, vol. 38, pp. 111–136, 1986.

[HOM 96] HOMER J., LONGSTAFF I., CALLAGHAN G., "High-resolution 3D SAR via multi-baseline interferometry", *IEEE International Geoscience and Remote Sensing Symposium (IGARSS'96)*, Lincoln, NE, no. 11, pp. 1145–1150, 1996.

[HOO 04] HOOPER A., ZEBKER H., SEGALL P., *et al.*, "A new method for measuring deformation on volcanoes and other natural terrains using InSAR persistent scatterers", *Geophysical Research Letters*, vol. 31, 2004.

[HOO 07] HOOPER A., SEGALL P., ZEBKER H., "Persistent scatterer interferometric synthetic aperture radar for crustal deformation analysis with application to Volcan Alcedo, Galapagos", *Journal of Geophysical Research*, vol. 112, no. B07407, 2007.

[HOO 08] HOOPER A., "A multi-temporal InSAR method incorporating both persistent scatterer and small baseline approaches", *Geophysical Research Letters*, vol. 35, 2008.

[HOO 12a] HOOPER A., BEKAERT D., SPAANS K., *et al.*, "Recent advances in SAR interferometry time series analysis for measuring crustal deformation", *Tectonophysics*, vol. 514–517, pp. 1–13, 2012.

[HOO 12b] HOOPER A., PRATA F., SIGMUNDSSON F., "Remote sensing of volcanic hazards and their precursors", *Proceedings of the IEEE*, vol. 100, pp. 2908–2930, 2012.

[HOR 10] HORION S., BERGAMINO N., STENUITE S., *et al.*, "Optimized extraction of daily bio-optical time series derived from MODIS/Aqua imagery for Lake Tanganika, Africa", *Remote Sensing of Environment*, vol. 114, pp. 781–791, 2010.

[HOU 09] HOUTEKAMER P., MITCHELL H., DENG X., "Model error representation in an operational ensemble Kalman filter", *Monthly Weather Review*, vol. 137, pp. 2126–2143, 2009.

[HOU 10] HOUSER P.R., DE LANNOY G.J.M., WALKER J.P., "Land surface data assimilation", in LAHOZ W., KHATTATOV B., MENARD R. (eds), *Data Assimilation*, Springer, Berlin, Heidelberg, pp. 549–597, 2010.

[IDE 97] IDE K., COURTIER P., GHIL M., *et al.*, "Unified notation for data assimilation operational, sequential and variational", *Journal of Meteorological Society of Japan*, vol. 75, pp. 181–189, 1997.

[IMB 99] IMBO P., SOUYRIS J.C., LOPES A., *et al.*, "Synoptic representation of the polarimetric information", *Proceedings of the CEOS SAR Calibration Workshop*, Toulouse, France, 1999.

[ING 04] INGLADA J., GIROS A., "On the possibility of automatic multisensor image registration", *IEEE Transactions on Geoscience and Remote Sensing*, vol. 42, no. 10, pp. 2104–2120, 2004.

[ING 07a] INGLADA J., MERCIER G., "A new statistical similarity measure for change detection in multitemporal SAR images and its extension to multiscale change analysis", *IEEE Transactions on Geoscience and Remote Sensing*, vol. 45, no. 5, pp. 1432–1446, May 2007.

[ING 07b] INGLADA J., MURON V., PICHARD D., *et al.*, "Analysis of artifacts in subpixel remote sensing image registration", *IEEE Transactions on Geoscience and Remote Sensing*, vol. 45, no. 1, pp. 254–264, 2007.

[IOR 11] IORDACHE M.D., BIOUCAS-DIAS J.M., PLAZA A., "Sparse unmixing of hyperspectral data", *IEEE Transactions on Geoscience and Remote Sensing*, vol. 49, no. 6, pp. 2014–2039, 2011.

[ISH 03] ISHIKAWA H., "Exact optimization for Markov random fields with convex priors", *IEEE Transactions on Pattern Analysis and Machine Intelligence*, vol. 25, no. 10, pp. 1333–1336, October 2003.

[JAC 81] JACKSON T., SCHMUGGE T., NICKS A., *et al.*, "Soil moisture updating and microwave remote sensing for hydrological simulation", *Hydrological Sciences Bulletin*, vol. 26, no. 3, pp. 305–319, 1981.

[JAC 90] JACQUEMOUD S., BARET F., "PROSPECT a model of leaf optical properties spectra", *Remote Sensing of Environment*, vol. 34, no. 2, pp. 75–91, 1990.

[JAC 09] JACQUEMOUD S., *et al.*, "PROSPECT and SAIL models: a review of use for vegetation characterization", *Remote sensing of environment*, vol. 113, pp. 56–66, 2009.

[JAN 04] JANSSEN V., GE L., RIZOSA C., "Troposheric corrections to SAR interferometry from GPS observations", *GPS Solutions*, vol. 8, pp. 140–151, 2004.

[JAR 08] JARLAN L., MANGIAROTTI S., MOUGIN E., *et al.*, "Assimilation of SPOT/VEGETATION NDVI data into a Sahelian vegetation dynamics model", *Remote Sensing of Environment*, vol. 112, no. 4, pp. 1381–1394, 2008.

[JOA 98] JOACHIMS T., "Text categorization with support vector machines: learning with many relevant features", *Proceedings of the European Conference on Machine Learning*, Springer, pp. 137–142, 1998.

[JOL 11] JOLIVET R., GRANDIN R., LASSERRE C., *et al.*, "Systematic InSAR tropospheric phase delay corrections from global meteorological reanalysis data", *Geophysical Research Letters*, vol. 38, 2011.

[JOL 13] JOLIVET R., LASSERRE C., DOIN M.P., *et al.*, "Spatio-temporal evolution of a seismic slip along the Haiyuan fault, China: implications for fault frictional properties", *Earth and Planetary Science Letters*, vol. 377–378, pp. 23–33, 2013.

[JUL 95] JULIER S., ULHMANN J., DURRANT H., *et al.*, "A new approach for filtering non linear systems", *Proceedings of the IEEE American Control Conference*, pp. 887–892, 1995.

[JUL 11] JULEA A., MÉGER N., BOLON P., *et al.*, "Unsupervised spatio-temporal mining of satellite image time series using grouped frequent sequential patterns", *IEEE Transactions on Geoscience and Remote Sensing*, vol. 49, no. 4, pp. 1417–1430, March 2011.

[JUL 12] JULEA A., MÉGER N., RIGOTTI C., *et al.*, "Efficient spatiotemporal mining of satellite image time series for agricultural monitoring", *Transactions on Machine Learning and Data Mining*, vol. 5, no. 1, pp. 23–45, 2012.

[KAL 60] KALMAN R., "A new approach to linear filtering and prediction problems", *Journal of Basic Engineering*, vol. 82, pp. 35–45, 1960.

[KAL 03] KALNAY E., *Atmospheric Modeling, Data Assimilation and Predictability*, Cambridge University Press, Cambridge, 2003.

[KAS 88] KASS M., WITKIN A., TERZOPOULOS D., "Snakes: active contours models", *International Journal of Computer Vision*, vol. 1, no. 4, pp. 321–331, 1988.

[KAU 76] KAUTH R.J., THOMAS G.S., "The tasseled Cap – a graphic description of the spectral-temporal development of agricultural crops as seen by LANDSAT", *Proceedings of the Symposium on Machine Processing of Remotely Sensed Data*, Purdue University of West Lafayette, IN, pp. 4B–41–4B–51, 1976.

[KER 95] KERGOAT L., FISHER A., MOULIN S., *et al.*, "Satellite measurements as a constraint on estimates of vegetation carbon budget", *Tellus*, vol. B47, pp. 251–263, 1995.

[KES 02] KESHAVA N., MUSTARD J.F., "Spectral unmixing", *IEEE Signal Processing Magazine*, vol. 19, no. 1, pp. 44–57, 2002.

[KNO 01] KNORR W., LAKSHMI V., "Assimilation of fAPAR and surface temperature into a land surface and vegetation model", in LAKSHMI V., ALBERTSON J., SHAAKE J. (eds), *Land Surface Hydrology, Meteorology and Climate: Observations and Modeling*, American Geophysical Union, Washington DC, 2001.

[KON 90] KONG J.A., *Electromagnetic Wave Theory*, John Wiley & Sons, Hoboken, NJ, 1990.

[KOS 02] KOSTER R.D., SUAREZ M.J., "Impact of land surface initialization on seasonal precipitation and temperature prediction", *Journal of Hydrometeorology*, vol. 4, pp. 408–423, 2002.

[LAB 00] LABEN C.A., BROWER B.V., "Process for enhancing the spatial resolution of multispectral imagery using pan-sharpening", US Patent No 6 011 875, 2000.

[LAP 72] LAPRADE G., "Stereoscopy–a more general theory", *Photogrammetric Engineering*, pp. 1177–1187, December 1972.

[LEC 82] LECLERC Y., "Constructing simple stable descriptions for image partitioning", *International Journal of Computer Vision*, vol. 3, pp. 73–102, 1982.

[LEE 94] LEE J.S., AINSWORTH T.L., GRUNES M.R. *et al.*, "Noise filtering of interferometric SAR images", *Image and Signal Processing for Remote Sensing, Rome*, vol. 2315, pp. 735-742, September 1994.

[LEE 00] LEE J.S., SCHULER D.L., AINSWORTH T.L., "Polarimetric SAR data compensation for terrain azimuth slope variation", *IEEE Transactions on Geoscience and Remote Sensing*, vol. 38, no. 5, pp. 2153–2163, 2000.

[LEE 08] LEE J.S., POTTIER E., *Polarimetric Radar Imaging: From Basics to Applications*, CRC Press, Boca Raton, FL, 2008.

[LEH 97] LEHEGARAT-MASCLE S., BLOCH I., VIDAL-MADJAR D., "Application of Dempster-Shafer evidence theory to unsupervised classification in multisource remote sensing", *IEEE Transactions on Geoscience and Remote Sensing*, vol. 35, no. 4, pp. 1018–1031, April 1997.

[LEH 09] LEHUREAU G., CAMPEDEL M., TUPIN F., *et al.*, "Combining SAR and optical features in a SVM classifier for man-made structures detection", *Proceedings of IEEE International Geoscience and Remote Sensing Symposium (IGARSS'09)*, CapeTown, South Africa, 2009.

[LEL 98] LELOGLU U.M., ROUX M., MAÎTRE H., "Dense urban dem with three or more high-resolution aerial images", *In GIS - Between Visions and Applications IAPRS*, vol. 32. no. 4, pp. 347–352, 1998.

[LEM 09] LEMEN C., Segmentation spatio-temporelle d'une séquence temporelle d'images satellitaires à haute résolution, PhD Thesis, Telecom ParisTech, 2009.

[LEN 93] LENOBLE J., *Atmospheric Radiative Transfer*, Deepak Publishing, Hampton, VA, 1993.

[LEN 02] LENNON M., Méthodes d'analyses d'images hyperspectrales. Exploitation du capteur aéroporté CASI pour des applications de cartographie agro-environnementale en Bretagne, PhD Thesis, University of Rennes I, 2002.

[LER 13] LEROUX D.J., KERR Y. H., *et al.*, "An approach to constructing a homogeneous time series of soil moisture using SMOS", *IEEE Transactions on Geoscience and Remote Sensing*, vol. 52, pp. 393–405, 2014.

[LET 93] LETOAN T., SMACCHIA P., SOUYRIS J., *et al.*, "On the retrieval of soil moisture from ERS-1 SAR data", *Proceedings of the 2nd ERS-1 Symposium: Space at the Service of Our Environment, ESA SP-361*, pp. 883–888, 1993.

[LI 06a] LI Z., CROSS P., ALBERT P., *et al.*, "Assessment of the potential of MERIS near-infrared water wapour products to correct ASAR interferometric measurements", *International Journal of Remote Sensing*, vol. 33, no. 2, pp. 349–365, 2006.

[LI 06b] LI Z., FIELDING E.J., CROSS P., *et al.*, "Interferometric synthetic aperture radar atmospheric correction: GPS topography-dependent turbulence model", *Journal of Geophysical Research*, vol. 111, 2006.

[LI 13] LI J., MARPU P.R., PLAZA A., *et al.*, "Generalized composite kernel framework for hyperspectral image classification", *IEEE Transactions on Geoscience and Remote Sensing*, vol. 51, no. 9, pp. 4816–4829, 2013.

[LIC 09] LICCIARDI G., PACIFICI F., TUIA D., *et al.*, "Decision fusion for the classification of hyperspectral data: outcome of the 2008 GRS-S data fusion contest", *IEEE Transactions on Geoscience and Remote Sensing*, vol. 47, no. 11, pp. 3857–3865, 2009.

[LIE 08] LIER P., VALORGE C., BRIOTTET X., *Imagerie spatiale*, CNES, Cepadues Editions, Paris, 2008.

[LIN 96] LINDE A., CLADWIN M., JOHNSTON M., *et al.*, "A slow earthquake sequence on the San Andreas fault", *Nature*, vol. 383, no. 6595, pp. 65–68, 1996.

[LIU 09] LIU G., BUCKLEY M., DING X., *et al.*, "Estimating spatiotemporal ground deformation with improved persistent scatterer radar interferometry", *IEEE Transaction on Geoscience and Remote Sensing*, vol. 47, no. 8, pp. 2762–2772, 2009.

[LIU 12] LIU Z., DEZERT J., MERCIER G., *et al.*, "Dynamical evidential reasoning for change detection in remote sensing images", *IEEE Transactions on Geoscience and Remote Sensing*, vol. 50, no. 5, pp. 1955–1967, May 2012.

[LIU 13] LIU G., JIA H., NIE Y., *et al.*, "Detecting subsidence in coastal areas by ultrashort baseline TCPInSAR on time series of high resolution TerraSAR-X images", *IEEE Transactions on Geoscience and Remote Sensing*, 2013.

[LOP 90] LOPES A., TOUZI R., NEZRY E., "Adaptative speckle filters and scene heterogeneity", *IEEE Transactions on Geoscience and Remote Sensing*, vol. 28, no. 6, pp. 992–1000, 1990.

[LOP 09] LOPEZ-QUIROZ P., DOIN M.P., TUPIN F., *et al.*, "Time series analysis of Mexico City subsidence constrained by radar interferometry", *Journal of Applied Geophysics*, vol. 69, no. 1, pp. 1–15, 2009.

[LOR 02] LORENA R.B., AO ROBERTO DOS SANTOS J., SHIMABUKURO Y.E., *et al.*, "A change vector analysis technique to monitor land use/land cover in SW Brazilian Amazon: Acre state", *ISPRS Symposium*, pp. 8, November 2002.

[LOR 03] LORENZ D.J., HARTMANN D.L., "Eddy-zonal flow feedback in the Northern Hemisphere winter", *Journal on Climate*, vol. 16, pp. 1212–1227, 2003.

[LOW 04] LOWE D., "Distinctive image features from scale invariant keypoints", *International Journal of Computer Vision*, vol. 60, no. 2, pp. 91–110, 2004.

[MAA 88] MAAS S., "Use of remotely sensed information in agricultural crop growth models", *Ecological Modelling*, vol. 41, pp. 247–268, 1988.

[MAC 67] MACQUEEN J., "Some methods for classification and analysis of multivariate observations", *Proceedings of 5th Berkeley Symposium on Mathematical Statistics and Probability*, 1967.

[MAE 97] MAES F., COLLIGNON A., VANDERMEULEN D., *et al.*, "Multimodality image registration by maximization of mutual information", *IEEE transactions on Medical Imaging*, vol. 16, pp. 187–198, 1997.

[MAH 91] MAHFOUF J.F., NOILHAN J., "Comparative study of various formulations of evaporations from bare soil using in situ data", *Journal of Applied Meteorology*, vol. 30, no. 9, pp. 1354–1365, 1991.

[MAI 01] MAÎTRE H., *Traitement des images de radar à synthèse d'ouverture*, Hermes, Paris, 2001.

[MAI 03] MAÎTRE H., *Le traitement des images*, Hermes–Lavoisier, Paris, 2003.

[MAI 04] MAIGNAN F., BRÉON F.M., LACAZE R., "Bidirectional reflectance of Earth targets: evaluation of analytical models using a large set of spaceborne measurements with emphasis on the hot spot", *Remote Sensing of Environment*, Photographer A. DERAMECOURT, vol. 90, no. 2, pp. 210–220, 2004.

[MAL 80] MALILA W.A., "Change vector analysis: an approach for detecting forest changes with Landsat", *Proceedings of 6th Annual Symposium on Machine Processing of Remotely Sensed Data*, pp. 326–335, June 1980.

[MAL 89] MALLAT S., "Theory for multiresolution signal decomposition: the wavelet representation", *IEEE Transactions on Pattern Analysis and Machine Intelligence*, vol. 11, no. 7, pp. 674–693, 1989.

[MAN 12] MANCONI A., CASU F., "Joint analysis of displacement time series retrieved from SAR phase and amplitude: impact on the estimation of volcanic source parameters", *Geophysical Research Letters*, vol. 39, no. 14, 2012.

[MAS 93] MASSONNET D., ROSSI M., CARMONA C., *et al.*, "The displacement field of the Landers earthquake mapped by radar interferometry", *Nature*, vol. 364, no. 6433, pp. 138–142, 1993.

[MAS 95] MASSONNET D., BRIOLE P., ARNAUD A., "Deflation of Mount Etna monitored by spaceborne radar interferometry", *Nature*, vol. 375, no. 6532, pp. 567–570, 1995.

[MAS 97] MASSONNET D., "Producing ground deformation maps automatically: the DIAPASON concept", *Proceedings of IEEE International Geoscience and Remote Sensing Symposium (IGARSS '97)*, vol. III, pp. 1338–1340, Singapore, 1997.

[MAS 98] MASSONNET D., FEIGL K.L., "Radar interferometry and its application to changes in the Earth's surface", *Review of Geophysics*, vol. 36, no. 4, pp. 441–500, 1998.

[MAS 08] MASSONNET D., SOUYRIS J., *Imaging with Synthetic Aperture Radar*, EPFL Press, Lausanne, 2008.

[MAU 01] MAURIS G., LASSERRE V., FOULLOY L., "A fuzzy approach for the expression of uncertainty in measurement", *International Journal of Measurement*, vol. 29, no. 3, pp. 165–177, 2001.

[MEG 11] MEGER N., JOLIVET R., LASSERRE C., *et al.*, "Spatiotemporal mining of ENVISAT SAR interferogram time series over the Haiyuan fault in China", *6th International Workshop on the Analysis of Multi-temporal Remote Sensing Images (Multi-Temp)*, pp. 137–140, 2011.

[MEI 81] MEIER S., "Planar geodetic covariance functions", *Review of Geophysics and Space Physics*, vol. 19, pp. 673–686, 1981.

[MEY 01] MEYZONNETTE J.L., LÉPINE T., *Bases de radiométrie optique*, Cepadues, Paris, 2001.

[MIC 99] MICHEL R., AVOUAC J.P., TABOURY J., "Measuring near field coseismic displacements from SAR images: application to the Landers earthquake", *Geophysical Research Letters*, vol. 26, no. 19, pp. 3017–3020, 1999.

[MIC 09] MICHEL J., INGLADA J., "Qualitative spatial reasoning for high-resolution remote sensing image analysis", *IEEE Transactions on Geoscience and Remote Sensing*, vol. 47, no. 2, pp. 599–612, 2009.

[MIL 86] MILLY P., "Integrated design of hydrological networks", (ed.) MOSS M.E., *International Association of Hydrological Sciences Publication*, vol. 158, pp. 201–211, 1986.

[MOR 90] MORCHIN W., *Airborne Early Warning Radar*, Artech House, London, 1990.

[MUM 89] MUMFORD D., SHAH J., *Image Understanding*, ed. ULLMAN S., RICHARDS W., Ablex Publishing corporation, Norwood, New Jersey, 1990.

[MUN 07] 40 MUNOZ-SABATER J., JARLAN L., CALVET J.-C. *et al.*, "From near-surface to root-zone soil moisture using different assimilation techniques", *Journal of Hydrometeorology*, vol. 8, no. 2, pp. 194–206, 2007.

[NAR 10] NARASIMHA R., Méthodes d'estimation de la profondeur par mise en correspondance stéréoscopique à l'aide de champs aléatoires couplés, PhD Thesis, University of Grenoble, September 2010.

[NAS 05] NASCIMENTO J., BIOUCAS-DIAS J., "Vertex component analysis: a fast algorithm to unmix hyperspectral data", *IEEE Transactions on Geosciences and Remote Sensing*, vol. 43, no. 8, pp. 898–910, 2005.

[NIC 77] NICODEMUS F.E., RICHMOND J.C., HSIA J.J., *et al.*, *Geometrical Considerations and Nomenclature for Reflectance*, vol. 160, National Bureau of Standards, October 1977.

[NIC 02] NICOLAS J.M., "Introduction aux statistiques de deuxième espèce: applications des log-moments et des log-cumulants à l'analyse des lois d'image radar", *Traitement du signal*, vol. 19, no. 3, pp. 139–167, 2002.

[NIE 98] NIELSEN A.A., CONRADSEN K., SIMPSON J.J., "Multivariate alteration detection (MAD) and MAF postprocessing in multispectral, bitemporal image data: new approaches to change detection studies", *Remote Sensing of Environment*, vol. 64, pp. 1–19, 1998.

[OFE 11] OFEIGSSON B.G., SIGMUNDSSON F., HOOPER A. *et al.*, "InSAR time series analysis at Hekla volcano, Iceland: inflation periods and crustal deformation associated with the 2000 eruption", *Journal of Geophysical Research*, 2011.

[OSM 11] OSMANOGLU B., DIXON T.H., WDOWINSKI S., *et al.*, "Mexico City subsidence observed with persistent scatterer InSAR", *International Journal of Applied Earth Observation and Geoinformation*, vol. 13, no. 1, pp. 1–12, 2011.

[OSU 97] OSUNA E., FREUND R., GIROSI F., "Training support vector machines: an application to face detection", *Proceedings of IEEE Computer Vision and Pattern Recognition Conference*, pp. 130–136, 1997.

[OTH 85] OTHA Y., KANADE T., "Stereo by intra- and inter-scanline search using dynamic programming", *IEEE Transactions on Pattern Analysis and Machine Intelligence*, vol. 7, no. 2, pp. 139–154, 1985.

[OTT 94] OTTLÉ C., VIDAL-MADJAR D., "Assimilation of soil moisture inferred from infrared remote sensing in a hydrological model over the HAPEX-MOBILHY region", *Journal of Hydrology*, vol. 158, no. 3, pp. 241–264, 1994.

[PAC 09] PACIFICI F., CHINI M., EMERY W., "A neural network approach using multi-scale textural metrics from very high-resolution panchromatic imagery for urban land-use classification", *Remote Sensing of Environment*, vol. 113, no. 6, pp. 1276–1292, 2009.

[PAG 07] PAGLI C., SIGMUNDSSON F., LUND B., *et al.*, "Glacio-isostatic deformation around the Vatnajókull ice cap, iceland, induced by recent climate warming: GPS observations and finite element modeling", *Journal of Geophysical Research*, vol. 112, no. B08405, 2007.

[PAP 00] PAPARODITIS N., THOM C., JIBRINI H., "Surface reconstruction in urban areas from multiple views of aerial digital frame cameras", *IAPRS*, vol. XXXIII, ISPRS, Amsterdam, 2000.

[PAR 92] PARRISH D.F., DERBER J.C., "The national meteorological center's spectral statistical-interpolation analysis system", *Monthly Weather Review*, vol. 120, pp. 1747–1763, 1992.

[PAR 09] PARCHARIDIS I., FOUMELIS M., KOURKOULI P., *et al.*, "Persistent scatterers InSAR to detect ground deformation over Rio-Antirio area (Western Greece) for the period 1992-2000", *Journal of Applied Geophysics*, vol. 68, pp. 348–355, 2009.

[PAT 06] PATHIER E., FIELDING E.J., WRIGHT T.J., *et al.*, "Displacement field and slip distribution of the 2005 Kashmir earthquake from SAR imagery", *Geophysical Research Letters*, vol. 33, no. 20, pp. 1–5, 2006.

[PAU 07] PAUWELS V., VERHOEST N., LANNOY J.D., *et al.*, "Optimization of a coupled hydrology. Crop growth model through the assimilation of observed soil moisture and leaf area index values using an ensemble Kalman filter", *Water Resources Research*, vol. 43, pp. 1–17, 2007.

[PEL 03] PELLENQ J., BOULET G., CHAPONNIÈRE A., *et al.*, "Soil moisture assimilation in a coupled water/energy scheme in a semi-arid region", *International Conference on the Hydrology in the Mediterranean and Semiarid Regions*, IAHS Publications, Red Books Series, pp. 9–14, 2003.

[PEL 04] PELLENQ J., BOULET G., "A methodology to test the pertinence of remote-sensing data assimilation into vegetation models for water and energy exchange at the land surface", *Agronomie*, vol. 24, no. 4, pp. 197–204, 2004.

[PEP 05] PEPE A., SANSOSTI E., BERARDINO P., *et al.*, "On the generation of ERS-ENVISAT DinSAR time-series via the SBAS Technique", *Geoscience and Remote Sensing Letters*, vol. 2, no. 3, pp. 265–269, 2005.

[PER 06] PERISSIN D., PRATI C., ENGDAHL M.E., *et al.*, "Validating the SAR wavenumber shift principle with the ERS-Envisat PS coherent combination", *IEEE Transactions on Geoscience and Remote Sensing*, vol. 44, pp. 2343–2351, 2006.

[PES 01] PESARESI M., BENEDIKTSSON J., "A new approach for the morphological segmentation of high-resolution satellite images", *IEEE Transactions on Geosciences and Remote Sensing*, vol. 39, no. 2, pp. 309–320, 2001.

[PET 12] PETITJEAN F., INGLADA J., GANAARSKI P., "Satellite image time series analysis under time warping", *IEEE Transactions on Geoscience and Remote Sensing*, vol. 50, no. 8, pp. 3081–3095, 2012.

[PHA 98] PHAM D., VERRON J., GOURDEAU L., "Singular evolutive Kalman filters for data assimilation in oceanography", *Compte rendu de l'académie des sciences*, vol. 326, no. 4, pp. 255–260, 1998.

[PIE 64] PIERSON W., MOSKOWITZ L., "A proposed spectral form for fully developped wind sea based on the similarity theory of S.A. KITAIGORODSKII", *Journal of Geophysical Research*, vol. 69, pp. 5181–5190, 1964.

[PIE 06] PIERROT-DESEILLIGNY M., PAPARODITIS N., "A multiresolution and optimization-based image matching approach: an application to surface reconstruction from SPOT5-HRS stereo imagery", *Proceedings of the ISPRS Conference Topographic Mapping From Space (With Special Emphasis on Small Satellites), ISPRS*, 2006.

[PIE 07] PIERROT-DESEILLIGNY M., "MicMac, un logiciel pour la mise en correspondance automatique d'images dans le contexte géographique", *Bulletin d'Information Scientifique et Technique de l'IGN*, vol. 77, 2007.

[PIN 07] PINEL V., SIGMUNDSSON F., STURKELL E., *et al.*, "Discriminating volcano deformation due to magma movements and variable surface loads: application to Katla subglacial volcano, Iceland", *Geophysical Journal International*, vol. 169, pp. 325–338, 2007.

[PIN 11] PINEL V., HOOPER A., DE LA CRUZ-REYNA S., *et al.*, "The challenging retrieval of the displacement field from InSAR data for andesitic stratovolcanoes: case study of Popocatepetl and Colima Volcano, Mexico", *Journal of Volcanology and Geothermal Research*, vol. 200, pp. 49–61, 2011.

[PLA 11] PLAZA A., MARTÍN G., PLAZA J., *et al.*, "Recent developments in spectral unmixing and endmember extraction", in BRUCE L., PRASAD S., CHANUSSOT J. (eds), *Optical Remote Sensing: Avances in Signal Processing and Exploitation*, Springer-Verlag, Berlin, Heidelberg, vol. 3, pp. 235–267, 2011.

[PON 07] PONS J.P., KERIVEN R., FAUGERAS O., "Multi-view stereo reconstruction and scene flow estimation with a global image-based matching score", *The International Journal of Computer Vision*, vol. 72, no. 2, pp. 179–193, April 2007.

[POU 11] POULAIN V., INGLADA J., SPIGAI M., *et al.*, "High-resolution optical and SAR image fusion for building database updating", *IEEE Transactions on Geoscience and Remote Sensing*, vol. 49, no. 8, pp. 2900–2910, 2011.

[PRE 84] PREVOT L., BERNARD R., TACONET O., *et al.*, "Evaporation from a bare soil evaluated using a soil water transfer model and remotely sensed surface soil moisture data", *Water Resources Research*, vol. 20, no. 2, pp.311–316, 1984.

[PRE 08] PREUSKER R., FISCHER J., BROCKMAN C., *et al.*, "Cloud screening and snow detection with MERIS", *ESA MERIS AATSR Meeting*, ESRIN, Italy, 2008.

[PRI 91] PRINCE S., "Satellite remote sensing of primary production: comparison of results for Sahelian grasslands 1981-1988", *International Journal of Remote Sensing*, vol. 12, no. 6, pp. 1301–1311, 1991.

[PRI 02] PRITCHARD M.E., SIMONS M., "A satellite geodetic survey of large-scale deformation of volcanic centres in the central Andes", *Nature*, vol. 418, pp. 167–171, 2002.

[PUY 07] PUYSSEGUR B., MICHEL R., AVOUAC J.P., "Tropospheric phase delay in interferometric synthetic apertur radar estimated from meteorological model and multispectral imagery", *Journal of Geophysical Research*, vol. 112, 2007.

[QUI 12] QUIN G., PINEL-PUYSSEGUR B., NICOLAS J.M., "Comparison of harmonic, geometric and arithmetic means for change detection in SAR time series", *9th European Conference on Synthetic Aperture Radar (EUSAR)*, pp. 255–258, 2012.

[RAD 97] RADARSAT Geology Handbook (RADARSAT International), 1997.

[RAM 86] RAMAPRIYAN H.K., STRONG J.P., HUNG Y., *et al.*, "Automated matching of pairs of SIR-B images for elevation mapping", *IEEE Transactions on Geoscience and Remote Sensing*, vol. 24, no. 4, pp. 462–472, 1986.

[RAS 13] RASI R., BEUCHLE R., BODART C., *et al.*, "Automatic updating of an object-based tropical forest cover classification and change assessment", *IEEE Journal of Selected Topics in Applied Earth Observations and Remote Sensing*, vol. 6, no. 1, pp. 66–73, 2013.

[REI 00] REIGBER A., MOREIRA A., "First demonstration of airborne SAR tomography using multibaseline L-band data", *IEEE Transactions on Geoscience and Remote Sensing*, vol. 38, no. 5, pp. 2142–2152, 2000.

[REI 01] REICHLE R.H., MCLAUGHLIN D.B., ENTEKHABI D., "Variational data assimilation of microwave radiobrightness observations for land surface", *Hydrology Applications*, vol. 39, no. 8, pp. 1708–1718, 2001.

[REI 02] REICHLE R.H., MCLAUGHLIN D.B., ENTEKHABI D., "Hydrologic data assimilation with the ensemble Kalman filter", *Monthly Weather Review*, vol. 130, pp. 103–114, 2002.

[REI 03] REICHLE R.H., KOSTER R.D., "Assessing the impact of horizontal error correlations in background fields on soil moisture estimation", *Journal of Hydrometeorology*, vol. 4, pp. 1229–1243, 2003.

[REI 04a] REICHLE R.H., KOSTER R.D., "Bias reduction in short records of satellite soil moisture", *Geophysical Research Letters*, vol. 31, pp. 2–5, 2004.

[REI 04b] REICHLE R., KOSTER R.D., DONG J., *et al.*, "Global soil moisture from satellite observations, land surface models and ground data implications for data assimilation", *Journal of Hydrometeorology*, vol. 5, pp. 30–442, 2004.

[REI 07] REICHLE R.H., KOSTER R.D., LIU P., *et al.*, "Comparison and assimilation of global soil moisture retrievals from the advanced microwave scanning radiometer for the Earth observing system (AMSR-E) and the scanning multichannel microwave radiometer (SMMR)", *Journal of Geophysical Research*, vol. 112, pp. 1–14, 2007.

[REI 08] REICHLE R.H., CROW W.T., KEPPENNE C.L., "An adaptive ensemble Kalman filter for soil moisture data assimilation", *Water Resources Research*, vol. 44, pp. 1–13, 2008.

[REM 03] REMY D., BONVALOT S., BRIOLE P., *et al.*, "Accurate measurements of tropospheric effects in volcanic areas from SAR interferometry data: application to Sakurajima volcano (Japan)", *Earth and Planetary Science Letters*, vol. 213, pp. 299–310, 2003.

[REM 05] REMER L.A., KAUFMAN Y.J., TANRÉ D., *et al.*, "The MODIS aerosol algorithm, products, and validation", *Journal of Atmospheric Sciences*, vol. 62, no. 4, pp. 947–973, 2005.

[ROD 92] RODRIGUEZ E., MARTIN J., "Theory and design of interferometric synthetic aperture radars", *IEEE Proceedings*, vol. 139, pp. 147–159, 1992.

[ROS 68] ROSENFIELD G.H., "Stereo radar techniques", *Photogrammetric Engineering*, vol. 34, no. 6, pp. 586–594, 1968.

[ROS 04] ROSEN P., HENSLEY S., PELTZER G., *et al.*, "Updated repeat orbit interferometry package released", *The Earth Observation System–Transactions, American Geophysical Union*, vol. 85, no. 5, 2004. Available at www.agu.org.

[ROU 01] ROUX M., MAÎTRE H., "Some more steps towards 3D reconstruction of urban areas from multiple views", (eds), BALKEMA PUBLISHERS, A.A. SWETS & ZEITLINGER B.V., LISSE, The Netherlands, *Automatic Extraction of Man-Made Objects from Aerial and Space Images (III)*, Ascona, Switzerland, pp. 135–147, June 2001.

[ROY 98] ROY S., COX I.J., "A maximum-flow formulation of the N-camera stereo correspondence problem", *IEEE Proceedings of International Conference on Computer Vision*, Bombai, January 1998.

[ROY 99] ROY S., "Stereo without epipolar lines: a maximum-flow formulation", *International Journal of Computer Vision*, vol. 34, no. 2–3, pp. 147–161, October 1999.

[RUC 70] RUCK G.T., BARRICK D.E., STUART W.D., *et al.*, "'Radar cross section handbook", Plenum, New York, 1970.

[RUD 92] RUDIN L., OSHER S., FATEMI E., "Nonlinear total variation based noise removal algorithms", *Physica D*, vol. 60, pp. 259–268, 1992.

[SCH 03] SCHMIDT D.A., BÜRGMANN R., "Time-dependent land uplift and subsidence in the Santa Clara valley, California, from a large interferometric synthetic aperture radar data set", *Journal of Geophysical Research*, vol. 108, no. B9, pp. ETG4.1–ETG4.13, 2003.

[SCH 05] SCHMIDT D.A., BÜRGMANN R., NADEAU R.M., *et al.*, "Distribution of aseismic slip rate on the Hayward fault inferred from seismic and geodetic data", *Journal of Geophysical Research*, vol. 110, no. B08406, pp. 1–15, 2005.

[SCH 06] SCHAEPMAN-STRUB G., SCHAEPMAN M., PAINTER T., *et al.*, "Reflectance quantities in optical remote sensing: definitions and case studies", *Remote Sensing of Environment*, vol. 103, no. 1, pp. 27–42, 2006.

[SCI 08] SCIPAL K.T.H., JEU R.D., NAEIMI V., *et al.*, "A possible solution for the problem of estimating the error structure of global soil moisture data sets", *Geophysical Research Letters*, vol. 35, pp. 2–5, 2008.

[SER 06] SERAFINO F., "SAR image coregistration based on isolated point scatterers", *IEEE Geoscience and Remote Sensing Letters*, vol. 3, no. 3, pp. 354–358, 2006.

[SHA 76] SHAFER G., *A Mathematical Theory of Evidence*, Princeton University Press, Princeton, 1976.

[SHU 82] SHUKLA J., MINTZ Y., "Influence of land-surface evapotranspiration on the earth climate", *Science*, vol. 215, no. 7, pp. 1498–1501, 1982.

[SIG 10] SIGMUNDSSON F., HREINSDOTTIR S., HOOPER A., *et al.*, "Intrusion triggering of the 2010 Eyjafjallajökull explosive eruption", *Nature*, vol. 468, pp. 426–430, 2010.

[SIM 02] SIMONETTO E., Extraction 3D de structures industrielles sur des images RAMSES haute résolution par radargrammétrie, PhD Thesis, University of Rennes I, 2002.

[SIM 07] SIMONS M., ROSEN P.A., "Interferometric synthetic aperture radar geodesy", in SCHUBERT G. (eds), *Treatise on Geophysics, Vol. 3*, Elsevier Press, Amsterdam, pp. 391–446, 2007.

[SIR 09] SIRMACEK B., UNSALAN C., "Urban-area and building detection using SIFT keypoints and graph theory", *IEEE Transactions on Geoscience and Remote Sensing*, pp. 1156–1167, 2009.

[SMA 06] SMARANDACHE F., DEZERT J., *Advances and Applications of DSmT for Information Fusion (Collected works)*, vol. 2, American Research Press, Rehoboth, MA, 2006.

[SME 94] SMETS P., KENNES R., "The transferable belief model", *Artificial Intelligence*, vol. 66, pp. 191–234, 1994.

[SMI 53] SMITH E.K., WEINTRAUB S., "The constants in the equation for atmospheric refractive index at radio frequencies", *Proceedings of the IRE*, vol. 41, pp. 1035–1037, 1953.

[SOI 04] SOILLE P., *Morphological Image Analysis*, Springer-Verlag, Heidelberg, 2004.

[SPA 12] SPARKS R., BIGGS J., NEUBERG J., "Monitoring volcanoes", *Science*, vol. 335, pp. 1310–1311, 2012.

[STA 10] STAGAKIS S., MARKOS N., SKYOTI O., *et al.*, "Monitoring canopy biophysical and biochemical parameters in ecosystem scale using satellite hyperspectral imagery: an application on a Phlomis fructicosa Mediterranean ecosystem using multisangular CHRIS/PROBA observations", *Remote Sensing Environment*, vol. 114, pp. 977–994, 2010.

[STR 02] STROZZI T., LUCKMAN A., MURRAY T., *et al.*, "Glacier motion estimation using SAR offset-tracking procedures", *IEEE Transactions on Geoscience and Remote Sensing*, vol. 40, no. 11, pp. 2384–2391, 2002.

[TAN 90] TANRÉ D., DEROO C., DUHAUT P., *et al.*, "Technical note description of a computer code to simulate the satellite signal in the solar spectrum: the 5S code", *International Journal of Remote Sensing*, vol. 11, no. 4, pp. 659–668, 1990.

[TIS 04] TISON C., NICOLAS J.M., TUPIN F., *et al.*, "A new statistical model of urban areas in high resolution SAR images for Markovian segmentation", *IEEE Transactions on Geoscience and Remote Sensing*, vol. 42, pp. 2046–2057, 2004.

[TIS 11] TISON C., MANENT C., AMIOT T., *et al.*, "Estimation of wave spectra with SWIM on CFOSAT – illustration on a real case", *(IEEE International Geoscience and Remote Sensing Symposium IGARSS'11)*, Vancouver, Canada, July 2011.

[TOU 88a] TOUZI R., LOPES A., BOUSQUET P., "A statistical and geometrical edge detector for SAR images", *IEEE Transactions on Geoscience and Remote Sensing*, vol. 26, no. 6, pp. 764–773, November 1988.

[TOU 88b] TOUZI R., LOPÈS A., BOUSQUET P., "A statistical and geometrical edge detector for SAR images", *IEEE Transactions on Geoscience and Remote Sensing*, vol. 26, no. 6, pp. 764–773, November 1988.

[TOU 00] TOUTIN T., GRAY L., "State of the art of elevation extraction from satellite SAR data", *ISPRS Journal of Photogrammetry and Remote Sensing*, vol. 55, pp. 13–33, 2000.

[TRO 96] TROUVÉ E., Imagerie interférentielle en radar à ouverture synthétique, PhD Thesis, Ecole Nationale Supérieure des Télécommunications, 1996.

[TRO 98] TROUVÉ E., NICOLAS J.M., MAÎTRE H., "Improving phase unwrapping techniques by the use of local frequency estimates", *IEEE Transaction on Geoscience and Remote Sensing*, vol. 36, no. 6, pp. 1963–1972, 1998.

[TSA 83] TSAI R.Y., "Multiframe image point matching and 3-D surface reconstruction", *IEEE Transactions on Pattern Analysis and Machine Intelligence*, vol. 5, no. 2, pp. 159–173, April 1983.

[TUI 09] TUIA D., PACIFICI F., KANEVSKI M. *et al.*, "Classification of very high spatial resolution imagery using mathematical morphology and support vector machines", *IEEE Transactions on Geoscience and Remote Sensing*, vol. 47, no. 11, pp. 3866–3879, 2009.

[TUI 10a] TUIA D., CAMPS-VALLS G., MATASCI G., *et al.*, "Learning relevant image features with multiple kernel classification", *IEEE Transactions on Geoscience and Remote Sensing*, vol. 48, no. 10, pp. 3780–3791, 2010.

[TUI 10b] TUIA D., RATLE F., POZDNOUKHOV A., *et al.*, "Multi-source composite kernels for urban image classification", *IEEE Geoscience and Remote Sensing Letters*, vol. 7, no. 1, pp. 88–92, 2010.

[TUP 98] TUPIN F., MAÎTRE H., MANGIN J.F., *et al.*, "Detection of linear features in SAR images: application to road network extraction", *IEEE Transactions on Geoscience and Remote Sensing*, vol. 36, no. 2, pp. 434–453, March 1998.

[TUP 99] TUPIN F., BLOCH I., MAÎTRE H., "A first step towards automatic interpretation of SAR images using evidential fusion of several structure detectors", *IEEE Transactions on Geoscience and Remote Sensing*, vol. 37, 3, pp. 1327–1343, 1999.

[TUP 05] TUPIN F., ROUX M., "Markov random field on region adjacency graphs for the fusion of SAR and optical data in radargrammetric applications", *IEEE Transactions on Geoscience and Remote Sensing*, vol. 43, no. 8, pp. 1920–1928, 2005.

[ULA 86] ULABY F.T., MOORE R.K., FUNG A.K., "Microwave Remote Sensing: Active and Passive, – Volume Scattering and Emission Theory, Advanced Systems and Applications", *Artech House, Inc., Dedham, Massachusetts*, vol. III, p. 1100, 1986.

[ULA 98] ULABY F., "SAR biophysical retrievals: lessons learned and challenges to overcome", *Proceedings of 2nd International Workshop on Retrieval of Biophysical Parameters from SAR Data for Land Applications*, pp. 21–28, 1998.

[USA 03] USAI S., "A least squares database approach for SAR interferometric data", *IEEE Transactions on Geoscience and Remote Sensing*, vol. 41, no. 4, pp. 753–760, 2003.

[VAN 96] VAN LEEUWEN P.J., EVENSEN G., "Data assimilation and inverse methods in terms of a probabilistic formulation", *Monthly Weather Review*, vol. 124, pp. 2898–2913, 1996.

[VAN 09] VAN LEEUWEN P.J., "Particle filtering in geophysical systems", *Monthly Weather Review*, vol. 137, no. 12, pp. 4089–4114, 2009.

[VAN 13] VANEGAS M.C., BLOCH I., INGLADA J., "Alignment and parallelism for the description of high-resolution remote sensing images", *IEEE Transactions on Geoscience and Remote Sensing*, vol. 51, no. 6, pp. 3542–3557, 2013.

[VAP 98] VAPNIK V., *Statistical Learning Theory*, John Wiley & Sons, NewYork, 1998.

[VAS 04] VASILE G., TROUVÉ E., CIUC M., *et al.*, "General adaptive-neighborhood technique for improving synthetic aperture radar interferometric coherence estimation", *Journal of Optical Society of America A*, vol. 21, no. 8, pp. 1455–1464, August 2004.

[VAS 08] VASILE G., TROUVÉ E., PETILLOT I., *et al.*, "High resolution SAR interferometry: estimation of local frequencies in the context of Alpine glaciers", *IEEE Transactions on Geoscience and Remote Sensing*, vol. 46, no. 4, pp. 1079–1090, 2008.

[VER 97] VERMOTE E., TANRE D., DEUZE J., *et al.*, "Second simulation of the satellite signal in the solar spectrum, 6S: an overview", *IEEE Transactions on Geoscience and Remote Sensing*, vol. 35, no. 3, pp. 675–686, 1997.

[VER 10] VERBESSELT J., HYNDMAN R., ZEILEIS A., *et al.*, "Detecting trend and seasonal changes in satellite image time series", *Remote Sensing of Environment*, vol. 114, no. 1, pp. 106–115, 2010.

[VIO 95] VIOLA P., WELLS W., "Alignment by maximization of mutual information", *5th International Conference on Computer Vision*, pp. 16–23, June 1995.

[VRU 06] VRUGT J.A., GUPTA H.V., NUALLAIN B.O., *et al.*, "Real-time data assimilation for operational ensemble streamflow forecasting", *Journal of Hydrometeorology*, vol. 7, pp. 548–565, 2006.

[WAL 01] WALKER J.P., WILLGOOSE G.R., KALMA J.D., "One-dimensional soil moisture profile retrieval by assimilation of near-surface observations. A comparison of retrieval algorithms", *Advances in Water Resources*, vol. 24, pp. 631–650, 2001.

[WAN 01] WAN E.A., VAN DER MERWE R., "The unscented Kalman filter", in HAYKIN S. (ed), *Kalman Filtering and Neural Networks*, Wiley, pp. 221–280, 2001.

[WAN 05] WANG Z., ZIOU D., ARMENAKIS C., *et al.*, "A comparative analysis of image fusion method", *IEEE Transactions on Geosciences and Remote Sensing*, vol. 43, no. 6, pp. 1391–1402, 2005.

[WAN 10] WANG M., HU F., LI J., "Epipolar arrangement of satellite imagery by projection trajectory simplification", *The Photogrammetric Record*, vol. 25, no. 132, pp. 422–436, December 2010.

[WEB 02] WEBLEY P.W., BINGLEY R.M., DODSON A.H., *et al.*, "Atmospheric water vapour correction to InSAR surface motion measurements on mountains: results from a dense GPS network on Mount Etna", *Physics and Chemistry of Earth*, vol. 27, pp. 363–370, 2002.

[WEG 05] WEGMULLER U., GAMMA IPTA processing example luxemburg, Technical report, *GAMMA Remote Sensing Research and Consulting AG*, 2005.

[WEI 95] WEIDNER U., FRSTNER W., "Towards automatic building extraction from high-resolution digital elevation models", *ISPRS Journal of Photogrammetry and Remote Sensing*, vol. 50, no. 4, pp. 38–49, 1995.

[WES 98] WESTON J., WATKINS C., Multi-class support vector machines, Report no. CSD-TR-98-04, Department of Computer Science, Royal Holloway, University of London, Egham, 1998.

[WRI 04] WRIGHT T.J., PARSONS B.E., LU Z., "Toward mapping surface deformation in three dimensions using InSAR", *Geophysical Research Letters*, vol. 31, no. L01607, pp. 1–5, 2004.

[WRI 06] WRIGHT T.J., EBINGER C., BIGGS J., *et al.*, "Magma-maintained rift segmentation at continental rupture in the 2005 Afar dyking episode", *Nature*, vol. 442, pp. 291–294, 2006.

[XIE 02] XIE H., PIERCE L., ULABY F., "Statistical properties of logarithmically transformed speckle", *IEEE Transactions on Geoscience and Remote Sensing*, vol. 40, no. 3, pp. 721–727, March 2002.

[YAN 11] YAN Y., Fusion de mesures de déplacement issues d'imagerie SAR: application aux modélisations séismo-volcaniques, PhD Thesis, University of Grenoble, 2011.

[YAN 12a] YAN Y., DOIN M.P., LOPEZ-QUIROZ P., *et al.*, "Mexico city subsidence measured by InSAR time series: joint analysis using PS and SBAS approaches", *IEEE Journal of Selected Topics in Applied Earth Observations and Remote Sensing*, vol. 5, no. 4, pp. 1312–1326, 2012.

[YAN 12b] YAN Y., MAURIS G., TROUVÉ E., *et al.*, "Fuzzy uncertainty representations of co-seismic displacement measurements issued from SAR imagery", *IEEE Transactions on Instrumentation & Measurement*, vol. 61, no. 5, pp. 1278-1286, 2012.

[YAN 12c] YAN Y., TROUVÉ E., PINEL V., *et al.*, "Fusion of D-InSAR and sub-pixel image correlation measurements for coseismic displacement field estimation: application to the Kashmir earthquake (2005)", *International Journal of Image and Data Fusion*, vol. 3, no. 1, pp. 71–92, 2012.

[YAN 13] YAN Y., PINEL V., TROUVÉ E., *et al.*, "Coseismic displacement field and slip distribution of the 2005 Kashmir earthquake from SAR amplitude image correlation and differential interferometry", *Geophysical Journal International*, vol. 193, no. 1, pp. 29–46, 2013.

[YOO 09] YOO H.Y., LEE K., KWON B.D., "Quantitative indices based on 3D discrete wavelet transform for urban complexity estimation using remotely sensed imagery", *International Journal of Remote Sensing*, vol. 30, no. 23, pp. 6219–6239, 2009.

[ZAD 78] ZADEH L.A., "Fuzzy sets as a basis for a theory of possibility", *Fuzzy Sets and Systems*, vol. 1, no. 1, pp. 3–28, 1978.

[ZAR 05] ZARCO-TEJADA P.J., BRERJÓN A., LÓPEZ-LOZANO R., *et al.*, "Assessing vineyard condition with hyperspectral indices: leaf and canopy reflectance simulation in a row-structured discontinuous canopy", *Remote Sensing of Environment*, vol. 99, pp. 271–287, 2005.

[ZEB 97] ZEBKER H.A., ROSEN P.A., HENSLEY S., "Atmospheric effects in interferometric synthetic aperture radar surface deformation and topographic maps", *Journal of Geophysical Research*, vol. 102, pp. 7547–7563, 1997.

[ZHA 05] ZHANG L., Automatic digital surface model (DSM) generation from linear array images, PhD Thesis, Swiss Federal Institute of Technology, Zurich, 2005.

[ZHA 08] ZHAO D., YUAN X., LIU X., "Epipolar Line Generation from IKONOS imagery based on rational function model", *ISPRS Archives, 21st ISPRS Congress, Technical Commission IV*, Beijing, China, 2008.

[ZHA 11] ZHANG L., DING X., LU Z., "Ground settlement monitoring based on temporarily coherent points between two SAR acquisitions", *ISPRS Journal of Photogrammetry and Remote Sensing*, vol. 66, no. 1, pp. 146–152, 2011.

List of Authors

Gilles BOULET
CESBIO
Toulouse
France

Olivier DE JOINVILLE
DGA
Bagneux
France

Olivier HAGOLLE
CESBIO CNES
Toulouse
France

Jordi INGLADA
CESBIO CNES
Toulouse
France

Lionel JARLAN
CESBIO
Toulouse
France

Grégoire MERCIER
Lab-STICC
Télécom Bretagne
Brest
France

Jean-Marie NICOLAS
LTCI
Télécom ParisTech
Paris
France

Virginie PINEL
ISTerre, IRD
University of Savoy
Chambéry
France

Michel ROUX
LTCI
Télécom ParisTech
Paris
France

Jean-Claude SOUYRIS
CNES
Toulouse
France

Florence TUPIN
LTCI
Télécom ParisTech
Paris
France

Emmanuel TROUVÉ
LISTIC
University of Savoy
Annecy
France

Flavien VERNIER
LISTIC
University of Savoy
Annecy
France

Devis TUIAS
LaSIG
EPFL
Lausanne
Switzerland

Yajing YAN
LISTIC/ISTerre
University of Savoy
Annecy/Chambéry
France

Index